Disaster Risk Reduction

Methods, Approaches and Practices

Series Editor

Rajib Shaw, Keio University, Shonan Fujisawa Campus, Fujisawa, Japan

Disaster risk reduction is a process that leads to the safety of communities and nations. After the 2005 World Conference on Disaster Reduction, held in Kobe, Japan, the Hyogo Framework for Action (HFA) was adopted as a framework for risk reduction. The academic research and higher education in disaster risk reduction has made, and continues to make, a gradual shift from pure basic research to applied, implementation-oriented research. More emphasis is being given to multi-stakeholder collaboration and multi-disciplinary research. Emerging university networks in Asia, Europe, Africa, and the Americas have urged process-oriented research in the disaster risk reduction field. With this in mind, this new series will promote the output of action research on disaster risk reduction, which will be useful for a wide range of stakeholders including academicians, professionals, practitioners, and students and researchers in related fields. The series will focus on emerging needs in the risk reduction field, starting from climate change adaptation, urban ecosystem, coastal risk reduction, education for sustainable development, community-based practices, risk communication, and human security, among other areas. Through academic review, this series will encourage young researchers and practitioners to analyze field practices and link them to theory and policies with logic, data, and evidence. In this way, the series will emphasize evidence-based risk reduction methods, approaches, and practices.

Kristoffer B. Berse · Juan M. Pulhin ·
Antonio G. M. La Viña
Editors

Climate Emergency in the Philippines

Impacts and Imperatives for Urgent Policy Action

Springer

Editors
Kristoffer B. Berse
National College of Public Administration and Governance
University of the Philippines Diliman
Quezon City, Philippines

University of the Philippines Resilience Institute
Quezon City, Philippines

Antonio G. M. La Viña
College of Law
University of the Philippines Diliman
Quezon City, Philippines

Juan M. Pulhin
College of Forestry and Natural Resources
University of the Philippines Los Baños
Los Baños, Philippines

ISSN 2196-4106 ISSN 2196-4114 (electronic)
Disaster Risk Reduction
ISBN 978-981-99-7803-8 ISBN 978-981-99-7804-5 (eBook)
https://doi.org/10.1007/978-981-99-7804-5

This Springer imprint is published by the registered company Springer Nature Singapore Pte Ltd.
The registered company address is: 152 Beach Road, #21-01/04 Gateway East, Singapore 189721, Singapore

Foreword

The setting for this compelling and timely volume from University of the Philippines' scholars is the undeniable reality that we are amid runaway climate change driven by human-caused greenhouse emissions. It is also the truth that the Philippines finds itself at the knife-edge of the climate catastrophe, both on account of its fragile and highly exposed ecology and its long history of environmental neglect and developmental vulnerability.

Carefully crafted by a team of researchers and scholars from the natural sciences, social sciences, engineering, humanities, and management, the book gives both the big picture to understand the magnitude of the problem that is upon this nation of 7,641 islands, and the nuts and bolts for concrete action in the immediate to long term. It is as much a briefer on how the climate emergency is unfolding in the Philippines, as it is a call to action to avert an existential crisis.

The climate conundrum is global and local. At the local level, the Philippines, like others in Southeast Asia, needs to undergo massive adaptation, as the die is already cast, and weather events to come will respond to the already accumulated emissions in the air. Equally, the global economy, including the Philippines', needs to take far-reaching steps in mitigation as all the adaptation in the world will not help unless the global economy makes a U-turn on carbon emissions.

The chapters in the book reveal a tapestry of sectoral realities that weave in steps in de-carbonization with the challenges of ensuring livelihoods and sustained socio-economic growth. The empirical evidence drawn from agriculture, food security, health systems, water resources, and energy systems stand out in this reader's mind. It leads invariably to needed actions, one of which is resource mobilization and climate financing, in the right place at the right time.

In a way, the signature feature of the book is the journey of local experiences across river basin ecosystems and coastal communities which are the hallmarks of lives and livelihoods in the Philippines. Rising sea levels in the face of land subsidence presents the most visible and tangible manifestation of climate change. It also showcases the symbiotic link between economic activities and protection of the environment.

It also pays homage to the role of various academic disciplines—including the creative arts—in building resilience and to the remarkable adaptability and resistance of the Filipino people. It is hoped that the lasting contribution of the book will be to provide a powerful reminder of this link and make its call to action one with traction as never before.

The other standout feature of the book is the spotlight on the strength of multi- and inter-disciplinary responses. Climate change is supremely cross-cutting in nature, and that calls for collaborative efforts across science, management, arts, economics, and other social sciences that have been lacking thus far. Top economics journals have hardly published research results on the science and economics of climate change—until very recently. Scientific journals, on the other hand, have stayed away from the economics of climate change which arguably is the root cause of global warming and the key to finding solutions.

Scientists, understandably, find economists' assumptions in their modeling bizarre. But greater collaboration between science and economics and a move to more multi-disciplinary analysis and action on climate change will be most helpful.

And then there is the call to action to confront the climate emergency with a sense of urgency. The actions are relevant and bold and include steps ranging from providing global leadership in climate negotiations to declaring a national climate emergency. Incremental steps will not be enough. Drawing on rigorous research and a deep sense of responsibility, the book makes the case for transformational change. This volume is more than a collection of chapters; it calls for collective and urgent action from the government, civil society, and the people to mitigate, if not completely avoid, the unprecedented impacts of a changing climate now and in the long future.

As I reflect on my recent writings, I am heartened to see a cadre of scholars from the University of the Philippines, the country's premier institution of higher learning, take up the mantle, and lead the way to a more sustainable world. This book is an example of the power of knowledge harnessed for the greater good, of how academia can be a driving force for change.

The response to the climate emergency needs all hands on deck. It demands a chorus of inter-disciplinary voices, and country, regional, and global movements. The Philippines, with its unique strengths and vulnerabilities, has a pivotal role in this journey. The book provides the essential evidence, guidance, and a game plan. Policymakers, practitioners, and citizens must take up the challenge of being better stewards of our planet.

Vinod Thomas
Former Senior Vice President, World Bank, and Director General, ADB
Author of "Climate Change and Natural Disasters" (2017), and
"Risk and Resilience in the Era of Climate Change" (2023)

Preface

The urgency of addressing the global Climate Emergency has reached unprecedented levels, reverberating through every corner of our planet. The Sixth Assessment Report of the Intergovernmental Panel on Climate Change (IPCC) has delivered a stark and unignorable message: we stand at the precipice of a climate catastrophe, and time is running out. This book stands as both an echo of that warning and a call to action.

To a highly vulnerable country like the Philippines, the IPCC's latest reports leave no room for doubt: the impacts of climate change are not abstract scenarios but tangible, immediate threats to our way of life. The rising average surface temperature and sea levels, intensified storms, extreme heat, and other manifestations of a changing climate put at risk the progress that we have made toward achieving the Sustainable Development Goals by 2030 and beyond.

As we navigate the complexities of this crisis, we recognize that addressing the Climate Emergency demands multi-, inter-, and trans-disciplinary approaches. Through the initiative of the University of the Philippines Resilience Institute (UP RI), we brought together experts from the physical and natural sciences, social sciences, arts and humanities, and management and law to take a closer look at how climate change is impacting key sectors and local communities, and to identify policy responses to address current and future challenges in such areas as disaster planning, public health, food and water security, education and risk communication, climate financing, and other critically affected areas. Our goal is not just to elucidate the extent of the climate crisis but to illuminate the paths forward.

This book underscores the indispensable role of academia in shaping our understanding of climate change's myriad facets and in informing the development of policies that can mitigate its impacts. The production of this book is a testament to the commitment of the University of the Philippines (UP) through UP RI to produce, apply, and share new knowledge, knowledge solutions, or know-how in managing climate and disaster risks in the country. Made possible through a small writing grant administered by UP RI's Research and Creative Work (RCW) Division, this volume represents the collective effort of leading climate change scholars and researchers from UP's various campuses, namely, UP Diliman, UP Manila, UP Los Baños, UP Baguio, UP Mindanao, UP Visayas, and UP Cebu.

As editors, we hope that this book will inform the programs, policies, and advocacy strategies of Philippine policymakers and stakeholders as the country responds to this global emergency. We have carefully selected the chapters to delve into the multifaceted nature of this crisis based on actual cases from select river basins, coastal and small island communities, and recent disasters. Yet the stories within these pages are not just about the devastating implications of climate change on human security and development; they are also about resilience, innovation, and the collective resolve of a community of scholars to bring climate science closer to policy and practice.

The chapters in this book will remind us that the Climate Emergency is not a distant threat but a present reality that casts a long shadow on the Philippines' future. Hence, it calls upon governments, civil society, the private sector, and academia to work hand in hand. We hope that vast wealth of knowledge and insights contained in each chapter can serve as powerful tools to inform the development of needed policies and inspire change now, not later.

This book itself is a product of the commitment of multiple institutions and individuals to collaborate and bring to fore the urgency for climate action. Thus, we wish to extend our heartfelt gratitude to the following entities, which have played integral roles in the publication of this book:

- The UP Office of the President under the successive leaderships of Atty. Danilo L. Concepcion (2017–2023) and Atty. Angelo A. Jimenez (2023–present) whose unwavering support, financial and otherwise, made this book project possible;
- The UPRI through its Executive Director Dr. Alfredo Mahar Francisco A. Lagmay whose indefatigable leadership ensured the availability of needed resources for the successful completion of this initiative;
- The UPRI's RCW Division Director and our co-editor, Dr. Kristoffer B. Berse, for birthing this ambitious vision and pursuing it amidst the unprecedented challenges of a global pandemic;
- Our esteemed chapter authors from UP's various constituent units throughout the country for their dedication and expertise to enrich this book with their research findings and insights, as well as for their patience and commitment throughout the revision process as they diligently addressed the feedback provided by both anonymous reviewers and the Springer staff;
- The anonymous reviewers whose meticulous assessment and constructive feedback greatly contributed to the refinement of the book's contents;
- Ms. Micah M. Nazal and Ms. Kirsten Lianne Mae Dedase of UPRI's RCW Division who tirelessly coordinated with the authors, editors, and Springer over the past years to ensure that the necessary chapter revisions and copyright requirements are complied with;
- Mr. Howard Luistro for his layout and graphics design assistance from the early stages to the eventual publication of this book;
- Dr. Rajib Shaw, foremost disaster and climate change scholar, for his encouragement and guidance in linking us with the publisher;

- Mr. Yosuke Nishida and Mr. Arulmurugan Venkatasalam of Springer Nature whose effective and prompt guidance and facilitation of the publication process have been indispensable to the success of this endeavor;

Thank you all for joining us in this critical and timely exploration of the Climate Emergency in the Philippines. May this book serve as both a wake-up call and a roadmap for a sustainable and resilient future for the Philippines and the world.

Quezon City, Philippines — Kristoffer B. Berse
Los Baños, Philippines — Juan M. Pulhin
Quezon City, Philippines — Antonio G. M. La Viña
February 2023

Contents

About the Editors

Kristoffer B. Berse Ph.D. is Associate Professor and Dean of the National College of Public Administration and Governance, University of the Philippines Diliman and previously served as Director for Research and Creative Work of the University of the Philippines Resilience Institute (UPRI).

Juan M. Pulhin Ph.D. is Professor and former Dean of the College of Forestry and Natural Resources, University of the Philippines Los Baños and a Lead Author of the United Nations Intergovernmental Panel on Climate Change (IPCC).

Antonio G. M. La Viña Ph.D. is Professorial Lecturer at the University of the Philippines College of Law and NCPAG, former Executive Director of the Manila Observatory, former Dean of the Ateneo School of Government, and was the lead negotiator for the Philippines in various international climate change issues.

Abbreviations

Chapter 1

ADB	Asian Development Bank
BAU	Business as Usual
CBMS	Community-Based Monitoring System
CDRA	Climate and Disaster Risk Assessment
CLRG	Center for Local and Regional Governance
CLUP	Comprehensive Land Use Plan
DPWH	Department of Public Works and Highways
DRMAPS	Disaster Risk Mitigation, Adaptation and Preparedness Strategies
DRR	Disaster Risk Reduction
GE	General Education
GHG	Greenhouse Gases
GIS	Geographic Information System
ICMSS	Information and Communication Management and Sharing System
IPCC	Intergovernmental Panel on Climate Change
LGU	Local Government Unit
ODK	Open Data Kit
PSF	People's Survival Fund
RCW	Research and Creative Work
RTMs	Risk Transfer Mechanisms
SCM	Science City of Muñoz
UP	University of the Philippines
UPRI	University of the Philippines Resilience Institute
UPV	University of the Philippines Visayas

Chapter 2

ARMM	Autonomous Region in Muslim Mindanao
BFAR	Bureau of Fisheries and Aquatic Resources
CCC	Climate Change Commission
CO_2	Carbon dioxide
CPUE	Catch per Unit Effort
DA	Department of Agriculture
DENR	Department of Environment and Natural Resources
DTI	Department of Trade and Industries
ENSO	El Niño-Southern Oscillation
HCO_3	Carbonic acid
IPCC	Intergovernmental Panel on Climate Change
KIG	Kalayaan Island Group
MPA	Marine Protected Area
NIPAS	National Integrated Protected Areas System
P	Precipitation
SLR	Sea Level Rise
SSH	Surface height
SST	Sea Surface Temperature
USAID	United States Agency for International Development
W	Wind data
WEP	Western Equatorial Pacific
WPS	West Philippine Sea

Chapter 3

CEPA	Communications, Education and Public Awareness
CLUP	Comprehensive Land Use Plan
DAO	DENR Administrative Order
DENR	Department of Environment and Natural Resources
DOST	Department of Science and Technology
EbA	Ecosystem-based Adaptation
EIA	Environmental Impact Assessment
ENSO	El Nino-Southern Oscillation
GHG	Greenhouse Gases
GMSL	Global Mean Sea Level
IPCC	Intergovernmental Panel on Climate Change
IPO	Interdecadal Pacific Oscillation
LGUs	Local Government Units
MGB	Mines and Geoscience Bureau
NIPAS	National Integrated Protected Areas System

PDO	Pacific Decadal Oscillation
PFDA	Philippine Fishery Development Authority
PPA	Philippine Ports Authority
RCP	Representative Concentration Pathway
RSLR	Relative Sea Level Rise
SLR	Sea Level Rise
SUCs	State University and Colleges
USA	United States of America
VIP	Verde Island Passage
WB	World Bank
WTP	Western Tropical Pacific

Chapter 4

CRED	Climate-Resilient Economic Development
DHF	Dengue Hemorrhagic Fever
ECA	Economics of Climate Adaptation
EM-DAT	Emergency Events Database
FNRI	Food and Nutrition Research Institute
GDP	Gross Domestic Project
HEMB	Emergency Management Bureau
IEC	Information, Education, and Communication
IEG	Independent Evaluation Group
NASA	National Aeronautics and Space Administration
NDCC	National Disaster Coordinating Council
NDRRMC	National Disaster Risk Reduction and Management Council
OFDA/CRED	Office of U.S. Foreign Disaster Assistance/Centre for Research on the Epidemiology of Disasters
PAGASA	Philippine Atmospheric, Geophysical, and Astronomical Services Administration
PTB	Pulmonary Tuberculosis
SARS	Severe Acute Respiratory Syndrome
SPEED	Surveillance for Post Extreme Emergencies and Disasters
UNISDR	United Nations International Strategy for Disaster Reduction
USAID	United States Agency for International Development
WHO	World Health Organization

Chapter 5

ARMM	Autonomous Region in Muslim Mindanao
BARMM	Bangsamoro Autonomous Region of Muslim Mindanao
CALABARZON	Cavite, Laguna, Batangas, Rizal, Quezon
CCC	Climate Change Commission
CO_2	Carbon dioxide
CSI	Climate Smartness Index
CSI	Coconut Scale Insect
DA	Department of Agriculture
DICT	Department of Information and Communications Technology
DILG	Department of the Interior and Local Government
DOST	Department of Science and Technology
DRR	Disaster Risk Reduction
DSSAT	Decision Support System for Agrotechnology Transfer
ECCP	European Chamber of Commerce of the Philippines
EPA	U.S. Environmental Protection Agency
FAW	Fall Army Worm
G7	Group of Seven
GCM	Global Climate Models
GDP	Gross Domestic Product
GHG	Greenhouse Gases
GIS	Geographic Information System
GVA	Gross Value Added
IFRI	International Food Policy Research Institute
IMPACT	International Model for Policy Analysis of Agricultural Commodities and Trade
IPCC	Intergovernmental Panel on Climate Change
LGU	Local Government Unit
MT	Megatonne
NEDA	National Economic and Development Authority
NDRRMC	National Disaster Risk Reduction and Management Council
NDVI	Normalized Difference Vegetation Index
OCD	Office of Civil Defense
PAGASA	Philippine Atmospheric, Geophysical and Astronomical Services Administration
PCAARRD	Philippine Council for Agriculture, Aquatic and Natural Resources Research and Development
PCIC	Philippine Crop Insurance Company
ppm	Parts per million
PRISM	Philippine Rice Information and Management System
PSA	Philippine Statistics Authority
R&D	Research and Development

SARAI	Smarter Approaches to Reinvigorate Agriculture as an Industry in the Philippines
SEAMS	SARAI-Enhanced Agricultural Monitoring System
SUCs	State Universities and Colleges
SWCC	System-wide Climate Change Office
WaNulCAS	Water, Nutrient and Light Capture in Agroforestry Systems
WIBI	Weather Index-Based Insurance
WMO	World Meteorological Organization

Chapter 6

ADB	Asian Development Bank
BWD	Baguio Water District
CAR	Cordillera Administrative Region
CCC	Climate Change Commission
GHG	Greenhouse Gases
HB	House Bill
ICLEI	International Council for Local Environmental Initiatives
IIED	Institute for Environment and Development
LGU	Local Government Unit
NDHS	National Demographic and Health Survey
NGWA	National Groundwater Association
NWRB	National Water Resources Board
ppm	Parts per million
PSA	Philippine Statistics Authority
PVC	Polyvinyl chloride
PWSSMP	Philippine Water Supply and Sanitation Master Plan
R&D	Research and Development
SWIM	Sustainable Water-Integrated Management and Governance
UN	United Nations
UNEP	International Institute for Environment and Development
USGSa	United States Geological Society
WASH	Water Access, Sanitation and Hygiene
WDM	Water Demand Management
WEF	World Economic Forum
WHO	World Health Organization
WRI	World Resources Institute

Chapter 7

ASEAN	Association of Southeast Asian Nations
AUUs	Abandoned, Underutilized, and Undeveloped
BFAR	Bureau of Fisheries and Aquatic Resources
BlueCARES	Comprehensive Assessment and Conservation of Blue Carbon Ecosystems and their Services in the Coral Triangle
CHED-PCARI	Commission on Higher Education - Philippine-California Advanced Research Institutes
DA	Department of Agriculture
DBM	Department of Budget and Management
DENR	Department of Environment and Natural Resources
DepEd	Department of Education
DILG	Department of the Interior and Local Government
DOH	Department of Health
DOST-PCIEERD	Department of Science and Technology-Philippine Council for Industry, Energy, and Emerging Technology Research and Development
DPWH	Department of Public Works and Highways
EIA	Philippine Environmental Impact Assessment
GR	General Register
FLAs	Fishpond Lease Agreements
FMB	Forest Management Bureau
FP	Fishpond
FSP	Fisheries Statistics of the Philippines
HEIs	Higher Education Institutions
ICRM	Integrated Coastal Resource Management
IAMBlueCECAM	Integrated Assessment and Monitoring of Blue Carbon Ecosystems for Conservation and Adaptive Management
IEC	Information, Education, and Communication
IUCRMP	Integrated Upland and Coastal Resource Management Planning
JICA-JST	Japan International Cooperation Agency-Japan Science and Technology Agency
KBA	Key Biodiversity Area
LGU	Local Government Unit
LLPCHEA	Las Piñas–Parañaque Critical Habitat and Ecotourism Area
LWUA	Local Water Utilities Administration
MBEMP	Manila Bay Environmental Management Project
MBSDMP	Manila Bay Sustainable Development Master Plan
MGRV	Mangrove
MMDA	Metropolitan Manila Development Authority
MWSS	Metropolitan Waterworks and Sewerage System
NAMRIA	National Mapping and Resource Information Authority

NCR	National Capital Region
NEDA	National Economic and Development Authority
NGO	Non-governmental Organization
OPMBCS	Operational Plan for Manila Bay Coastal Strategy
PCARI	Philippine-California Advanced Research Institutes
PCG	Philippine Coast Guard
PEMSEA	Partnerships in Environmental Management for the Seas of East Asia
PNP-MG	Philippine National Police Maritime Group
PPA	Philippine Ports Authority
PRA	Philippine Reclamation Authority
PRB	Pampanga River Basin
RCW	Research and Creative Writing
SuitMaPs	Suitability mapping for guiding mangrove replanting
TWG-RRA	Technical Working Group for Refined Risk Assessment
UPRI	University of the Philippines Resilience Institute

Chapter 8

2D	Two dimensional
CDOR	Cagayan de Oro River
DPWH	Department of Public Works and Highways
GEV	General Extreme Value
IPCC	Intergovernmental Panel on Climate Change
JICA	Japan International Cooperation Agency
MSL	Mean Sea Level
NHRC	National Hydraulic Research Center
TS	Tropical Storm

Chapter 9

ADSDPP	Ancestral Domain Development and Sustainable Protection Plan
ASL	Above Sea Level
Brgys	Barangays
CADT	Certificate of Ancestral Domain Title
CC	Climate Change
CDRRMO	City Disaster Risk Reduction and Management Office
COA	Commission on Audit
DA	Department of Agriculture
DCDRRMC	Davao City Disaster Risk Reduction and Management Council

DENR	Department of Environment and Natural Resources
DEWS	Deployment of Early Warning Systems
DILG	Department of the Interior and Local Government
DOLE	Department of Labor and Employment
DPWH	Department of Public Works and Highways
DRB	Davao River Basin
DREAM	Disaster Risk Exposure and Assessment for Mitigation
DRiver	Davao River
DRRMC	Disaster Risk Reduction and Management Council
EI	Exposure Index
ENSO	El Niño-Southern Oscillation
Estd	Estimated
FPIC	Free Prior Informed Consent
GIS	Geographic Information Systems
HELP	Hydrology for Environment Life and Policy
ICHARM	International Centre for Water Hazard and Risk Management
ICMSS	Information and Communication Management and Sharing System
IDIS	Interfacing Development Interventions for Sustainability
IHP	Intergovernmental Hydrological Programme (of the UNESCO)
IMS	Information Management System
IP	Indigenous Peoples
IPCC	Intergovernmental Panel on Climate Change
IPO	Indigenous Political Office
IPRA	Indigenous Peoples Rights Act
IPS	Indigenous Political Structure
IT	Information Technology
ITC	Information Technology and Communication [or] Information Technology Center
IWRM	Integrated Water Resource Management
JICA	Japan International Cooperation Agency
LCCAP	Local Climate Change Action Plan
LGU	Local Government Unit
MinLand	Mindanao Land Foundation
NCIP	National Commission on Indigenous Peoples
NDRRMC	National Disaster Risk Reduction and Management Council
NGO	Non-governmental Organization
NGP	National Greening Program
PAGASA	Philippine Atmospheric, Geophysical and Astronomical Services Administration
PD	Presidential Decree
PDAT	Purok Disaster Action Team
Phil-LIDAR	Philippines-Light Detection and Ranging
PhP	Philippine Peso
PO	People's Organization
RA	Republic Act

S&T	Science and Technology
SDGs	Sustainable Development Goals
SEC	Securities and Exchange Commission
SETI	Science Engineering and Technology Innovation
SIR	Slum Improvement Resettlement
SUGAR	Scaling Up Governance and Resiliency
UN	United Nations
UNESCO	United Nations Educational, Scientific and Cultural Organization
UPMin	University of the Philippines Mindanao
UP TCAGP	University of the Philippines Training Center for Applied Geodesy and Photogrammetry
WWII	World War II

Chapter 10

CDP	Center for Disaster Preparedness
CR	Comfort Room
ECLAC	Economic Commission for Latin America and the Caribbean
FGDs	Focus Group Discussions
HUC	Highly Urbanized City
IDB	Inter-American Development Bank
IDP	Internally Displaced Person
INGO	International Non-government Organization
IPCC	Intergovernmental Panel on Climate Change
LGBT	Lesbian, Gay, Bisexual, and Transgender
NEDA	National Economic and Development Authority
NGO	Non-government Organization
NHA	National Housing Authority
NSO	National Statistics Office
PROVIA	Programme of Research on Climate Change Vulnerability, Impacts and Adaptation
PWD	Persons with Disabilities
ST	Super Typhoon
TRRP	Tacloban Recovery and Rehabilitation Plan

Chapter 11

AIP	Annual Investment Program
AWS	Automated Weather System
BDC	Barangay Development Council
BDRRMC	Barangay Disaster Risk Reduction and Management Committee
BFAR	Bureau of Fisheries and Aquatic Resources
BMIS	Barangay Management Information Systems
CA	Christian Aid
CCA	Climate Change Adaptation
CCC	Climate Change Commission
CCDRR	Child-Centered Disaster Risk Reduction
CCPRA	Child-Centered Participatory Risk Assessment
CCS	Coastal CORE Sorsogon
CD	Community Development
CO	Community Organizing
COMSCA	Community-Managed Savings and Credit Associations
DBM	Department of Budget and Management
DDR	Disaster Risk Reduction
DENR	Department of Environment and Natural Resources
DPWH	Department of Public Works and Highways
DOST	Department of Science and Technology
DOT	Department of Tourism
DRRM	Disaster Risk Reduction and Management
ECM	Evacuation Management
FPE	Foundation for the Philippine Environment
GIS	Geographic Information System
ICBDRR	Inclusive Community-Based Disaster Risk Reduction
ICODE	Iloilo Caucus of Development Non-government Organizations, Inc.
IRA	Internal Revenue Allotment
ISATU	Iloilo Science and Technology University
ISDA	Island Sustainable Development Alliance
GFDRR	Global Facility for Disaster Reduction and Recovery
GIDA	Geographically Isolated and Disadvantaged Areas
HSA	Hot Spots Assessment
LGU	Local Government Unit
MDRRMO	Municipal Disaster and Risk Reduction Management Office
MPDO	Municipal Planning and Development Office
NAPOCOR	National Power Corporation
NIACDEV	Northern Iloilo Alliance for Coastal Development
PAGASA	Philippine Atmospheric, Geophysical and Astronomical Services Administration
PAR	Participatory Action Research
PCG	Philippine Coast Guard

PCVA	Participatory Capacity and Vulnerability Analysis
PhP	Philippine Peso
PWD	Person with Disability
RA	Republic Act
RDRRMC	Regional Disaster Risk Reduction and Management Council
RISE	Rehabilitation for Island Sustainability and Empowerment
SHINe	School Hydrometeorological Information Network
SIDS	Small Island Developing States
SK	Sangguniang Kabataan
SSA	School Safety Assessment
SURGE	Scaling Up Resilience in Governance
SWM	Solid Waste Management
UMPA	Unified Marine Protected Area
UNDP	United Nations Development Programme
UNGA	United Nations General Assembly
UNICEF	United Nations Children's Fund
UNISDR	United Nations Office for Disaster Risk Reduction
UP CIDS	University of the Philippines Center for Integrative and Development Studies
UPV	University of the Philippines Visayas
UPVFI	University of the Philippines Visayas Foundation, Inc.
VAWC	Violence against Women and Children
WASAR	Water Search and Rescue
WVCA	Western Visayas Caving Association
ZSL	Zoological Society of London

Chapter 12

BFAR	Bureau of Fisheries and Aquatic Resources
CBA	Community-Based Adaptation
CCA	Climate Change Act
CCA	Climate Change Adaptation
CDP	Comprehensive Development Plan
CLUP	Comprehensive Land Use Plan
CNN	Cable News Network, Inc.
CPDO	City Planning Development Officer
CPPDO	Cebu Province Planning and Development Office
DILG	Department of Interior and Local Government
DENR	Department of Environment and Natural Resources
DENR-MGB	Department of Environment and Natural Resources-Mines and Geosciences Bureau
DND	Department of National Defense

DOST	Department of Science and Technology
DPWH	Department of Public Works and Highways
DRRM	Disaster Risk Reduction and Management
DSWD	Department of Social Welfare and Development
ECVE	Extreme Climate Variable Events
ESA	Emergency Shelter Assistance
IPCC	Intergovernmental Panel on Climate Change
IRA	Internal Revenue Allotment
LGA	Local Government Academy
LGU	Local Government Unit
MC	Memorandum Circular
MSL	Mean Sea Level
MSWD	Municipal Social Welfare and Development
NASA	National Aeronautics and Space Administration
NDRRMC	National Disaster Risk Reduction and Management Council
NGO	Non-government Organization
No.	Number
OPARR	Office of the Presidential Assistant for Rehabilitation and Recovery
PHP	Philippine Peso
Project NOAH	Project Nationwide Operational Assessment of Hazards
PSA	Philippine Statistics Authority
RA	Republic Act
SDG	Sustainable Development Goals
Sec.	Section
SLR	Sea Level Rise
USAID	United States Agency for International Development
USD	United States Dollar

Chapter 13

CBMS	Community-Based Monitoring System
CDRA	Climate and Disaster Risk Assessments
CLUP	Comprehensive Land Use Plans
COA	Commission on Audit
DOST	Department of Science and Technology
DOST-PCIEERD	Department of Science and Technology-Philippine Council for Industry, Energy, and Emerging Technology Research and Development
DREAM	Disaster Risk Exposure and Assessment for Mitigation
DRRM	Disaster Risk Reduction and Management
GIS	Geographic Information System

GPS	Global Positioning System
HEI	Private Higher Education Institutions
LDRRMO	Local Disaster Risk Reduction and Management Offices
LGU	Local Government Unit
LiDAR	Light Detection and Ranging Technology
NAPC	National Anti-poverty Commission
NOAH	UP Nationwide Operational Assessment of Hazards
ODK	Open Data Kit
PADO Inc.	Philippine Augustinians Development Office Inc.
PAGASA	Philippine Atmospheric, Geophysical and Astronomical Services Administration
PHIL-LIDAR	UP Light Detection and Ranging Technology
Project DATOS	Remote Sensing and Data Science
PSA	Philippine Statistics Authority
PWD	Persons with Disabilities
SNAF	Sto. Nino de Cebu Augustinian Social Development Foundation
SRR	Search, Rescue, and Retrieval
SUC	State University and College
UP Cebu CENVI	UP Cebu Center for Environmental Informatics

Chapter 14

CA	Commonwealth Act
CCA-DRR	Climate Change Assessment and Disaster Risk Reduction
CCC	Climate Change Commission
CDRA	Climate and Disaster Risk Assessment
CLSU	Central Luzon State University
CLUP	Comprehensive Land Use Plan
CPDO	City Planning and Development Office
CRMF	Climate Risk and Management Framework
DENR	Department of Environment and Natural Resources
DILG	Department of Interior and Local Government
EO	Executive Order
EP	Ecological Profile
GAM	Goal Achievement Matrix
GIS	Geographic Information System
GOTS	Goals, Objectives, Targets, and Strategies
HLURB	Housing and Land Use Regulatory Board
IPCC	Intergovernmental Panel on Climate Change
LCCAP	Local Climate Change Action Plans
LDRRMP	Local Disaster Risk Reduction and Management Plan

LGBTQ	Lesbian, gay, bisexual, transgender, and queer
LGC	Local Government Code
LGU	Local Government Unit
NCCAP	National Climate Change Action Plan
NEDA	National Economic and Development Authority
NFPP	National Framework for Physical Planning
NOAH	UP Nationwide Operational Assessment of Hazards
NSFCC	National Strategic Framework for Climate Change
PIG	Power-Interest Grid
PLUP	Provincial Land Use Plan
PPFP	Provincial Physical Framework Plan
RA	Republic Act
RCP	Representative Concentration Pathway
SCM	Science City of Muñoz
SGLG	Seal of Good Local Governance
SP	Sangguniang Panglunsod
SWOT	Strengths, Weaknesses, Opportunities, and Threats
TWG	Technical Working Group
UDHA	Urban Development and Housing Act
UPRI	University of the Philippines Resilience Institute
VRG	Vision-Reality Gap
ZO	Zoning Ordinance

Chapter 15

ARMM	Autonomous Region of Muslim Mindanao
CCA	Climate Change Assessment
CCC	Climate Change Commission
CCET	Climate Change Expenditure Tagging
CDRA	Climate & Disaster Risk Assessment
CLRG	Center for Local and Regional Governance
COA	Commission on Audit
CPEIR	Climate Public Expenditure and Institutional Review
CRI	Climate Risk Index
DA	Department of Agriculture
DAE	Direct Access Entity
DBCC	Development Budget Coordination Committee
DBM	Department of Budget and Management
DBP	Development Bank of the Philippines
DENR	Department of Environment and Natural Resources
DOF	Department of Finance
DPWH	Department of Public Works and Highways

DRR	Climate and Disaster Risk Reduction
DRRM	Disaster Risk Reduction and Management
DSWD	Department of Social Welfare and Development
e.g.	Example
ENTF	El Niño Task Force
EO	Executive Order
et.al.	And others
etc.	Et cetera
FY	Fiscal Year
GAA	General Appropriations Act
GCF	Green Climate Fund
GHG	Greenhouse Gas
GSIS	Government Service Insurance System
IPCC	International Panel on Climate Change
JMC	Joint Memorandum Circular
LBP	Land Bank of the Philippines
LDRRM	Local Disaster Risk Reduction and Management
LGU	Local Government Unit
MINDER	Mesoscale Insurance for Disaster Readiness and Recovery
MOOE	Maintenance and Other Operating Expenses
NCCAP	National Climate Change Action Plan
NDRRM	National Disaster Risk Reduction and Management
NDRRMC	National Disaster Risk Reduction and Management Council
NDRRMP	National Disaster Risk Reduction and Management Plan
NEDA	National Economic Development Authority
NGA	National Government Agency
NIA	National Irrigation Administration
ODA	Official Development Assistance
OXFAM	Oxfam International
PAGASA	Philippine Atmospheric, Geophysical and Astronomical Services Administration
PCIC	Philippine Crop Insurance Corporation
PSF	People's Survival Fund
QRF	Quick Response Funds
RA	Republic Act
RAIN	Roadmap to Address the Impacts of El Niño
RTM	Risk Transfer Mechanisms
Sec	Section
SES	Socio-ecological Systems
TSI	Technical Support Instrument
UN-OCHA	United Nations Office for the Coordination of Humanitarian Affairs
vs.	Versus

Chapter 16

AB	Bachelor of Arts
ASEAN	Association of the Southeast Asian Nations
BA	Bachelor of Arts
Dr.	Doctor
DRMAPS	Disaster Risk Mitigation, Adaptation and Preparedness Strategies
GP	General Practitioner
MA	Master of Arts
UN	United Nations
UP	University of the Philippines

Chapter 17

3D	Three Dimensional
ANSI	American National Standard Institute
AR	Augmented Reality
CC	Climate Change
CCA	Climate Change Act
COVID-19	Coronavirus disease-2019
DIY	Do-It-Yourself
DRMAPS	Disaster Risk Mitigation, Adaptation, and Preparedness Strategies
DRR	Disaster Risk Reduction
DRRM	Disaster Risk Reduction Management
EIDR	Emerging Interdisciplinary Research
et.al.	And others
etc.	Et cetera, and other similar things
FGD	Focused Group Discussion
GE	General Education
GIS	Geographic Information System
MECQ	Modified Enhanced Community Quarantine
PAGASA	Philippine Atmospheric Geophysical Astronomical Services Administration
PDRRM	Philippine Disaster Risk Reduction and Management
QC	Quezon City
SAVIMA	San Vicente Mangrove Association
UNISDR	United Nations International Strategy for Disaster Reduction Secretariat
URL	Uniform Resource Locator

Chapter 18

BFAR	Bureau of Fisheries and Aquatic Resources
CDRA	Climate and Disaster Risk Assessment
cf.	Compare
CLUP	Comprehensive Land Use Plans
DENR	Department of Environment and Natural Resources
DND	Department of National Defense
DRM	Disaster Risk Management
DRR-CCAM	Disaster Risk Reduction-Climate Change Adaptation Mitigation
et.al.	And others
FLA	Fishpond Lease Agreements
LGU	Local Government Unit
NDRRMC	National Disaster Risk Reduction and Management Council
NECIP	National Evacuation Center Investment Program
NEDA	National Economic Development Authority
NPTE	National Panel of Technical Experts
OCD	Office of Civil Defense
OP	Office of the President
PDRRM	Philippine Disaster Risk Reduction and Management
RA	Republic Act
UK	United Kingdom
UP	University of the Philippines
UPRI	University of the Philippines Resilience Institute

Chapter 1
Climate Emergency in the Philippines: Sectoral Impacts, Local Experiences, and Multidisciplinary Responses

Kristoffer B. Berse, Antonio G. M. La Viña, Juan M. Pulhin, and Micah M. Nazal

Abstract This chapter provides an overview of the climate emergency as a global movement and as a local reality in the Philippines. Citing the different sections of the book, it discusses how the changing climate drastically impacts multiple sectors in different geographical settings. It also lays down the imperative for urgent action, drawing from the multi-disciplinary responses discussed in the volume. It underscores the need to treat the climate emergency as an ongoing crisis that requires a whole-of-government and whole-of-society intervention at the policy and governance levels.

Keywords Climate emergency · Sectoral impacts · Local experiences · Multidisciplinary responses · Philippines

In August 2018, 15-year-old Greta Thunberg skipped school and sat outside the Swedish Parliament, holding up a "School Strike for Climate" sign and handing out leaflets that declared: "I am doing this because you adults are shitting on my future". In no unequivocal terms, she was demanding politicians to "focus on the climate and treat it like a crisis" (Crouch 2018). What began as a two-week protest on cobblestones in Stockholm ignited a global #FridaysForFuture movement, as calls for decisive and drastic climate action from authorities resonated across oceans (Woodward 2020). Young Filipino climate activists have since responded to the challenge (Abad 2019), with various groups officially joining the Global Climate Strike a year later (Madarang 2019). Thunberg, in a worldwide student walkout

K. B. Berse (✉) · A. G. M. La Viña
University of the Philippines, Diliman, Quezon City, Philippines
e-mail: kbberse@up.edu.ph

K. B. Berse · M. M. Nazal
University of the Philippines Resilience Institute, Quezon City, Philippines

J. M. Pulhin
University of the Philippines, Los Baños, Los Baños, Philippines

K. B. Berse et al. (eds.), *Climate Emergency in the Philippines*, Disaster Risk Reduction, https://doi.org/10.1007/978-981-99-7804-5_1

in March 2019, would sum up the youth's overwhelming source of anxiety in the following words:

> We have only been born into this world, we are going to have to live with this [climate] crisis our whole lives. So will our children and grandchildren and coming generations.... We are not going to accept this. We are striking because we want a future and we are going to carry on. (Westbrook and Fraser 2019)

This growing global outrage among the youth is not without scientific basis. In 2018, Sweden experienced their hottest recorded summer in 262 years, as large parts of the country experienced heatwaves, drought, and wildfires (Anderson and Cowell 2018; McElwee 2018) with fatal consequences (Åström et al. 2019).

Over the years, undeniable impacts of climate change have become more apparent as these continue to intensify and imperil people's lives and livelihoods, and endanger development gains at the community and national levels. Global warming, sea level rise, ocean acidification, extreme weather events, and unprecedented hydrometeorological hazard-induced disasters such as droughts, winter storms, hurricanes, flooding, and wildfires are no longer sporadic events as they become normal scenarios in this century. In the recent Intergovernmental Panel on Climate Change (IPCC) Special Report (2018), anthropogenic activities like the burning of fossil fuels, deforestation, and environmental degradation were identified as the main culprits for the global warming of 1.0 °C from pre-industrial levels. Despite efforts to abate climate change, global commitments still seem to be insufficient in reducing climate polluting emissions. The IPCC report states that the typical Business as Usual (BAU) response of most countries could actually enhance the global warming increase of more than 4 °C above the pre-industrial levels by the end of the century.

In the IPCC's comparison between the 1.5 and 2 °C global warming increase, the likelihood that the Arctic Ocean turns into a complete bare of sea ice in summer may occur once a century at 1.5 °C, but it may take place once a decade at 2 °C. Hence, this will result in a sea level increase of 100 cm higher at 2 °C than at 1.5 °C. Although both will exacerbate the vulnerability of coastal communities, 1.5 °C will only affect 6 million people but with 2 °C, 10 million more people will be endangered at the end of the century. Similarly, the frequency and intensity of droughts, storms, and extreme weather events will most likely increase as the warming intensifies.

Yet, striving to limit the increase within the 1.5 °C, in spite of its already serious impacts, is crucial because every fraction of additional warming beyond this point will bring worse impacts that will most likely exacerbate prevailing concerns on lives, livelihoods, and economies (IPCC 2018).

For countries like the Philippines, a 1.5 °C increase would already cause a significant increase in the number of extremely hot days, greater risks from droughts and precipitation deficits, and tremendous devastating disasters such as droughts and dry spells. As such, the country may progress into entirely new climate regimes because of the frequent occurrence of unprecedented weather extremes. Predictions on the upward trends in annual mean precipitation translate in the increase in the frequency and intensity of heavy rainfall events that result in severe flooding scenarios (Asian Development Bank [ADB] 2017; IPCC 2018). Within the Paris Agreement, the sea level rise could increase beyond 1 m over the centuries which could trigger an increased wave height of storm surges by 10–15%. Thus, the probability of more intense and catastrophic hydrometeorological disasters and potential submergence of many densely populated low-lying cities in the country (Brecht et al. 2012). A rough estimation by Filipino climate change experts, some of whom belong to the IPCC, indicates that as much as 28 out of 81 provinces in the country are at high risk to coastal inundation due to a 1-m sea level rise (Comiso et al. 2014), which could seriously disrupt the lives and livelihoods of millions of Filipinos living in coastal cities and municipalities (Perez et al. 1999).

On the other hand, water shortages are expected to be the usual scenario due to the increasing demand from rapid population and economic growth. Agriculture and food security are highly threatened as the climate crisis ushers in complex challenges brought about by higher temperatures, drier conditions, sea level rise, and flooding, along with the threat in the potential effects of carbon dioxide fertilization. Likewise, the situation will greatly exacerbate the overwhelming concerns on declining soil productivity, groundwater depletion, reduced water availability, pest incidence, and water salinity that directly affect food security in the area. Severe coral bleachings are also expected to cause the massive dieback which will insecure coastal livelihoods like fisheries and tourism as the fish stock decrease because of impaired reef ecosystem (ADB 2017; IPCC 2018).

Consequently, the impaired ecosystem results in a chain reaction and multiplied threats that lead to unfavorable impacts on vulnerable population (Huntjens and Nachbar 2015; IPCC 2018). These potential changes can amplify human insecurity. As early as 2020, projections of climate change and its adverse impacts have never been more real as the Philippines experienced a series of intensified typhoons, severe flooding and landslide incidences claiming precious lives, properties, and infrastructures, all occurring in the middle of a global pandemic.

1.1 What is Climate Emergency?

In 2019, more than 11,000 scientists across 153 countries declared that "clearly and unequivocally, [the] planet Earth is facing a climate emergency" (Ripple et al. 2020: 1). This was on the heels of Thunberg's call for drastic action among political leaders (United Nations News 2019).

Climate emergency is a situation in which urgent action is required to reduce or halt climate change and avoid potentially irreversible environmental damage resulting from it (Oxford Dictionary n.d.). Crises like these are considered complex emergencies because they have interdependent and multi-scalar political, environmental, economic, and social impacts that are dynamic throughout the emergency. They can affect multiple scales, systems, and sectors simultaneously and involve a significant degree of uncertainty in terms of their potential magnitude, consequences, frequency, and extremes. Oftentimes, the risk of these emergencies is unprecedented, hence the difficulty over its prediction, preparation, and quantification (Handmer and Proudley

2005; Handmer and Dovers 2013). While there are existing studies and projections of future climate conditions, the severity of its impacts is yet to unfold fully as the world progresses into the next crucial decade that could define the climate condition throughout the century (Steffen et al. 2018).

As of 2019, the global surface temperature continues to rise because of the increase in the three most abundant atmospheric Greenhouse Gases (GHG), namely, carbon dioxide, methane, and nitrous oxide. Likewise, global ice has been rapidly disappearing as proven by the decline in the minimum summer Arctic sea ice, Greenland and Antarctic ice sheet and glacier thickness worldwide. Ocean heat content, ocean acidity, sea level, deforestation and forest degradation, and the damage costs from extreme weather events are also trending upward rapidly (IPCC 2018). More importantly, there is an overwhelming risk that the biogeophysical feedback processes within the Earth System espoused with direct human degradation could result in a trajectory beyond the planetary threshold. Although the planetary threshold is still uncertain, if crossed, the increased temperature will further lead to the activation of other tipping elements in a domino-like cascade that could further raise global temperature. If unabated, this could lead to the irreversible "Hothouse Earth" pathway that amplifies the difficulty in the existence of human and other life forms regardless of the fulfillment of the Paris Agreement (Ripple et al. 2020; Steffen et al. 2018).

Declarations of climate emergency are hoped to be the loudest call that could catalyze the shift from the business-as-usual climate management to a new emergency mode of climate governance (Sutton 2017). This mode demands a radical and urgent mobilization of economic and social resources beyond the normal level of intensity and scale to better address the climate crisis and ensure a safe climate (Spratt 2019). In this age of intensified climate crisis, the stewardship role of this generation requires a deliberate and sustained action to implicate an integral, adaptive part of the Earth system dynamics that could create feedbacks which can keep the system on stabilized earth conditions (Steffen et al. 2018). Although the declaration is the first major step to leverage on the climate crisis, response toward it greatly requires well-thought out, long-term, and more flexible plans and policies anchored on strong leadership and coordination, with a particular emphasis on adaptation and recovery (Aldunce et al. 2016). Countries must invest in effective mitigation measures that could greatly decarbonize their economies while adaptation measures should be implemented to protect the most vulnerable population (ADB 2017). From the business-as-usual scheme, this is now the opportunity to turn empty words and false promises into real actions that could effectuate change leading to a stabilized and resilient Earth.

By elaborating the tangible realities and consequences of the climate crisis on the country's essential sectors, this book aims to capture the severity of its impact on the vulnerable ecosystem and population. It highlights the need to not only declare climate emergency but the urgency of implementing sound policies buttressed on science and technological advances and strengthened and sustained by good governance. This book is divided into three main parts. Part 1 consists of chapters that highlight the impacts of the climate emergency on selected sectors, namely, coastal communities and the health, food, and water sectors. Part 2 delves on stories and local experiences of communities in recent years. Lastly, Part 3 showcases multidisciplinary responses to the climate emergency in the Philippines, encompassing prospects and challenges related to mapping, planning, financing, education, and communication.

1.2 Sectoral Realities of Climate Emergency in the Philippines

Following this introductory chapter, David et al. bring to fore the impacts of climate change on the safety and food security of coastal communities. Compared to the rest of the world, Filipinos consume twice the amount of seafood per capita per year, while fisheries and related industries provide direct and indirect livelihood for 30 million Filipinos. Furthermore, 60% of the Philippine population reside in coastal municipalities. The country's dependence on the marine system for food and livelihood, its high exposure to hazards, and projected impacts of increasing sea surface temperature, extreme events of rain and storms, and rising sea levels make it vulnerable to climate change, compelling decision-makers to take action. In Chap. 2, David et al. propose early adaptation strategies and policy interventions to build marine resilience. They call for the identification of the most vulnerable localities to tailor-fit corresponding interventions and highlight the need for international support in order to prepare for future climate change scenarios.

In Chap. 3, Siringan and Sta. Maria warn that beaches and coasts are in danger of being lost due to climate change impacts of sea level rise and coastal erosion. The Philippines is particularly vulnerable to these threats due to its natural location in the Western Tropical Pacific and the typhoon belt, and due to subsidence that results from human activities such as groundwater extraction. Sea level rise, which is caused by the melting of ice sheets and glaciers, pushes waves farther toward land, eroding coasts and sediments thereby narrowing beaches. While this occurs due to natural and human causes, climate change speeds up the process. The authors propose that policymakers revise the Environmental Impact Assessment guidelines to include longshore currents and geomorphic compartments, improve easement guidelines, and redesign and rethink the structure of piers and ports. The authors argue that to build the resilience of coastal communities, national and local governments must also address the human activities that endanger Philippine coasts such as beach and offshore mining and strictly enforce existing laws.

From the coastal communities sector, we shift to the topic of health in Lu's Chap. 4. A changing climate does not only bring about more intense and frequent natural hazards that could result in injury and loss of life, but it could directly and indirectly affect public health with the emergence of epidemics, diseases, and health emergencies. For instance, extreme thermal conditions such as heat stress could cause cardiovascular and respiratory problems or even death. Dengue fever cases and the increased incidence of transmission of infectious diseases and epidemic outbreaks are among such associated effects of climate change. On the other hand, ecological mediations, air pollution, vector-borne and marine-borne diseases, and illnesses borne from poor sanitation practices, and water shortages during evacuation are among the indirect effects to health. It is also important to recognize how disasters affect peoples' mental health. The author discusses the link between epidemics, diseases, and health emergencies to climate change and disasters and proposes recommendations for policymakers to strengthen and prepare the health sector to face the climate emergency.

From the health sector, we turn to the agriculture sector. Climate change impacts on the agriculture sector do not only disrupt economic growth and development but they affect the lives and livelihoods of Filipino farmers and food security for all. In Chap. 5, Ortega-Espaldon and Medina discuss the challenges and impacts of the climate change on the agriculture sector. Crop production is projected to have a general decline in the face of different climate scenarios from 2030 to 2050. The authors argue that the status quo and "farming as usual" are no longer an option in the face of a climate emergency. It is imperative to reevaluate traditional knowledge and agricultural practices and adopt science and technology for monitoring and forecasting in order to prepare for predicted climate change impacts. The authors present a brief review and profile of Philippine agriculture and the "Filipino farmer" in the context of agricultural systems in the country and adaptation approaches to manage climate risks. The chapter seeks to inform and guide policymakers, however daunting the task to transform ingrained mindsets and ways among farmers, in order to ensure their well-being and to forward a resilient agricultural sector.

The next chapter focuses on water security. In Chap. 6, Ciencia et al. argue that the water supply crisis that hits Manila in March 2019 is indicative that the Philippines is facing the urgent problem of water insecurity that occurs when there is dwindling water supply and rapid population growth. Climate change exacerbates water insecurity, manifesting in the form of unpredictable occurrences of heavy rainfall and severe flooding or drought, and threatening the quality and quantity of available water sources. Since water is a valuable resource essential for human life and survival, national and local leaders must address the problem of water insecurity. The authors discuss the issues and challenges in the management of the water resources sector of the Philippines and recommend that a single holistic and integrated framework to regulate all aspects of the sector is necessary. National legal frameworks and regulations must be revisited and improved, with consideration to the impacts of climate change. They recommend that the government invest in research and development in the areas of water desalination, water filtration, water purification, water recycling, rainwater harvesting, wastewater treatment, etc. as well as the construction and maintenance of water sewerage systems and wastewater treatment plants all over the country as part of the government's infrastructure program. Local governments must faithfully implement water-related regulations in monitoring the quality of water resources and in the granting of permits for water extraction activities.

Apart from the above, a wide range of equally important sectors are also greatly affected by climate change such as the various forest ecosystems. Severity of deforestation and forest degradation have been multiplied through time as driven by various proximate and underlying causes (Geist and Lambin 2002). Despite its importance, human-induced activities such as the conversion of forestlands to agricultural areas, logging and other resource extraction, increased fire incidence, hunting and wildlife trade, and global interconnectivity clearly affected the forests and their functions (Malhi et al. 2014; Steffen et al. 2011). These activities often lead to the impairment of the forest's capacity to produce ecosystem services such as carbon storage and wood products due to anthropogenic activities and environmental changes (Thompson et al. 2013). In particular, tropical forests are important sources and sinks of carbon. For this reason, every decrease in the forest cover translates to the release of carbon dioxide in the atmosphere which in turn contributes to global warming as they become a carbon source.

Since the deforestation in the fifteenth century until the twentieth century, the Philippines' deforestation of almost 20.9 M ha has contributed to the 3.7 Pg (1015 g) of C to the atmosphere with the 2.6 Pg accounted for the twentieth century alone (Lasco and Pulhin 2000). Similarly, forest cover loss of 157,000 ha from 2000 to 2005 has an estimated aggregate emission of 0.12587 Gt Co2 (FAO 2005). With today's remaining 7.014 million ha(FMB 2019), the Philippine forests can sequester between 7.1 and 27.2 Tg C annually (Lasco and Pulhin 2000). On the contrary, concerns of deforestation and degradation continue to persist and intensify despite the existing policies, reforestation programs, and other initiatives. The country, which is one of the 17 most megadiverse countries of the world, continues to lose its rich biodiversity resources from human activities and is now one of the world's most threatened hotspots (Cruz et al. 2017). Therefore, responsive policies and programs anchored on good governance principles coupled with the mobilization of the citizens are necessary to save the forest and mitigate climate change.

In terms of the built environment ecosystems, they are likewise affected by the consequences of global warming yet, they too, emit a disproportionate amount of GHG largely due to the concentration of human activities in urban areas. Burgeoning population accompanied by its demand on energy sources and natural resources often lead to entangling problems on poverty, conflicts, and migration. Climate change impacts continue to undermine the vulnerable condition of the marginalized population in the built environments. Potential infrastructure failures are also expected because of the continued reliance on unsustainable fossil fuels and energy imports, and intermittent performance of hydropower plants. This is mainly caused by the future instabilities on water discharge and reduced capacity of thermal power plants due to the increasing scarcity of cooling water (ADB 2017; IPCC 2018). Similarly, concerns on sea level rise continue to increase the risk of intense and more frequent storm surges that could result in the devastation of the coastal cities. In the analysis of Brecht et al. (2012), Manila, Philippines ranked first in the cities with the highest affected population if wave height of possible storm surges will be increased by 10–15%. Further, it was followed by six other cities: Taguig City, Caloocan City, Davao City, Butuan City, and Malabon and Iloilo City (Brecht et al. 2012). This study emphasized the increased vulnerability of these built environments as climate change intensifies. Thus, it is imperative that urban planning of all the cities across the country should anticipate these future events that could imperil lives and incur great loss.

Other sectors not covered in this book similarly needs research in order to assess the severity of the climate change impacts that will aid in addressing the concerns accordingly. Local context of climate change experiences enriches the mitigation and adaptation that could support regional and national plans as the country progresses into the next crucial decades of this century.

1.3 Local Experiences from Selected River Basin Ecosystems and Coastal Communities

The second part of this book immerses readers in the stories and experiences of local communities in the Philippines. Chapters 7, 8, and 9 take the readers to Manila Bay in Luzon and to two major river basins in Mindanao. Chapters 10, 11, and 12 present narratives that shed light on the wide array of environmental, policy, and socio-political challenges experienced by communities in Tacloban North, Gigantes, Iloilo, and Daanbantayan, Cebu. It is essential that those most affected by disaster and climate change impacts be given the opportunity to participate and decide on solutions affecting their future, hence, Part 2 of this book aims to bring these voices to the fore.

Due to various environmental challenges plaguing Manila Bay, the government has launched programs to clean, rehabilitate, and preserve its waters and coastal ecosystems. Chapter 7, revisits the changes in mangrove occurrence and distribution relative to the spatio-temporal changes and history of the 200-km coastline of the entire Manila Bay. In reconstructing the Bay's history, it was found that there were about 75,000–90,000 ha of mangroves in the 1990s comprising 10% of the river basin, but, at present, mangroves now only comprise less than 1%. This massive deforestation is the result of conversion into brackish water fishponds, salt pans, and rice paddies, and development into residential and industrial settlements. In Chap. 7, Rollon et al. argue that the Manila Bay story seems to be a microcosm of national and global trends, exemplifying how we deal with development projects which transform our coastal ecosystems into other uses. Specific policy recommendations for national and local governments are recommended in setting a clear rehabilitation roadmap for Manila Bay Coastal Ecosystems considering present and projected climate scenarios.

From Manila Bay in Luzon, we travel to two major river basins in Mindanao, the Cagayan De Oro River Basin and the Davao River Basin. The next chapters discuss disaster and climate change strategies focusing on flood hazards. In Chap. 8, readers will learn about flood hazard mapping which is vital for forecasting, early warning, and planning for flood disasters. Flood inundation modeling and simulation is conducted to describe the spatial extent and distribution of flood waters over time, for a given hydrologic design flood. Engineers use the "return period" or probability of occurrence as the basis for designing flood management schemes or control structures to protect vulnerable communities. However, with the changing climate and urbanization, the flood frequency analysis must be performed in the framework of nonstationary extreme value analysis rather than treating the river flow process as stationary in time which is the usual practice. In Chap. 8, Tabios applies flood hazard mapping with nonstationary extreme values under a changing climate for the Cagayan de Oro River Basin which was devastated during Typhoon Sendong in December 2011. Tabios proposes two key policy recommendations: first, for the engineering practice of DPWH to utilize nonstationary extreme value analysis over stationary analysis and second, for policies and decisions to be based on actual trends or shifts from observed data rather than just relying on climate change design parameters and projections that depend on scenarios, storylines, and global climate models that may not be appropriate for Philippine climate conditions. Tabios points that a strategic policy that should be adopted is piecewise engineering or optimal project sequencing and staging that considers the evolutionary nature and uncertainty of climate change.

For the past 20 years, the frequency and magnitude of the flash floods and landslides in the Davao River Basin have wrought destruction in lives, properties, and public and private investments in development, and the physical and mental health of surviving families in affected barangays. In Chap. 9, Acosta and Gamboa look into the experiences of the Davao River Basin communities who are vulnerable to flood hazards and disasters. Twenty-one (21) of the 36 barangays in the city are prone to flooding while five barangays are most vulnerable to recurring floods and landslides that are a direct result of climate change impacts of extreme weather events. Chapter 9 discusses climate change strategies to address the climate emergency in the Davao River Basin, one of which is the recommendation to create an Information and Communication Management and Sharing System (ICMSS) that is continuously populated, open access and interactive, and staffed by a technical unit. It is imperative that disaster risk and climate information are mainstreamed and communicated to the vulnerable communities by engaging with and learning from Indigenous peoples (Obu-Manuvu, Matigsalug-Manobo, Ata, Ata-Manobo, and Bagobo-Klata), and partnering with civil society organizations and the academe in Davao City.

The succeeding chapters take us from Mindanao to Visayas where we look into the experiences of three communities vulnerable to hydrometeorological hazards. Chapters 10 and 12 tell the narratives of resettled communities still recovering from the loss and damage brought about by Super Typhoon Yolanda (internationally named Haiyan) that struck the Philippines in November 2013. Chapter 11 discusses the different risks and vulnerabilities of small island communities that must be considered in building their capacities and resilience.

Super Typhoon Yolanda destroyed the informal settlements located along Tacloban City's coastline hence, the government relocated thousands of families to Tacloban North. Their experiences are narrated in Chap. 10. Three years after Typhoon Yolanda, these resettled survivors are still recovering but constrained and burdened by poverty, livelihood issues, and insubstantial housing. Relocatees maintained that the dwelling units are poorly constructed, overcrowded, and lack water and sanitation facilities. In Chap. 10, Mangada discusses the factors that undermine the ability of the resettled communities to prepare and adapt to predicted climate change impacts such as poor governance and lack of participation in decision-making. Mangada recommends that policymakers at the national level must review mass relocation as a strategy to protect survivors, improve contractor-driven housing to be disaster-proof, and strengthen communication and inter-agency cooperation. Local governments must conduct community-based risk assessments, the results of which must be integrated in development plans and communicated to the public in the local language, especially the vulnerable communities.

Isolated communities located in the open sea are exposed to different kinds of hydrometeorological and climatological hazards, such as typhoons, which have proven to be the most destructive, monsoon winds, dry spells, and micro-climate anomalies. Chapter 11, presents the experiences of the small island communities of Barangay Asluman, Gabi, Granada, and Lantangan, of the municipality of Carles, in the province of Iloilo. In Chap. 11, Ebay et al. demonstrate how these communities are made even more vulnerable because they are isolated from basic services, compounded by insecurity over land tenure, economic deprivation, and cultural inferiority. This chapter discusses the public service initiative of the University of the Philippines Visayas (UPV) carried out in partnership with the local communities of Gigantes Islands to mainstream disaster risk reduction and climate change adaptation into local plans and practices by way of community development. Three objectives linked the programmatic and multi-year effort: (1) to expand community and local government knowledge about small island community vulnerabilities and risks; (2) to establish functional links between the islands and the mainland government services; and (3) to enhance local capacities to gain recognition and voice and mainstream small island community concerns into local planning and policy. Local governments and small island communities around the Philippine archipelago may adopt the "Small Island Community Resilience Framework" developed by the UPV team, and the good practices and recommendations discussed in this chapter.

Finally, Chap. 12, marks the final stop for Part 2 of the book. The municipality of Daanbantayan, Cebu was among the most hard-hit by Super Typhoon Yolanda in 2013, and just like many coastal communities in the country, it is highly exposed to climate change impacts of extreme weather events and sea level rise. In Chap. 12, Corro discusses the experiences and perceptions of former coastal residents of Daanbantayan, on the dangers of residing in coastal areas, their reasons for relocation, and the modes of assistance given in the aftermath of Typhoon Yolanda. This chapter examines perceptions of former coastal residents in Daanbantayan, Cebu, who were hard-hit by Super Typhoon Haiyan, on the danger of residing in coastal areas, their reasons for relocating, and the different modes of assistance they received to cope with the impact of the super typhoon. This chapter also covers problems encountered by the LGU in relation to the classification of coastal residents whether they are residing in an unsafe or danger zone vis-à-vis the guidelines on the implementation of the government's Emergency Shelter Assistance program. Recommendations are made using the Building Sustainable and Resilient Coastal Communities Framework, to apply governance and technological adaptive measures, which can be mainstreamed to address climate change problems.

1.4 Multidisciplinary Responses, Prospects and Challenges

Part 3 of this book presents multidisciplinary responses, prospects, and challenges posed by the climate emergency in the Philippines that aim to help readers and decision-makers not only to take concrete action but to rethink the way they frame and approach problems and solutions. Part 3 covers topics on mapping, planning, financing, education, and communication in the context of disasters and the climate emergency.

The changing climate is projected to lead to frequent and more intense natural hazards and disasters making prevention, mitigation, and preparedness efforts vital to save lives. Hazard maps are important tools for disaster risk reduction and management but at status quo, Philippine hazard maps have large gaps in data further exacerbating vulnerabilities particularly to flooding and landslides. In Chap. 13, Sinogaya et al. discuss how data voids limit evidence-based planning and constrain response efforts to save lives and recommend policies for building the technical capacities of local governments. The climate emergency adds to the urgency to improve data collection and risk assessments that could be done through funding local governments and state universities and colleges, encouraging collaborations with the academe, and enhancing climate action planning, disaster risk management, and land use planning. Chapter 13 discusses the utilization of the Open Data Kit (ODK) approach where the Community-Based Monitoring System (CBMS) helps to fill the gap of providing demographic information essential for disaster risk assessments in the country. This chapter recommends the integration of CBMS with hazard maps and capacity building of local government's DRR and GIS offices.

As the country addresses the climate emergency, it is vital for local governments to mainstream climate and disaster risk information into their plans to be able to adapt to the impacts of climate change. The 1987 Constitution and Local Government Code of 1991 provides for the sustainable management, development, allocation, and use of land resources and has devolved environmental responsibilities and planning to local governments through the Comprehensive Land Use Plan (CLUP). In Chap. 14, Lagmay et al. discuss the process on how to formulate the CLUP with an integrated climate and disaster risk assessment (CDRA) through the experience of the Science City of Muñoz (SCM) in the province of Nueva Ecija. Chapter 14 presents how CDRA is mainstreamed in every step and describes the benefit of using a participatory approach among stakeholders to strengthen understanding of the city's risks and vulnerabilities, specifically to floods. Their CLUP was executed by UPRI from 2018 to 2019 as part of its main goal to build resilient communities. This chapter is a practical guide and resource for policymakers, practitioners, and planners. Without a scientific, systematic and participatory climate and disaster risk assessment, local governments will not be able to anticipate future hazards that are predicted to be bigger than what is covered by local knowledge and historical records.

In the wake of a climate emergency, an important question to ask is whether Philippine policies on climate and disaster financing are efficient and effective in addressing climate impacts. In Chap. 15, Alampay and dela Torre examine such policies by integrating outcomes of climate finance-related projects of the Center for Local and Regional Governance (CLRG) in the past 5 years. This includes cases of El Niño 2015–2016 and the problem of funding slow-onset climate disasters, as well as assessments in the aftermath of Typhoons Ondoy, Pepeng, and Yolanda. Chapter 15 also looks at the issues and challenges hounding the implementation of climate and disaster risk reduction funds such as the People's Survival Fund (PSF), risk transfer mechanisms (RTMs), and national DRR funds, especially with respect to budget execution. Lastly, it looks at the case of insurance as a strategy for poor LGUs as a strategy to transfer risk and recommends urgent climate financing reforms as part of "transformative governance" to address the climate emergency, including the use of a climate finance policy framework.

The last chapters of Part 3 invite the readers to look inward and consider how they think about the "climate emergency" and their important role as risk communicators. Chapter 16 poses the question, "If artists, art historians, and scholars were to respond to the climate emergency, how would they confront this crisis?".

Climate change is not only a scientific problem, but also a cultural one. In this chapter, Datuin argues that the climate emergency must be understood in all its multiple and interdisciplinary dimensions, recognizing that there are deeply ingrained and unconscious beliefs that shape ecology and how people understand their relationship with the environment. Drawing on case studies from the General Education course—DRMAPS (disaster risk mitigation, adaptation and preparedness strategies)—of the University of the Philippines, this chapter recommends initiating conversations across disciplines and the conduct of training of trainers at national and local levels to reframe "climate change". This can be done through the creation of curricular and education modules for teachers to better explain the science of climate change and underscore the value of the arts and humanities in shaping culture, action, and transformation.

In a country beset with natural hazards and disasters that are projected to worsen due to climate change, there is an urgency for policymakers, practitioners, and educators to understand their role and become better skilled risk communicators. In Chap. 17, Rosete and Mendoza pose the question "What is so special about risk communication?" While communication is considered an art and a science, risk communication comes with greater burden in the sense that it seeks to deliver correct information that the viewer cannot interpret in any other way, elicit the desired response, and incite action with urgency. It is also imperative to communicate messages within the context and "realm" of the audience, otherwise it will be lost. Capturing the attention of the viewer with technical information is a creative challenge. Chapter 17 presents an overview of theoretical foundations of communication; discusses risk communication standards in texts and graphic requirements, accepted conventions, language and symbols for specific groups; and provides concrete applications of communication strategies and new approaches. The authors give concrete recommendations on how to design and create risk communication materials to help and equip the reader who is also a communicator.

Chapter 18, builds on key lessons from the preceding chapters and presents a set of changes that need to happen to effectively confront the #ClimateEmergency. La Viña et al. call for the mainstreaming of nature-based solutions and institutionalization of an integrated natural resources and environment management in DRR efforts. The authors also brought to fore the need to invest in improved infrastructure, including safer and durable evacuation centers; enhance inter-agency collaboration for early warning and response; promote inter-LGU cooperation in planning, response, and recovery; strengthen the national focal agency for disaster risk management; enact an Open Data Law to enhance disaster risk assessment; and ultimately declare a nationwide climate emergency.

1.5 About This Volume

As of this writing, globally, 1,863 jurisdictions in 33 countries have declared a climate emergency. In the Philippines, some local government units (LGUs), such as Cebu City, Bacolod City, and Tolosa, have taken the same tack by declaring a climate emergency in their respective territories (Aidt 2020). At the national level, in November 2020, the Philippines' House Committee on Disaster Resilience adopted House Resolution No. 535, authored by Albay Representative Joey Salceda, declaring a national "disaster and climate emergency". The UPRI together with various policymakers, civil society organizations, and concerned sectors supported this call.

This project is part of UPRI's mandate to produce, apply, and share new knowledge, knowledge solutions, or knowhow on reducing and managing climate and disaster risks in the Philippines. This initiative was made possible through a small writing grant administered by UPRI's Research and Creative Work (RCW) Division. The authors approach the issue of climate emergency from a wide array of disciplines, from physical sciences, social sciences, arts and humanities, management and law.

As previously mentioned, this book is divided into three main parts. The first section zeroes in on climate-vulnerable sectors and communities in the Philippines to introduce readers to the current and pressing impacts of the #ClimateEmergency. The second part presents case studies of local experiences from Luzon, Visayas, and Mindanao, and aims to situate the reader in different environments and contexts affected by the changing climate. The third part features transdisciplinary responses to the climate emergency, including salient challenges in relation to planning, financing, communication, and education.

Each chapter comes with an abstract, a list of key points, and a Summary for Policymakers. The latter, in particular, is intended to provide clear guidance for action for both local and national policymakers. A concluding chapter highlights eight areas for action to tackle the climate emergency.

This book aims to provide a snapshot of the manifestations of #ClimateEmergency in the Philippines. This compilation puts together studies conducted by UP researchers and faculty across the whole UP System, from UP Baguio to UP Mindanao, to provide policymakers and the public a clear picture of why we need to confront climate change with a sense of desperation and urgency. The 18 chapters in this volume highlight the need for a multidisciplinary approach to effectively tackle the challenges posed by a changing climate now, not later.

References

Abad M (2019) 'Kapag may isa, maraming sumusunod': leading a Philippine youth climate strike. https://specials.rappler.com/newsbreak/profiles/241268-philippine-youth-climate-strike-leaders/index.html. Accessed 23 Sep 2020

Aidt M (2020) Climate emergency declarations in 1,769 jurisdictions and local governments cover 820 million citizens. https://climateemergencydeclaration.org/climate-emergency-declarations-cover-15-million-citizens/. Accessed 25 Sep 2020

Aldunce P, Handmer J, Beilin R, Howden M (2016) Is climate change framed as 'business as usual' or as a challenging issue? The practitioners' dilemma. Environ Plan C Gov Policy 34:999–1019

Anderson C, Cowell A (2018) Heat wave scorches Sweden as wildfires rage in the Arctic circle. https://www.nytimes.com/2018/07/19/world/europe/heat-wave-sweden-fires.html. Accessed 3 Nov 2020

Asian Development Bank (2017) A region at risk: the human dimensions of climate change in Asia and the Pacific. https://doi.org/10.22617/TCS178839-2

Åström C, Bjelkmar P, Forsberg B (2019) Ovanligt många dödsfall i Sverige sommaren 2018 - Drygt 600 kan ha dött till följd av värmeböljan [High mortality during the 2018 heatwave in Sweden]. Lakartidningen. 2019 May 21; 116: FLFH. Swedish. PMID: 31192425

Brecht H, Dasgupta S, Laplante B, Murray S, Wheeler D (2012) Sea-level rise and storm surges: high stakes for a small number of developing countries. J Environ Dev 21:120–138. https://doi.org/10.1177/1070496511433601

Comiso et al (2014) Changing Philippine climate: impacts on agriculture and natural resources. UP Press, Quezon City

Crouch D (2018) The Swedish 15-year-old who's cutting class to fight the climate crisis. https://www.theguardian.com/science/2018/sep/01/swedish-15-year-old-cutting-class-to-fight-the-climate-crisis. Accessed 21 Sep 2020

Cruz RVO, Aliño PM, Cabrera OC, David CPC, David LT, Lansigan FP, Lasco RD, Licuanan WRY, Lorenzo FM, Mamauag SS, Peñaflor EL, Perez RT, Pulhin JM, Rollon RN, Samson MS, Siringan FP, Tibig LV, Uy NM, Villanoy CL (2017) 2017 Philippine climate change assessment: impacts, vulnerabilities and adaptation. The Oscar M. Lopez Center for Climate Change Adaptation and Disaster Risk Management Foundation, Inc. and Climate Change Commission

Food and Agricultural Organization (FAO) of the United Nations (2005) Global forest resources assessment

Forest Management Bureau (FMB) (2019) Philippine forestry statistics 2019. DENR, Quezon City

Geist HJ, Lambin EF (2002) Proximate causes and underlying driving forces of tropical deforestation. Bioscience 52(2):143–150. https://doi.org/10.1641/0006-3568(2002)052[0143:PCAUDF]2.0.CO;2

Handmer J, Proudley B (2005) The science of surprise: implementing post-normal science for managing complex unbounded problems: report for EMA. Environmental Management Australia, Canberra

Handmer JW, Dovers S (2013) Handbook of disaster policies and institutions: improving emergency management and climate change adaptation, 2nd edn. Routledge, New York

Huntjens P, Nachbar K (2015) Climate change as a threat multiplier for human disaster and conflict policy and governance recommendations for advancing climate security climate change as a threat multiplier for current and future conflict policy and governance recommendations for advance. http://www.thehagueinstituteforglobaljustice.org/working-paper-9

Intergovernmental Panel on Climate Change (2014) Climate Change 2014: mitigation of climate change. Contribution of Working Group III to the fifth assessment report of the IPCC. Cambridge University Press, Cambridge, UK

Intergovernmental Panel on Climate Change (IPCC) (2018) Summary for policymakers. In: Masson-Delmotte V, Zhai P, Pörtner HO, Roberts D, Skea J, Shukla PR, Pirani A, Moufouma-Okia W, Péan C, Pidcock R, Connors S, Matthews JBR, Chen Y, Zhou X, Gomis MI, Lonnoy E, Maycock T, Tignor M, Waterfield T (eds) Global warming of 1.5°C. An IPCC special report on the impacts of global warming of 1.5°C above pre-industrial levels and related global greenhouse gas emission pathways, in the context of strengthening the global response to the threat of climate change, sustainable development, and efforts to eradicate poverty. World Meteorological Organization, Geneva, Switzerland, 32 pp

Lasco RD, Pulhin FB (2000) Forest land use change in the Philippines and climate change mitigation. Mitig Adapt Strat Glob Change 5(1):81–97. https://doi.org/10.1023/A:1009629220978

Madarang CRS (2019) Millennials and Gen Z stage events in Philippines as part of the global protest on climate change. https://interaksyon.philstar.com/trends-spotlights/2019/09/19/155075/millennials-and-gen-z-stage-events-in-philippines-as-part-of-the-global-protest-on-climate-change. Accessed 3 Nov 2020

Malhi Y, Gardner TA, Goldsmith GR, Silman MR, Zelazowski P (2014) Tropical forests in the Anthropocene. Annu Rev Environ Resour 39(1):125–159. https://doi.org/10.1146/annurev-environ-030713-155141

McElwee R (2018) Sweden: wildfires, heatwave and drought. https://www.aljazeera.com/news/2018/7/19/sweden-wildfires-heatwave-and-drought. Accessed 3 Nov 2020

Oxford Dictionary. (n.d.) Climate emergency. In: Oxford online dictionary. https://www.oxfordlearnersdictionaries.com/climateemergency

Perez R et al (1999) Climate change impacts and responses in the Philippine coastal sector. Clim Res J 12:97–107

Ripple WJ, Wolf C, Newsome TM, Barnard P, Moomaw WR (2020) World scientists' warning of a climate emergency. Bioscience 70(1):8–12. https://doi.org/10.1093/biosci/biz088

Spratt D (2019) Understanding climate emergency & local government. In: Breakthrough. https://www.breakthroughonline.org.au/guides

Steffen W, Rockström J, Richardson K, Lenton TM, Folke C, Liverman D (2018) Trajectories of the earth system in the Anthropocene. Proc Natl Acad Sci 115(33):8252–8259. https://doi.org/10.1073/pnas.1810141115

Steffen W, Grinevald J, Crutzen P, Mcneill J (2011) The Anthropocene: conceptual and historical perspectives. https://doi.org/10.1098/rsta.2010.0327

Sutton P (2017) Local-first implementation: why a strong climate declaration is needed at the local government level and what it can do. Research and Strategy for Transition Initiation Inc. http://www.green-innovations.asn.au/RSTI/Local=first-implementation_local-govt.pdf

Thompson ID, Guariguata MR, Okabe K, Bahamondez C, Nasi R, Heymell V, Sabogal C (2013) An operational framework for defining and monitoring forest degradation. Ecol Soc 18(2). https://doi.org/10.5751/es-05443-180220

United Nations (UN) News (2019) Greta Thunberg tells world leaders 'you are failing us', as nations announce fresh climate action. UN Department of Economic and Social Affairs, Youth

Westbrook T, Fraser A (2019) 'Worse than Voldemort': photos show students around the world skipping school to protest climate change inaction. https://www.businessinsider.com/students-protest-climate-change-inaction-photos-2019-3. Accessed 3 Nov 2020

Woodward A (2020) Greta Thunberg turns 17 today. Here's how she started a global climate movement in just 18 months. https://www.businessinsider.com/greta-thunberg-bio-climate-change-activist-2019-9. Accessed 3 Nov 2020

Kristoffer B. Berse Ph.D. is Associate Professor and Dean of the National College of Public Administration and Governance (NCPAG), University of the Philippines Diliman and previously served as Director for Research and Creative Work of the University of the Philippines Resilience Institute (UPRI).

Antonio G. M. La Viña Ph.D. is Professorial Lecturer at the University of the Philippines College of Law and NCPAG, former Executive Director of the Manila Observatory, former Dean of the Ateneo School of Government, and was the lead negotiator for the Philippines in various international climate change issues.

Juan Pulhin Ph.D. is Professor and former Dean of the College of Forestry and Natural Resources, University of the Philippines Los Baños and a Lead Author of the United Nations Intergovernmental Panel on Climate Change (IPCC).

Micah M. Nazal served as a senior researcher at UPRI and currently works as a communications consultant at an international development organization.

Part I
Sectoral Realities of Climate Emergency in the Philippines

Chapter 2
Climate Change Impacts on Coastal Communities

Laura T. David, Rhodora Azanza, Talna dela Cruz, Roselle Borja-del Rosario, Hannah Eunice Amihan, Ma. Simeona Martinez, and Eileen Peñaflor

Abstract One of the primary resources that Filipinos depend on for food and livelihood is marine biodiversity. On the other hand, with the advent of climate change, there have been unprecedented environmental shifts, characterized by sea level rise and increasing ocean temperatures. In this regard, this chapter will delve into the potential ramifications of the shifts to the coastal habitats, with a subsequent effect on food security. To comprehensively examine the extent of the impact, distribution maps of habitats, fisher population, and fisher density were used in identifying the most susceptible provinces. Findings have shown that those include the provinces of Maguindanao, Lanao del Sur, and Tawi-Tawi (ARMM), alongside South Cotabato, Cotabato City (Region XII), South Cotabato, Cotabato City (Region XII), and Bulacan and Pampanga (Region III). Additionally, the chapter delved into the assessment of coastal inundation vulnerability through storm surge and sea level rise simulations, with emphasis on the municipalities within these vulnerable provinces that require urgent attention. In light of the findings, implementing protective measures for all three habitats coupled with the provision of alternative livelihood for fisher folks is one of the ways to address these challenges. The study also looks into leveraging the capacity of mangrove and seagrass ecosystems for carbon sequestration. It suggests that the Philippines should explore debt trading to preserve its coastal habitats, resulting in better carbon sequestration and marine resources.

Keywords Climate change impacts · Coastal communities · Marine habitats · Food security · Tiered community planning

---

L. T. David (✉) · R. Azanza · T. Cruz · R. B. Rosario · H. E. Amihan · Ma. S. Martinez · E. Peñaflor
University of the Philippines Diliman, Quezon City, Philippines
e-mail: ltdavid@msi.upd.edu.ph

K. B. Berse et al. (eds.), *Climate Emergency in the Philippines*, Disaster Risk Reduction, https://doi.org/10.1007/978-981-99-7804-5_2

Key Points

- Filipinos rely heavily on food fish for protein resource; as such we are vulnerable to any change that happens in the marine environment.
- Fishers have the highest risk since they rely on the ocean for food and livelihood, and typically live within areas exposed to sea level rise and storm surges.
- Climate change impacts are site specific, and therefore adaptation strategies will have to be tailored to each locality.
- Adaptation is necessary for our survival but it can also provide opportunities so that the Philippines can thrive amid climate change.

Summary for Policymakers

Food Security and Fisher Livelihood

The problem of food security in terms of food catch and production is a direct result of climate change impacts.

- Fisheries and associated livelihoods are one of the main economic sectors of the Philippine society providing direct and indirect livelihood to more than 30 million Filipinos.
- Filipinos are most vulnerable since we consume more than twice the amount of seafood per capita per year compared to global average.
- The Philippines is deemed to be of high climate exposure with projected increase in sea surface temperature, locally prolonged drought, intense episodic rainfall, more intense storms and sea level rise that is three to five times the global average.
- This reliance on the marine system combined with the high climate exposure makes the Philippines highly vulnerable to climate change.

Since the Philippines is a low carbon emitter country, any carbon we sequester or keep sequestered is more than what is needed to balance our production. It is possible to "sell" this positive balance to countries that are big producers but are otherwise not able to compensate this production with their mitigation measures.

To address this, policymakers at the local level should enable the implementation of effective adaptation strategies so that collectively we can survive the projected change through:

- Inclusion of all coastal habitats (mangrove, seagrass, coral reefs) in making marine-protected areas.
- Implementing a no-build buffer zone on the landward side of mangrove forests to allow for natural migration as sea level rises.
- Providing appropriate alternative livelihood for the fisher families so that they are not reliant on fisheries as their sole source of income.

To address this, policymakers at the national level must institute appropriate strategies that will allow us to fund all necessary adaptation strategies.

We suggest that the national government should make full use of the capacity of our coastal resources to sequester carbon. Since the Philippines is a low carbon emitter country, producing less than 1% of global carbon (USAID 2016), any carbon we sequester or keep sequestered is actually more than what is needed to balance our production. It is possible to "sell" this positive balance to countries that are big producers but are otherwise not able to compensate this production with their mitigation measures. There have been several studies and attempts for Brazil trading its national debt in protection of the Amazon forests, its oxygen contribution to the planet, and all its biodiversity (e.g., US-Brazil debt for nature swap). Similarly, the Philippines can study doing a debt trade in the protection of its coastal habitats for its carbon sequestration and its marine biodiversity.

Coastal Community Safety

The problem of coastal community survival is a direct result of climate change impacts.

- Filipinos are most vulnerable since all of our big cities can be found right beside the coast.
- Sixty percent of the population live within coastal municipalities.
- The Philippines experiences 20 typhoons a year which are becoming stronger with climate change.

Policymakers must revisit the law of easement from the coast and base it on height from sea level instead of distance from shore.

To address this, policymakers at the local level must:

- Plan for future development of essential structures (e.g., hospitals, evacuation centers, schools) to be located higher than the highest simulated storm surge heights (i.e., 4–8 m above mean sea level).
- Assist in the adaptation of current coastal dwellers either by slowly relocating them away from the coast or by giving them technical assistance to adapt locally (e.g., houses on stilts, above-ground closed septic systems).

To address this, policymakers at the national level must revisit the law of easement from the coast and base it on height from sea level instead of distance from shore.

2.1 Introduction

The average protein requirement for adult Filipinos is estimated at 49–57 g per day (DOST-FNRI 2015) with seafood recognized being the principal source of animal protein. People near the coast are reported to consume 43 kg of seafood per year while those away from the coast still consume 23 kg of seafood per year (FAO 2000; Pedro et al. 2007). Additionally, fisheries and associated livelihoods are one of the main economic sectors of the Philippine society. This sector currently provides livelihood directly to 1 million Filipinos and associated industries estimated to more than 30 million Filipinos. This reliance on the marine system combined with the high climate exposure makes the Philippines highly vulnerable to climate change.

Projections of the CMIP5 models (IPCC5) highlight that in the near-term climate change scenario (2016–2035), the Western Equatorial Pacific (WEP) region shall experience significant increase in sea surface temperature (SST) (mean ocean surface change of 0.5–0.75 °C). There is a projected seasonal mean percentage precipitation change of up to 10%. More importantly, locally prolonged drought and intense episodic rainfall with increase in variability is the likely scenario. Models also agree that storms will become more intense but the frequency will either decrease or remain unchanged. Global sea level rise (SLR) is projected at 4–6 cm per decade with the WEP region likely experiencing the higher of these global estimates. Ocean pH is globally projected to decrease by 0.1 unit in the near term. These impacts are additional pressure on top of the many anthropogenic pressures which the marine ecosystem is already experiencing. Although the Philippines's high biodiversity can help reduce overall vulnerability, urgent actions are needed to build marine food resiliency. A "business-as-usual" approach would compromise 25–75% of our marine food resources and place our coastal communities in harm's way.

2.2 Global Ocean Climate Projections and the Philippines

The coast is a dynamic environment. It is continuously exposed to changing sea level due to daily tides. It is also exposed to different wave energies due to changing monsoon winds and intermittent storms. Depending on the latitude, coastal areas are also exposed to changing seasonal temperature. It is an environment that is always adapting to change. Most of these changes, however, either have limits that the coastal system is already used to (e.g., tides and seasonal winds and temperature) or is infrequent in nature (e.g., occasional storm event). What is referred to as climate change in the coastal and marine environment pertains to changes that affect the frequency, predictability, and intensity of these natural cycles or changes beyond these natural cycles. Climate and ocean change include changes in intensity and pattern of sea surface temperature (SST), water budget, storms, sea level, and ocean acidification.

2.2.1 Temperature

Surface water found along the equator always flow toward the western boundary of oceans. Consequently, as the expanse of the Pacific Ocean warms, it is the Western Equatorial Pacific (WEP) region which will experience significantly higher increase

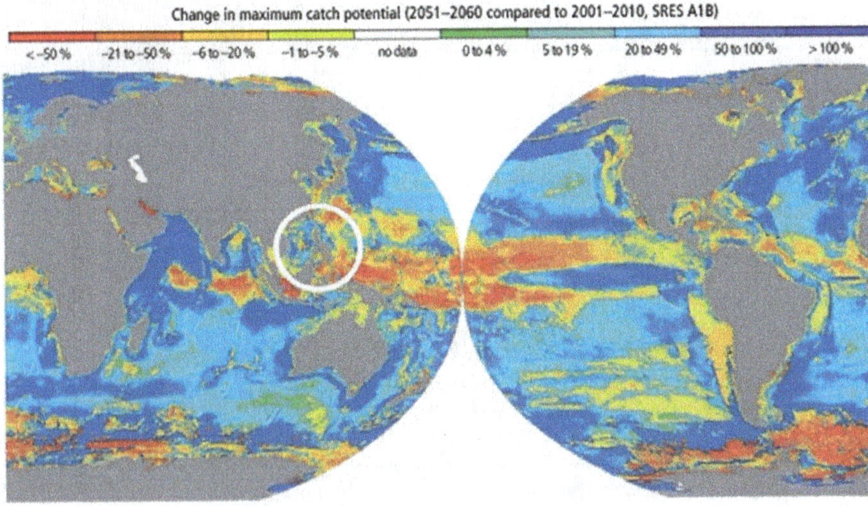

Fig. 2.1 Projected change in maximum catch potential between 2001–2010 and 2051–2060 under IPCC SRES A1B temperature scenario but without taking to consideration any additional pressure from overfishing (IPCC AR5 WG2 2014)

in sea surface temperature (SST). Projections of the CMIP5 models (IPCC5) highlight that in the near-term climate change scenario (2016–2035) the mean ocean surface change in this region is 0.5–0.75 °C.

For the pelagic environment, a short-term increase in temperature can result in migration of fish to deeper or cooler waters. Changes in fish distribution and availability can lead to decrease in catch per unit effort (CPUE) and consequently increase in fish prices. A prolonged warming could disrupt reproduction cycles of target species. Recruitment failures for the less mobile species are also likely.

Shown in Fig. 2.1 is the projected temperature-driven global redistribution of maximum catch potential of ~1000 exploited marine fish and invertebrate species. The waters surrounding the Philippines mostly show a decrease of 6–50% with the southern part being more drastically affected. Only the areas that are known to be upwelling sites seem to fare well. These are sites where colder water from the deeper portions of the water column is seasonally driven to the surface such as the area off of the Bicol shelf or the sub-surface upwelling site west of Luzon.

2.2.2 Sea Level Rise (SLR)

Ocean dynamic topography is typically highest in the western boundary of oceans. In the same manner, as the Pacific Ocean warms and expands, it is the Western Equatorial Pacific (WEP) region which will experience significantly higher increase in SLR. Global SLR in the near future is projected to be higher at 4–6 cm per decade

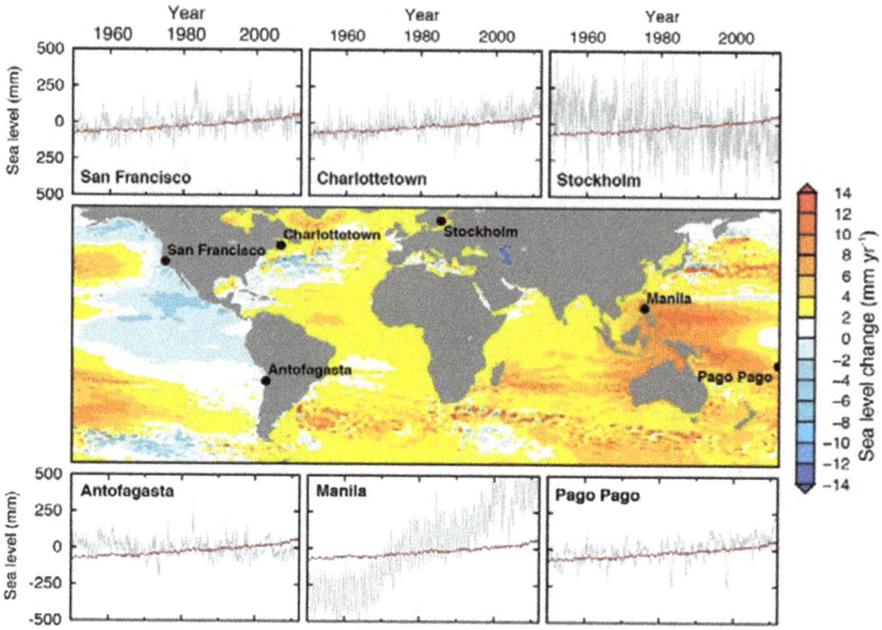

Fig. 2.2 Sea level trends derived from satellite measurements (1993–2012) and actual in situ historical data (1950–2012) (IPCC AR5 WG1 2013)

(4-6mm/yr) with the WEP region likely experiencing the higher of these global estimates.

Shown in Fig. 2.2 are the historical sea level trends for different parts of the world using satellite-derived altimetric mean sea surface data. It can be seen from the figure that the western boundary of the Pacific Ocean is one of the regions already experiencing relatively higher SLR. In the same figure, tide gauge records over the period 1950–2012 show the relative higher increase in Manila Port.

Mangrove forest of the Pacific Islands is predicted to decline by 13% using upper projections of global SLR by 2100 (Gilman et al. 2006). Similarly, a 10–15% mangrove loss is projected globally (Alongi 2008). The main effect is on the establishment of propagules and growth of seedlings. Seedlings in general have been found to thrive better in lower saline environments (3–17 psu) during their first 4 to 5 months. Sea level rise would expose more of them to higher saline environments which have been known to affect their survival rates, cumulative shoot height, mean growth rates, and mean total leaf area, and mean dry weight. In addition, mangrove seedlings thrive when they are above sea surface during daytime. Prolonged submergence has been observed to cause a significant decline in photosynthetic rates.

2.2.3 *Extreme Events: Rain and Storm*

In the tropics and sub-tropics, majority of heavy rainfall events are associated with storm events. Ocean temperature, on the other hand, influence the generation and strength of storms. The warmer the ocean, the more powerful storms.

Multiple studies have associated change in storm intensity to human influence (Fig. 2.3). IPCC5 different scenario models agree that storms will become more intense but the frequency will either decrease or remain unchanged. A more recent study using data from 155 countries showed that an increase of 1°C in global temperature could result in the increase of Category 3 events by seven times in the WEP. Another climate model estimates that if carbon dioxide (CO_2) concentrations double, then the frequency of Category 5 storms may triple (Anderson and Bausch 2006).

Extreme rainfall and storm events over a denuded watershed can lead to sedimentation—one of the highest stressors to the marine coastal environment. The more denuded the watershed and the heavier the rainfall at any given time, the higher the potential sedimentation. Persistent seagrass burial due to sedimentation of 2–19.5 cm has been seen to result in 50% shoot mortality and most likely lead to regression and complete destruction of the seagrass meadow (Cabaco et al. 2008).

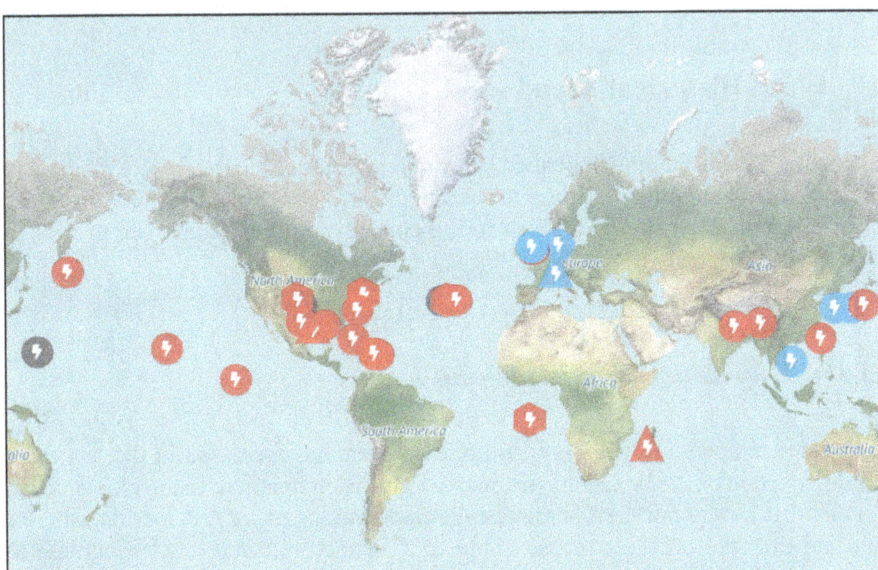

Fig. 2.3 Literature review on the attribution of changes in storm events to human influence shows positive correlation in 20 (shown in RED) out of the 28 studies (Carbon Brief 2020—https://www.carbonbrief.org/mapped-how-climate-change-affects-extreme-weather-around-the-world/)

2.2.4 Ocean Acidification

An increase in carbonic acid (HCO_{3-}) in ocean water is expected as atmospheric CO_2 increases. A higher (HCO_{3-}) environment is not conducive for building or keeping the integrity of calcified structures such as the shells of invertebrates and skeletal base of coral reefs. Hence, these marine creatures will be negatively affected. There are studies, however, that show a non-detrimental or even positive effect for some marine species that can make use of (HCO_{3-}) in photosynthesis (e.g., seagrass) (Burnell et al. 2014; Maberly et al. 2022) and those that do not form hard structures (e.g., jelly fish) (Du et al. 2018; Enrique-Navarro et al. 2021). At current rates of ocean pH trajectory, the global projection is a decrease by 0.1 unit in the near term and up to 0.3 units by 2100.

Aragonite saturation state is commonly used to monitor carbonate ion concentration. Plankton, shellfish, corals, and other hard structure-forming invertebrates are stressed when aragonite saturation state falls below 3. Currently, the aragonite saturation in the waters surrounding the Philippines is still adequate. In the medium term (2040–2060), the Pacific Seaboard is overall expected to have low but still acceptable carbonate ion concentration, while the West Philippine Sea (WPS) is expected to fall below 3 (Johnson et al. 2016).

2.3 Local Historical Trends and Typology

The Pacific Ocean waters are pushed toward the western boundary due to the mean northern Pacific circulation. Being located in the western boundary, the Philippines is hence particularly highly exposed to the changes in temperature, sea level, and storm events occurring in the Pacific Ocean.

2.3.1 Temperature and Bleaching Events

Sea surface temperature (SST) in Philippine waters has increased in the last three decades at a rate of 0.12–0.20 °C per decade. This is comparable to the global ocean average of 0.13 and northern hemisphere ocean average of 0.19 °C per decade, for the same time period. This is also about 25% of the projected rate of increase in temperature for this part of the world in the next two decades. The question therefore is if the Philippines can cope with a rate four times faster than what it has been experiencing so far?

Aside from this slow-onset temperature increase, the waters of the Philippines have also experienced severe episodic warming of 1 degree or higher lasting for at least 4 weeks. Shown below (Fig. 2.4) is the analysis for the 1998 warming episode that happened all over the Coral Triangle. The pie graph gives the percent coral cover

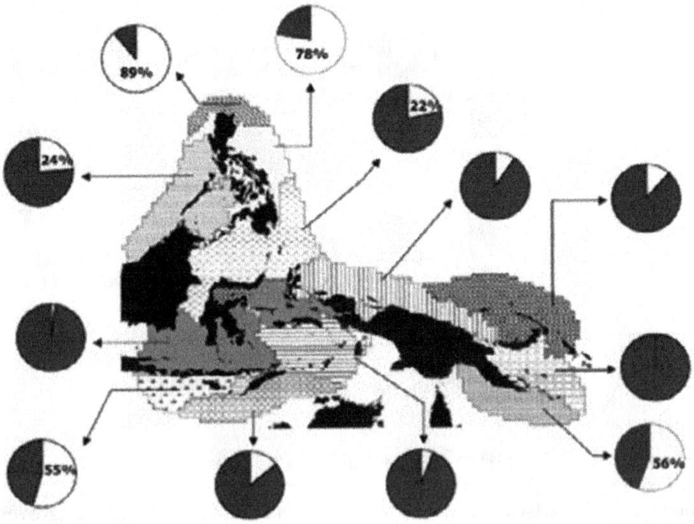

Fig. 2.4 Pie chart emphasizes percent of coral pixels that experienced severe warming in 1998 (derived from Peñaflor et al. 2009)

that experienced severe warming within each cluster and highlights the higher occurrence in the waters surrounding the Philippines. This episode caused the massive coral bleaching event documented in 1998.

2.3.2 Sea Level Rise (SLR) and Coastal Erosion

Using weekly data from 1992 to 2008 for the surrounding and inner waters of the Philippines, David et al. (2015) analyzed that changes in sea level ranged from 5.4 to 8.3 cm per decade. During the same time period, the global average was 3.2 cm per decade. This means that the rate of SLR is twice (in the Pacific seaboard) to almost three times (in the rest of the Philippine waters) that of the world.

This high rate exposes more of our low-lying coasts to daily wave action resulting in more pronounced coastal erosion. Shown below is a schematic of how a few centimeters of SLR can result to extensive coastal erosion (Fig. 2.5).

Sea level rise coupled with unregulated groundwater extraction may also increase the salinity intrusion into pumped groundwater thus negatively affecting coastal agriculture. Unregulated groundwater extraction can also lead to land subsidence further exacerbating the effect of SLR. Figure 2.6 shows how SLR and groundwater extraction have affected the areas around Manila Bay.

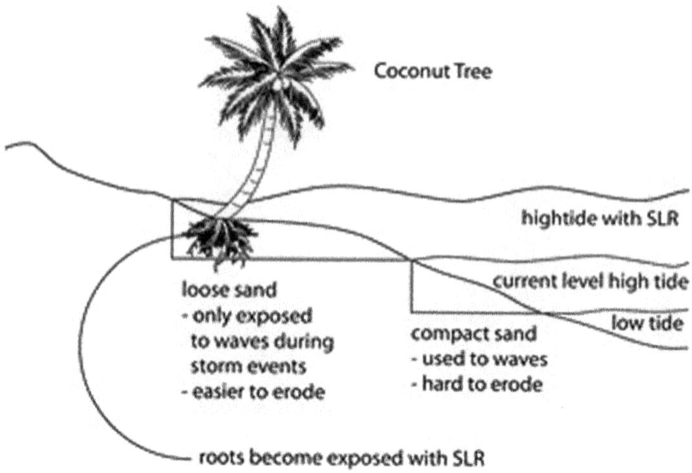

Fig. 2.5 Schematic of increased exposure to higher salinity environment of mangroves seedlings to sea level rise. Likewise, the potential increase in erosion due to exposure of loose sediment to daily waves

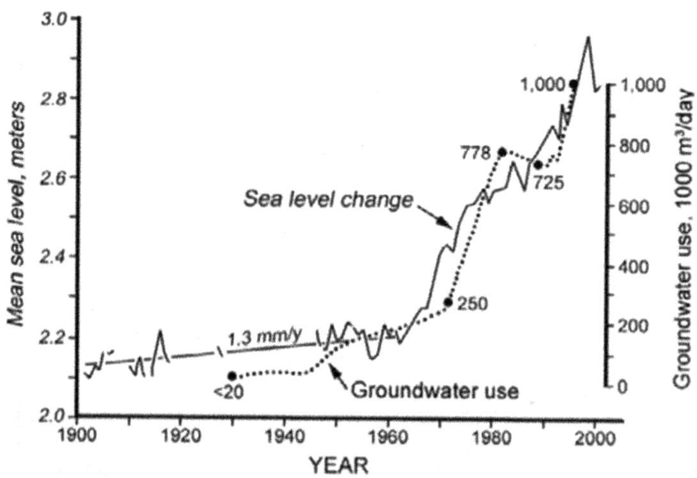

Fig. 2.6 Sea level rise recorded at Manila's South Harbour from 1902–2000 and groundwater use in Metro Manila over the same period (Rodolfo and Siringan 2006)

2.3.3 *Extreme Rainfall Events and Sedimentation*

Aside from storm events, easterly winds and the monsoons dictate the rainfall patterns over the Philippines. The strength of the easterly winds influences the general position of rain clouds over the tropical Pacific. If the easterlies are strong, the rains end up over the Philippines similar to La Niña situations; if they are weak, the rains remain over the central Pacific similar to El Niño times. In the short term, it seems that the intensity and frequency of ENSO events are increasing. However, the data are too short to say how ENSO activity will change in the coming decades.

Monsoons, on the other hand, are dictated by the difference in temperature between land masses and oceans. Over the Philippines, during winter months of December, January, and February, when the land masses in the northern hemisphere are colder compared to the oceans, surface dry and cold air rushes from the northeast. During June, July, and August when the land masses in the northern hemisphere are warm, there is reversal of wind patterns, and moist surface air rushes from the relatively cooler southwest seas causing an increase in precipitation on the western side of the country. With lesser difference in temperature due to climate change, patterns of monsoons are becoming less predictable. In some areas, this is predicted to result in periods of reduced rainfall in certain regions.

2.3.4 *Overall Typology*

Cluster analysis on satellite-derived sea surface temperature (SST), sea surface height (SSH), wind data (W), and precipitation (P) naturally divided the archipelagic waters of the Philippines into 11 distinct air-sea climate exposure clusters (Fig. 2.7). The northern and eastern waters (clusters II, I, X, IX, and VIII) are most prone to extreme temperature while the eastern cluster has the highest rate of increase in average ocean temperature. The entire Philippines suffer twice to three times the magnitude of the global sea level rise (SLR). In terms of precipitation hazards, western Luzon is the only cluster that has suffered extreme rainfall events. Western Luzon all the way down to northern Palawan has suffered from unpredictability of seasonal rainfall. This analysis highlights the need for different adaptation approaches for the different coastal and marine areas of the Philippines.

2.3.5 *Exposure and Sensitivity*

Filipinos are vulnerable to any change in our oceans since up to 20–40% of our food protein comes from aquatic and marine resources. The sector that will be most vulnerable is the fishers. Not only do they rely on the oceans for food, but also for their livelihood.

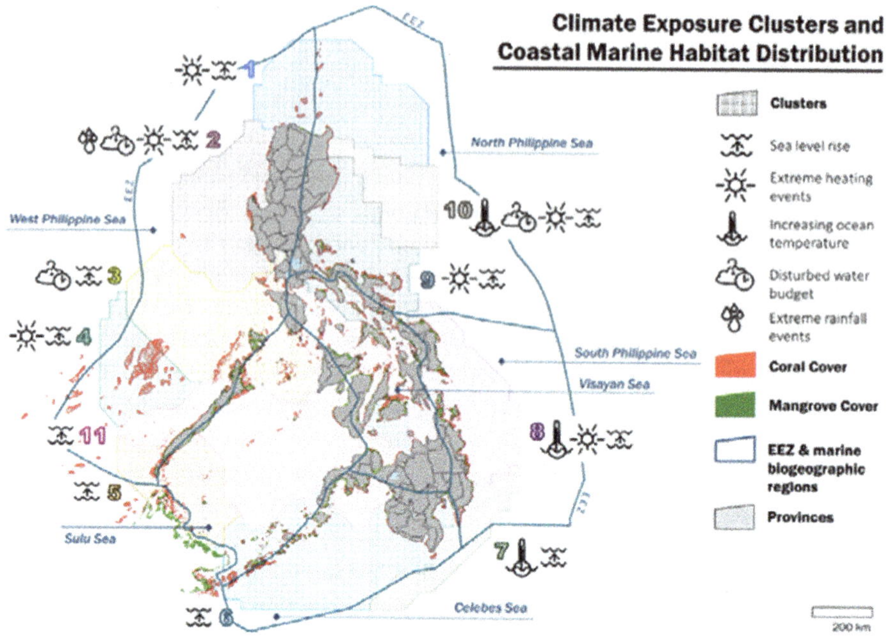

Fig. 2.7 Climate exposure clusters determined from historical satellite-derived data (1980–2010) (derived from David et al. 2015)

2.3.6 *Fisher Livelihood*

According to the 2015 statistics, there are a total of 1,856,500 fishers in the Philippines. Eighty-seven percent (87%) are fishing in municipal waters; 1% are in commercial fishing operations; and 12% are farming fishes, mollusks, seaweeds, and other aquatic organisms. They are also considered among the poorest of our poor along with farmers (Table 2.1). The 2000 census showed that, in general, fishersfolks are more likely to live in makeshift houses with less access to safe water, sanitary toilets, and electricity. The average households of fishers had higher dependents, and had lower education levels. All these translates to a low capacity to adapt to changes.

Data aggregated at the regional scale indicated that highest number of fishers are in the ARRM, composing 13% of the country's fisher population. MIMAROPA (IV-B), the Bicol region (V), and the Visayan regions (VI, VII and VIII) have 7–9% each. Together these regions provide 60% of our municipal fisheries. Province-wise Palawan, Maguindanao, Cebu, Sulu, Leyte, Bohol, Tawi-Tawi, Cagayan, Masbate, and Iloilo have 3–4% each. Negros Occidental, Quezon, Western Samar, Batangas, Pangasinan, Camarines Sur, Lanao del Sur, Zamboanga del Sur, Isabela, Eastern Samar, and Zamboanga del Norte have 2% each.

Aside from the number of fishers, the density of fishers is also important because it gives a handle on how much resource is available per capita. The higher the density,

Table 2.1 Poverty incidence for farmers and fishermen: 2006, 2009, 2012, and 2015 (PSA 2017)

Sector	2006		2009		2012		2015		Increase/Decrease		
	Poverty incidence (%)	CV	Poverty incidence (%)	CV	Poverty incidence (%)	CV	Poverty incidence (%)	CV	2006–2009	2009–2012	2012–2015
Philippines	**26.6**	**1.9**	**26.3**	**2.0**	**25.2**	**2.1**	**21.6**	**3.5**	**(0.3)**	**(1.1)**	**(3.6)**
Farmers	38.5	2.1	38.0	2.1	38.3	2.5	34.3	3.0	(0.5)	0.2	(4.0)
Fishermen	41.2	4.6	41.3	4.0	39.2	4.7	34.0	5.8	0.1	(2.1)	(5.2)

the greater is the competition between fishers. The number of fishers per kilometer of coastline (fisher/km) is a good proxy for competition in habitat-based fisheries, while the number of fishers per square kilometer of municipal waters (fisher/km^2) is a good proxy for pelagic-based fisheries (Fig. 2.8).

Using these proxies, the areas with the highest competition in habitat-based fisheries are South Cotabato, Bulacan, and Cotabato City with more than 500 fishers/km, followed by Maguindanao, Pampanga, Cavite, Lanao Del Sur, Manila, and Davao del Sur with more than 150 fishers/km. The areas with the highest competition in pelagic-based fisheries are Pampanga and South Cotabato with more than 100 fisher/

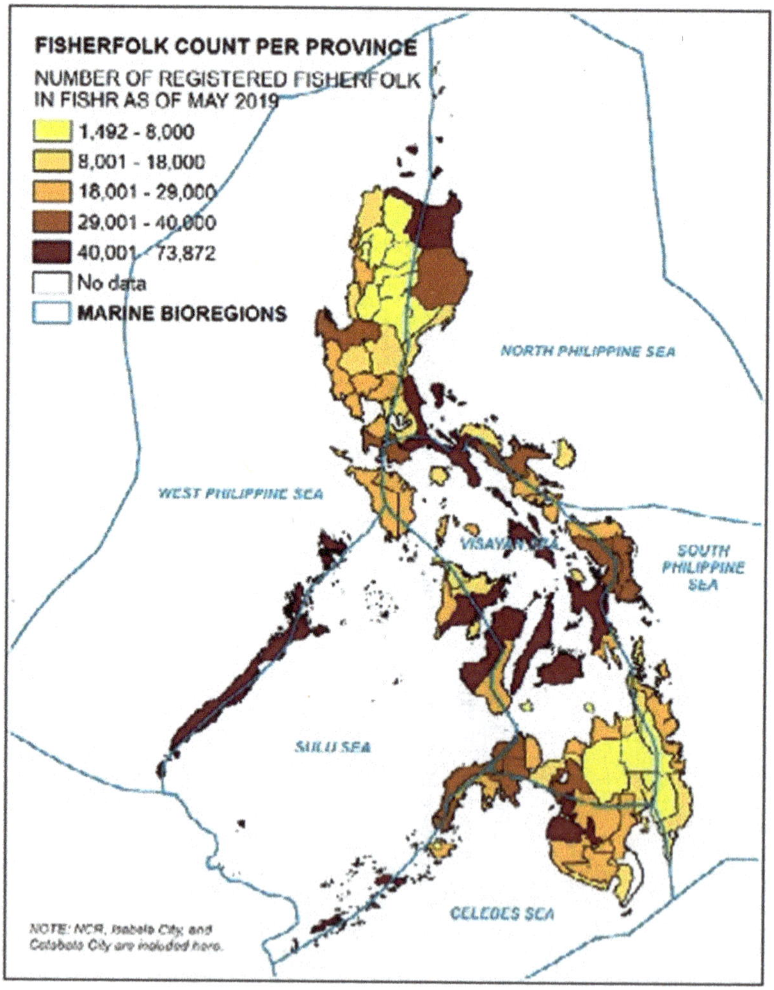

Fig. 2.8 Fisherfolk count and fisherfolk density per square kilometer of municipal waters (Fisherfolk count from BFAR 2019; Fisherfolk density analyzed for this paper)

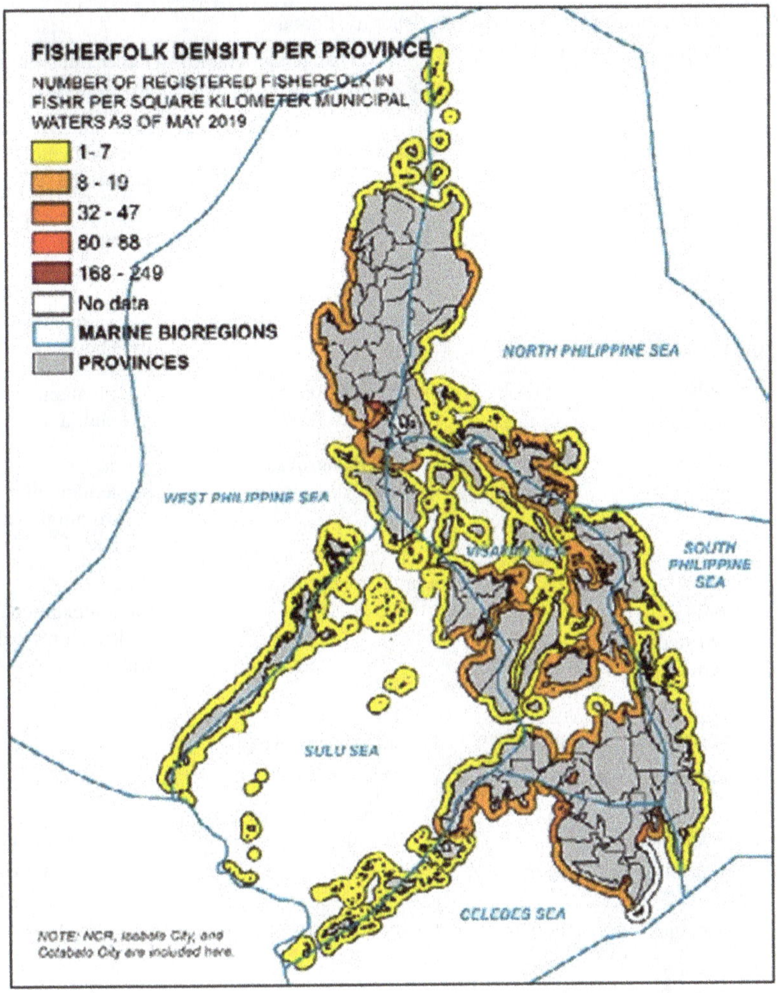

Fig. 2.8 (continued)

km^2. Followed by Cotabato, Bulacan, Lanao Del Sur, Maguindanao, and Manila with around 50 fisher/km^2 or more.

The high fisher population and the high competition for resources contribute to the fishers having a low capacity to adapt.

Taking into consideration the exposure to climate change, the observed and projected vulnerability of fisheries of the different habitats, and the fishers' low adaptive capacity, we were able to identify areas where fisher livelihood to be at high risk due to climate change (Table 2.2).

ARMM is seen to be at highest risk with four of its provinces highlighted. Sea level rise (SLR) is their major climate change concern and hence mangrove forests will

Table 2.2 Identified high-risk sites vis-à-vis fisher livelihood

Region	Province	Fisherfolk count	Climate change vulnerable habit that needs protection/management	Other interventions in addition to protection
ARMM	Maguindanao	72,143	Mangroves	Supplemental livelihood
	Lanao del Sur	37,510		Supplemental livelihood
	Sulu	65,259		
	Tawi-Tawi	54,772		
XII	South Cotabato	27,369	Mangroves	Supplemental livelihood
	Cotabato City	6,976		
III	Bulacan	25,747	Seagrass & mangroves	Supplemental livelihood
	Pampanga	26,344		
VII	Bohol	54,953	Corals, mangroves + offshore	Strict implementation of 15-km municipal waters
	Cebu	70,361		
VIII	Leyte	55,511	Corals, mangroves + offshore	Strict implementation of 15-km municipal waters
	Samar (Western Samar)	39,533		
	Eastern Samar	33,571		
V	Masbate	48,822	Corals & mangroves	
	Camarines Sur	37,713		
VI	Negros Occidental	48,007	Corals, seagrass, mangroves + offshore	Strict implementation of 15-km municipal waters
	Iloilo	48,577		
IV-A	Batangas	39,030	Corals, seagrass & mangroves	Supplemental livelihood
	Quezon	47,614		
	Cavite	21,962		
IV-B	Palawan	73,872	Corals & mangroves	
IX	Zamboanga del Norte	33,527	Mangroves	
	Zamboanga del Sur	37,246		
I	Pangasinan	37,898	Corals, seagrass & mangroves	
II	Isabela	33,945	Corals & mangroves	

need significant protection and forward-looking management. Reforestation effort will need to take into consideration the SLR projections to assure that planting of saplings will not be done in areas that will eventually be permanently flooded. Maguindanao and Lanao Del Sur will also need additional assistance in terms of supplemental livelihood because of the very high competition on available resources. If the resources are compromised, there will be many fisher families that will be affected. Supplemental or alternative livelihood need not necessarily target the fishers themselves but at least someone within the family. If the wife or children are given skill-appropriate livelihoods, then there will be another source of income for the family giving them a higher capacity to adapt.

The analysis is similar for fishers of South Cotabato and Cotabato City of Region XII, and Bulacan and Pampanga of Region III. In addition, Region III is also exposed to disturbed water budgets which could result in increased sedimentation over seagrass beds. Management interventions include direct protection and also mangrove reforestations to stabilize the coast and keep the sediment from going into the water column.

Regions VII, VIII, and VI are exposed to SLR, extreme heating events, and increasing ocean temperature. Mass coral bleaching has been documented historically. In order to lessen the risk for the fishers in Bohol, Cebu, Leyte, and Western and Eastern Samar, there is a need to reduce other stress on the coral reefs (e.g., destructive fishing, overfishing) to increase their resilience to climate change. In addition, the high rate of increase in surrounding oceans is projected to impact the pelagic fisheries. To lessen the pressure, there is a need to strictly implement the 15 km municipal water restriction on all commercial-sized fisheries.

Habitat protection through integrated MPAs, incorporating mangrove, seagrass, and corals would be most effective to protect the livelihood of fishers in all the highlighted provinces and contribute to the food security of the nation.

2.3.7 Coastal Communities

The highly populated cities of the Philippines are mostly found near the coast. Co-located with this urban setting are much of the industries associated with the marine environment such as tourism; fisheries; manufacturing; shipping; and the electricity, gas, and water utilities (PEMSEA 2016). This makes the population and economy of the Philippines vulnerable to coastal inundation brought about by storm surge or sea level rise (SLR). Stronger storms passing through the region are given as a higher probability in the IPCC5 report. Stronger storms would result in bigger storm surges. Based on simulations done for the Philippine setting, a category 5 typhoon similar to Haiyan (Yolanda) could generate a storm surge of up to 4–7 m in the country (Lapidez et al. 2015) (Table 2.3). Additionally, the persistent slow-onset hazard SLR in the Northwestern Pacific is about three times higher than the global mean, reaching around 12 cm per decade.

Table 2.3 Top 30 provinces with the highest simulated storm surge heights (Lapidez et al. 2015)

Rank	Province	Max surge height (m)	Population density (per km^2)
1	Samar	7.45	100–250
2	Leyte	6.84	>1000
3	Palawan	6.71	25–100
4	Iloilo	6.29	500–1000
5	Biliran	6.26	100–250
6	Camarines Sur	6.17	500–1000
7	Quezon	5.86	100–250
8	Masbate	5.45	100–250
9	Southern Leyte	5.32	100–250
10	Bataan	5.04	500–1000
11	Dinagat Islands	5.00	100–250
12	Surigao del Norte	5.00	100–250
13	Cebu	4.77	250–500
14	Pampanga	4.76	250–500
15	Bohol	4.45	250–500
16	Bulacan	4.42	100–250
17	Negros Occidental	4.41	250–500
18	Guimaras	4.41	100–250
19	Albay	4.36	250–500
20	Negros Oriental	4.05	100–250
21	Capiz	4.04	250–500
22	Metro Manila	3.90	>1000
23	Eastern Samar	3.87	100–250
24	Surigao del Sur	3.72	25–100
25	Camarines Norte	3.69	100–250
26	Maguindanao	3.65	500–1000
27	Lanao del Sur	3.65	<25
28	Zamboanga del Sur	3.59	25–100
29	Sulu	3.46	100–250
30	Marinduque	3.39	100–250

According to Kulp and Strauss (2019), SLR (of 1 m) alone can already permanently flood four to seven million people, while an additional episodic storm surge (of 4–7 m) can affect as much as 24–31 million (Table 2.4).

In general, we therefore recommend coastal cities to proactively employ a tiered community approach (Fig. 2.9). Coastal spatial planning should limit the location of critical facilities such as hospitals, schools, and evacuation areas above the projected

Table 2.4 Simulation of number of people (in millions) that will be affected by different inundation scenarios using SRTM and the higher resolution CoastalDEM data (Kulp and Strauss 2019)

Place; DEM used	Water heights (meters)										
	<1	1	2	3	4	5	6	7	8	9	10
Philippines; SRTM	0.6	1.2	2.3	4.1	6.5	9.6	13	17	20	22	24
Philippines; CoastalDEM	4.3	7.5	12	17	22	24	29	31	33	35	36
World; SRTM	28	63	110	190	280	370	460	550	630	710	780
World; CoastalDEM	110	230	400	550	680	770	840	900	950	1,000	1,040

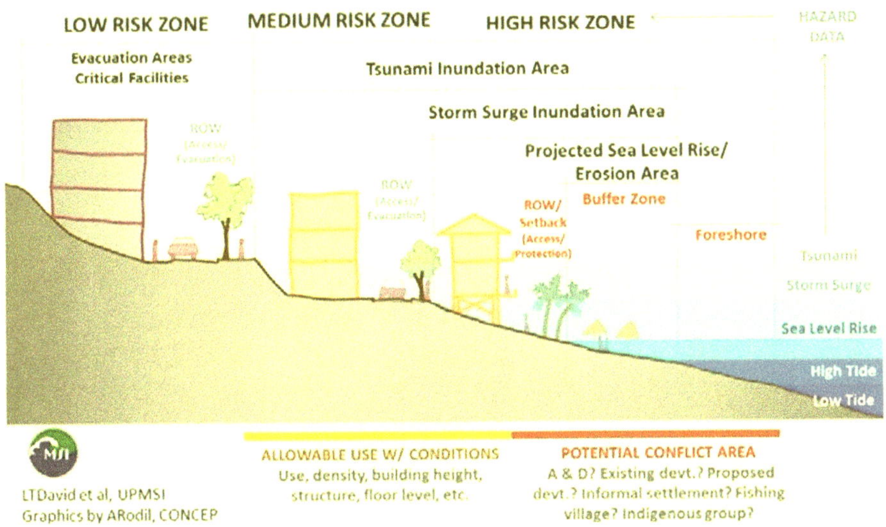

Fig. 2.9 Tiered community planning taking into consideration the different causes of coastal inundation (David et al.)

combined height of the SLR and storm surges. Other infrastructure should at the very least be located above projected sea level rise (SLR) of 2050. Only fishing villages are recommended to be situated lower on the coast but still behind the 40 m buffer zone measured from high tide level.

Isolating the K14 probabilistic projection for sea level and moderate flooding by 2050 for our identified vulnerable fisher population, we were able to pinpoint the municipalities where fishers might need relocation assistance (Table 2.5). Fishers from Cotabato City (Region XII) and the province of Bulacan (Region III) are at highest risk since a significant portion of the coastal and low-land areas are shown to be annually flooded (Fig. 2.10). The other areas of concern are the coastal municipalities and small low-lying islands of Sulu and Tawi-Tawi (AARM), Camarines Sur (V), Negros Occidental and Iloilo (VI), Quezon (IV-A), and Pangasinan (I).

Table 2.5 Sea level rise (Kopp et al. 2014) and flooding projections using the Climate Central data analyzed for our a priori identified provinces

Region	Province	Sea Level Rise (SLR) + Moderate flooding	Municipalities affected
ARMM	Sulu	High	Small islands & Omar, Kalingalan Caluang
	Tawi-Tawi	High	Small islands & Bongao, Languyan
XII	Cotabato City	Very high	Entire municipality of Cotabato City
III	Bulacan	Very high	All coastal areas of Bulacan
VII	Bohol	Medium	Inabanga, Cadijay, Mahanay Island
	Cebu	Medium	Balamban, Cebu City, Mandaue City, Mactan
VIII	Eastern Samar	Medium	Guiuan, Salcedo, Gen. McArthur, Arteche, Oras, Dolores
V	Masbate	Medium	Aroroy, Milagros, Cawayan, Uson, Placer
	Camarines Sur	High	Cabusao, Canaman, San Jose
VI	Negros Occidental	High	Ilog, Pulupandan, Victorias, Manapla, Sagay
	Iloilo	High	Kalibo, Roxas, San Jose, Iloilo City, Dumangas, Concepcion, Numancia, New Washington
IV-A	Batangas	Medium	Nasugbu, San Juan, Batangas City, Balayan
	Quezon	High	Small islands, General Nakar, Infanta, Calauag, Macalelon
	Cavite	Medium	Cavite City, Noveleta, Naic
IV-B	Palawan	Medium	Taytay, Puerto Princesa City, Narra
IX	Zamboanga del Norte	Medium	Dapitan City, Dipolog City
	Zamboanga del Sur	Medium	Dinas, Pagadian City, Dimataling, Zamboanga City
I	Pangasinan	High	Bani, Lingayen, Dagupan, Mangaldan, San Fabian
II	Isabela	Medium	Palanan, Divilacan

2.4 Adaptation and Opportunities

2.4.1 Adaptation Options at the Local Scale

The above assessment shows that the risk of fishers is site specific. Those that live on the northern and eastern side will face intermittent and long-term changes in temperature affecting their coral reef-associated fisheries (e.g., snappers/"maya-maya") as well as pelagic ones (e.g., jacks/"talakitok"). Those living on the western side will

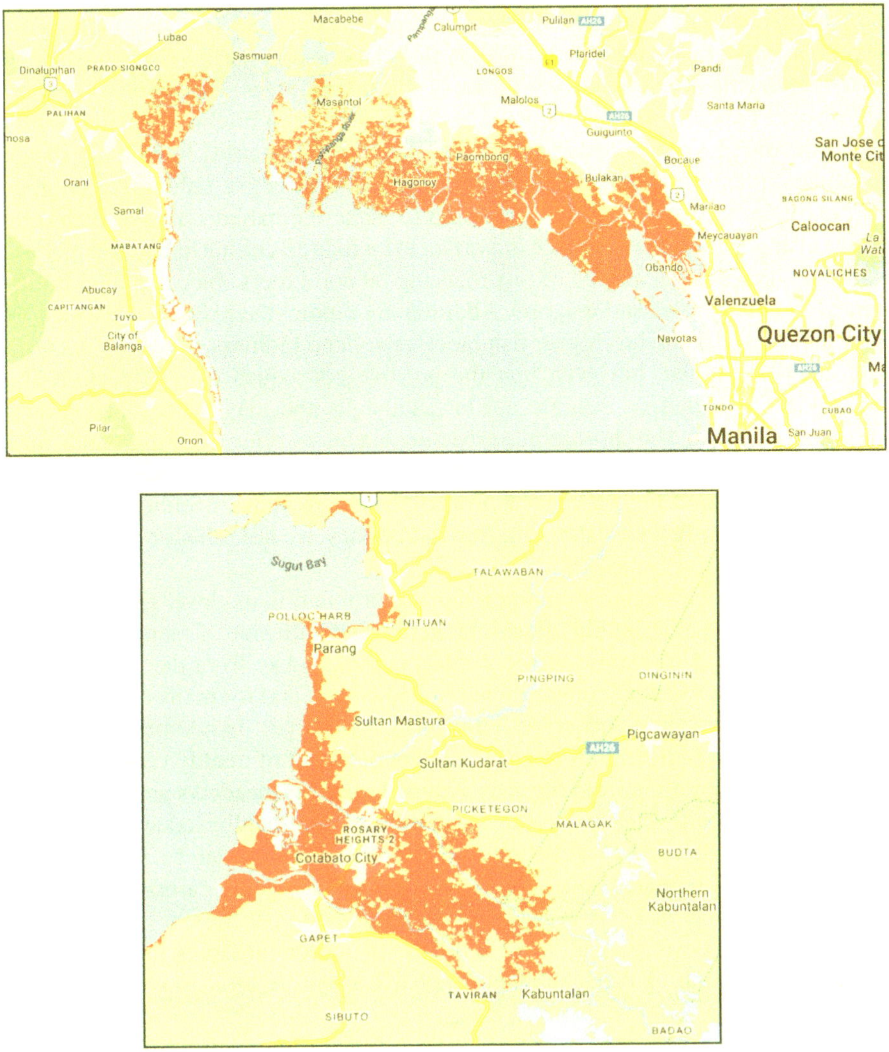

Fig. 2.10 Bulacan in the north and the city municipality of Cotabato in the south are shown to have the highest exposure to coastal inundation due to sea level rise and annual flooding projection for 2050. The simulation makes use of CoastalDEM, SRTM, JAXA and MERITDEM elevation datasets, for present-day sea level (Climate Central 2020—https://coastal.climatecentral.org/map/11/124.2445/7.2226/?theme=water_levelmap_type=water_level_above_mhhw)

experience extreme and long-term changes in rainfall. This will have impact on seagrass fisheries, like that of rabbit fish, or what is locally known as "danggit". Fishers from all over the country will have to deal with sea level rise (SLR) at rates twice to five times that of the mean global SLR. This will impact mangrove-fisheries (e.g., mangrove crab/"alimango"). Moreover, the high rate of SLR will include the risk of episodic or perennial inundation of their land and homes.

Adaptation hinges on the pragmatic use of coastal resources. This involves sustainable practices such as harvesting just enough fish to keep marine habitats healthy and productive. Leaving enough fish to keep the habitat healthy. Regulating illegal human activities around different habitats, such as destructive fishing or activities that contribute to water pollution, will help to make the habitats able to withstand climate change more. It will also be important to include all coastal habitats in making marine-protected areas instead of just focusing on coral reefs since our food is also sourced from mangrove and seagrass. Additionally, studies have shown that a healthy intact multi-habitat site has higher fish biodiversity and biomass.

It should be noted, however, that the positive consequence of protection and management of a marine area are not instantaneous and may in fact take several years to a decade. In the short term, restricting access to prior fishing grounds may initially result in decreased catch. Therefore, it is necessary to provide appropriate alternative livelihood for the fisher families. These alternative livelihoods can also have the positive effect on fisher families so that they are not reliant on fisheries as their sole source of income.

In addition, since the most common climate problem is sea level rise, there is a need to implement a no-build buffer zone on the landward side of mangrove forests to allow for natural migration of succeeding saplings away from the rising sea. A healthy mangrove forest not only provides a nursery ground for many economically important food fish, they also act as buffer for storm surges. In addition, mangrove forests act to hold sediment, decreasing erosion potential of muddy coasts and also decreasing sedimentation from the coasts toward seagrass meadows and coral reefs.

Healthy growth of seagrass and mangroves, also continually sequesters carbon. More than 40% of carbon sequestered in ocean sediments can be found within coastal habitats. Compared to tropical forests, the soil organic carbon of seagrass and mangroves contain 2.5–8 times more carbon dioxide per hectare (Fig. 2.11). If these habitats are destroyed, all these sequestered carbon dioxide will be released. Dedicated protection of these habitats is needed to ensure their role as long-term carbon sinks.

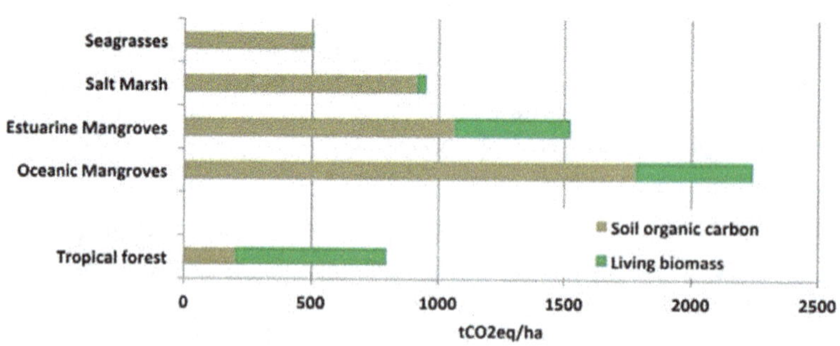

Fig. 2.11 Total carbon sequestered per hectare habitat (Murray et al. 2011)

2.4.2 National Opportunities

The protection of our coastal habitats is typically delegated to our Local Government Units (LGU) under the Fisheries Code (RA 8550). In addition, there is also a focus on mangroves and corals within the National Integrated Protected Areas System or NIPAS Act (RA 7586) under the Department of Environment and Natural Resources (DENR). Due to its productivity, its health is also of interest to the Department of Agriculture-Bureau of Fisheries and Aquatic Resources (DA-BFAR). We suggest that the health of mangroves and seagrasses should also be of interest to the Climate Change Commission (CCC) as the corresponding sequestered carbon can be included in our National Accounting (IPCC 2013 Supplement to the 2006 Guidelines for National Greenhouse Gas Inventories: Wetlands). Further, we suggest that the capacity of our coastal resources to sequester carbon should also be an interest to the Department of Trade and Industries (DTI).

Since the Philippines is a low carbon emitter country, producing less than 1% of global carbon (USAID 2016), any carbon we sequester or keep sequestered is actually more than what is needed to balance our production. It is possible to "sell" this positive balance to countries that are big producers but are otherwise not able to compensate this production with their mitigation measures. There have been several studies and attempts for Brazil trading its national debt in protection of the Amazon forests, its oxygen contribution to the planet, and all its biodiversity (e.g., US-Brazil debt for nature swap). Similarly, the Philippines can study doing a debt trade in the protection of its coastal habitats for its carbon sequestration and its marine biodiversity.

Finally, there is a need to take into consideration the future sea level rise scenarios in planning for future development of essential structures along coastal communities. Presently, the easement law from bodies of water prescribes "three (3) meters in urban areas, twenty (20) meters in agricultural areas and forty (40) meters in forest areas". Under the climate change lens, it is much more pragmatic to prescribe a height above sea level/projected sea level. Essential structures (e.g., hospitals, evacuation centers, schools) should be located higher than the highest simulated storm surge heights (i.e., 4–8 m above mean sea level). Coastal communities that will be impacted by the projected sea level rise should likewise be slowly relocated away from the coast. Only fisher communities should be allowed near the coast but should also be given technical assistance to adapt locally (e.g., houses on stilts, above-ground closed septic systems).

2.5 Conclusion

Healthy coral reefs, seagrass meadows, and mangrove forests provide food, livelihood, and protection. Filipinos are particularly reliant on these resources and are therefore vulnerable to climate impacts on these habitats. Either locally or with international assistance, the protection of our coastal resources and the early adaptation to highly probable future scenarios are necessary. Doing nothing will compromise our food security and place some coastal communities in harm's way. The resilience of the fisheries sector, the safety of our coastal communities, and the security of our food fish resources ultimately rely on this action.

References

Anderson J, Bausch C (2006) Climate change and natural disasters: scientific evidence of a possible relation between recent natural disasters and climate change—brief number 02a/2006, Policy brief for the EP environment committee, IP/A/ENVI/FWC/2005-35. European Parliament—Policy Department Economy and Science

Alongi DM (2008) Mangrove forests: resilience, protection from tsunamis, and responses to global climate change. Estuar Coast Shelf Sci 76(1):1–13. https://doi.org/10.1016/j.ecss.2007.08.024

BBC News (2010) US-Brazil debt for nature swap to protect forests, August 12, 2010. https://www.bbc.com/news/world-latin-america-10958695

BFAR (2019) Municipal Fisherfolk Registry (FishR) dataset of provincial level fisherfolk count as of May 6, 2019

Brito B (2017) Potential trajectories of the upcoming forest trading mechanism in Pará State, Brazilian Amazon. PLoS One 12(4):e0174154. https://doi.org/10.1371/journal.pone.0174154

Burnell OW, Connell SD, Irving AD, Watling JR, Russell BD (2014) Contemporary reliance on bicarbonate acquisition predicts increasedgrowth of seagrass Amphibolis antarctica in a high-CO_2 world. Conserv Physiol 2(1):cou052

Cabaco S, Santos R, Duarte CM (2008) The impact of sediment burial and erosion on seagrasses: A review. Estuarine, Coastal and Shelf Science 79(2008):354–366.

Carbon Brief (2020) Attributing extreme weather to climate change. https://www.carbonbrief.org/mapped-how-climate-change-affects-extreme-weather-around-the-world

Christensen JH, Kanikicharla KK, Aldrian E, An S-I, Cavalcanti IFA, de Castro M, Dong W, et al (2013) Climate phenomena and their relevance for future regional climate change. In: Stocker TF, Qin D, Plattner G-K, Tignor MMB, Allen SK, Boschung J, Nauels A, Xia Y, Bex V, Midgley PM (eds) Climate change 2013: the physical science basis. Contribution of working group I to the fifth assessment report of the intergovernmental panel on climate change. Cambridge University Press, Cambridge, United Kingdom and New York, NY, USA

Church JA, Clark PU, Cazenave A, Gregory JM, Jevrejeva S, Levermann A, Merrifield MA, et al (2013) Sea level change. In Stocker TF, Qin D, Plattner G-K, Tignor MMB, Allen SK, Boschung J, Nauels A, Xia Y, Bex V, Midgley PM (eds) Climate change 2013: the physical science basis. Contribution of working group I to the fifth assessment report of the intergovernmental panel on climate change. Cambridge University Press, Cambridge, United Kingdom and New York, NY, USA

Ciais P, Sabine C, Bala G, Bopp L, Brovkin V, Canadell J, Chhabra A, et al (2013) Carbon and other biogeochemical cycles. In: Stocker TF, Qin D, Plattner G-K, Tignor MMB, Allen SK, Boschung J, Nauels A, Xia Y, Bex V, Midgley PM (eds) Climate change 2013: the physical science basis. Contribution of working group I to the fifth assessment report of the intergovernmental panel on climate change. Cambridge University Press, Cambridge, United Kingdom and New York, NY, USA

Climate Central (2020) Land projected to be below annual flood level in 2050. https://sealevel.climatecentral.org/maps/

David LT, Rosario RB-D, Peñaflor EL, Cordero-Bailey K, Villanoy CL, Aliño PM (2015) Developing a Philippine climate-ocean typology as input to national vulnerability assessments. In: 36th Asian conference on remote sensing 2015 (ACRS 2015): fostering resilient growth in Asia, vol 1. Asian Association on Remote Sensing (AARS) and Curran Associates, Inc, pp 34–38

Department of Science and Technology-Food and Nutrition Research Institute (DOST-FNRI) (2015) Philippine dietary reference intakes 2015: summary tables.
Donelson JM, Munday PL, McCormick MI, Pankhurst NW, Pankhurst PM (2010) Effects of elevated water temperature and food availability on the reproductive performance of a coral reef fish. Marine Ecol Progress Ser 401:233–243. https://doi.org/10.3354/meps08366
Du X, Dong C, Sun SC (2018) Octopicola huanghaiensis n. sp. (Copepoda: Cyclopoida: Octopicolidae), a new parasitic copepod of the octopuses Amphioctopus fangsiao (d'Orbigny) and Octopus minor (Sasaki) (Octopoda: Octopodidae) in the Yellow Sea. Systematic Parasitology 95(8–9):905–912
Enrique-Navarro A, Huertas IE, León Cobo MJ, Prieto L (2021) Impact of ocean warming and ocean acidification on asexual reproduction and statolith formation of the symbiotic jellyfish Cotylorhiza tuberculata. PLoS ONE 16(8):e0254983
Field CB, Barros VR, Mach KJ, Mastrandrea MD, van Aalst MK, Neil Adger W, Arent DJ, et al (2014) Technical summary. In: Field CB, Barros VR, Jon Dokken D, Mastrandrea MD, Mach KJ, Eren Bilir T, Chatterjee M, et al (eds) Climate change 2014: impacts, adaptation, and vulnerability. Part A: global and sectoral aspects. contribution of working group II to the fifth assessment report of the intergovernmental panel on climate change. Cambridge University Press, Cambridge, United Kingdom and New York, NY, USA
Food and Agriculture Organization of the United Nations (FAO) (2000) The state of world fisheries and aquaculture 2000. Food and Agriculture Organization of the United Nations, Rome
Gilman E, Van Lavieren H, Ellison J, Jungblut V, Wilson L, Areki F, Brighouse G, et al (2006) Pacific Island mangroves in a changing climate and rising sea—UNEP regional seas reports and studies no 179. United Nations Environment Programme, Regional Seas Programme, Nairobi, Kenya
Grinsted A, Moore JC, Jevrejeva S (2013) Projected Atlantic hurricane surge threat from rising temperatures. Proc Natl Acad Sci 110(14):5369–5373. https://doi.org/10.1073/pnas.1209980110
Hilomen VV (2017) Effects of climate change on fisheries. Presented at the Visayas Regional Scientific Meeting (RSM), April 19
Hoegh-Guldberg O, Cai R, Poloczanska ES, Brewer PG, Sundby S, Hilmi K, Fabry VJ, Jung S (2014) The ocean. In: Barros VR, Field CB, Jon Dokken D, Mastrandrea MD, Mach KJ, Eren Bilir T, Chatterjee M, et al (eds) Climate change 2014: impacts, adaptation, and vulnerability. Part B: regional aspects. Contribution of working group II to the fifth assessment report of the intergovernmental panel on climate change. Cambridge University Press, Cambridge, United Kingdom and New York, NY, USA
IPCC (2013) Annex II: climate system scenario tables. Prather M, Flato G, Friedlingstein P, Jones C, Lamarque J-F, Liao H, Rasch P (eds). In: Stocker TF, Qin D, Plattner G-K, Tignor MMB, Allen SK, Boschung J, Nauels A, Xia Y, Bex V, Midgley PM (eds) Climate change 2013: the physical science basis. Contribution of working group I to the fifth assessment report of the intergovernmental panel on climate change. Cambridge University Press, Cambridge, United Kingdom and New York, NY, USA
IPCC (2014) 2013 supplement to the 2006 IPCC guidelines for national greenhouse gas inventories: Wetlands. In: Hiraishi T, Krug T, Tanabe K, Srivastava N, Jamsranjav B, Fukuda M, Troxler T (eds). IPCC, Switzerland
Johnson J, Bell J, Sen Gupta A (2016) Pacific Islands ocean acidification vulnerability assessment. SPREP, Apia, Samoa
Kirtman B, Power SB, Adedoyin AJ, Boer GJ, Bojariu R, Camilloni I, Doblas-Reyes F, et al (2013a) Near-term climate change: projections and predictability. In: Stocker TF, Qin D, Plattner G-K, Tignor MMB, Allen SK, Boschung J, Nauels A, Xia Y, Bex V, Midgley PM (eds) Climate change 2013: the physical science basis. Contribution of working group I to the fifth assessment report of the intergovernmental panel on climate change. Cambridge University Press, Cambridge, United Kingdom and New York, NY, USA

Kirtman B, Power SB, Adedoyin AJ, Boer GJ, Bojariu R, Camilloni I, Doblas-Reyes F, Fiore AM, et al (2013b) Chapter 11—near-term climate change: projections and predictability. In: Climate change 2013: the physical science basis. IPCC working group I contribution to AR5. IPCC. Cambridge University Press, Cambridge

Kopp RE, Horton RM, Little CM, Mitrovica JX, Oppenheimer M, Rasmussen DJ, Strauss BH, Tebaldi C (2014) Probabilistic 21st and 22nd century sea-level projections at a global network of tide-gauge sites. Earth's Fut 2(8):383–406. https://doi.org/10.1002/2014ef000239

Kulp SA, Strauss BH (2019) New elevation data triple estimates of global vulnerability to sea-level rise and coastal flooding. Nat Commun 10(1):1–12. https://doi.org/10.1038/s41467-019-12808-z

Lapidez JP, Tablazon J, Dasallas L, Anne Gonzalo L, Cabacaba KM, Ramos MMA, Kenneth Suarez J, Santiago J, Francisco Lagmay AM, Malano V (2015) Identification of storm surge vulnerable areas in the Philippines through the simulation of Typhoon Haiyan- induced storm surge levels over historical storm tracks. Nat Hazards Earth Syst Sci 15(7):1473–1481. https://doi.org/10.5194/nhess-15-1473-2015

Maberly SC, Stott A W and Gontero B (2022) The differential ability of two species of seagrass to use carbon dioxide and bicarbonate and their modelled response to rising concentrations of inorganic carbon. Front. Plant Sci. 13:936716

Mangora MM, Mtolera MSP, Björk M (2014) Photosynthetic responses to submergence in mangrove seedlings. Mar Freshw Res 65(6):497–504. https://doi.org/10.1071/mf13167

Murray BC, Pendleton L, Aaron Jenkins W, Sifleet S (2011) Green payments for blue carbon: economic incentives for protecting threatened coastal habitats. Duke Nicholas Institute for Environmental Policy Solutions

Nurse LA, McLean RF, Agard J, Briguglio LP, Duvat-Magnan V, Pelesikoti N, Tompkins E, Webb A (2014) Small islands. In: Barros VR, Field CB, Jon Dokken D, Mastrandrea MD, Mach KJ, Eren Bilir T, Chatterjee M, et al (eds) Climate change 2014: impacts, adaptation, and vulnerability. Part B: regional aspects. Contribution of working group II to the fifth assessment report of the intergovernmental panel on climate change. Cambridge University Press, Cambridge, United Kingdom and New York, NY, USA

Pankhurst NW, Munday PL (2011) Effects of climate change on fish reproduction and early life history stages. Mar Freshw Res 62(9):1015–1026. https://doi.org/10.1071/mf10269

Pedro MaRA, Barba CVC, Benavides-de Leon R (2007) Nutrition transition in the Philippines. Philipp Popul Rev 6(1). https://doi.org/10.3860/ppr.v6i1.264

Peñaflor EL, Skirving WJ, Strong AE, Heron SF, David LT (2009) Sea-surface temperature and thermal stress in the Coral Triangle over the past two decades. Coral Reefs 28(4):841–850. https://doi.org/10.1007/s00338-009-0522-8

Philippine Statistics Authority (PSA) (2017) Farmers, Fishermen and Children consistently posted the highest poverty incidence among basic sectors—PSA, June 30, 2017. https://psa.gov.ph/content/farmers-fishermen-and-children-consistently-posted-highest-poverty-incidence-among-basic

Rodolfo KS, Siringan FP (2006) Global sea-level rise is recognised, but flooding from anthropogenic land subsidence is ignored around northern Manila Bay, Philippines. Disasters 30(1), 118–139. https://doi.org/10.1111/j.1467-9523.2006.00310.x

The New York Times (1989) Brazil's debt can save the Amazon, February 3, 1989, sec. opinion. https://www.nytimes.com/1989/02/03/opinion/brazil-s-debt-can-save-the-amazon.html

Trenberth KE, Jones PD, Ambenje P, Bojariu R, Easterling D, Tank AK, Parker D, et al (2007) Observations: surface and atmospheric climate change. In: Solomon S, Qin D, Manning M, Marquis M, Averyt K, Tignor MMB, LeRoy Miller H, Jr, Chen Z (eds) Climate change 2007: the physical science basis. Contribution of working group I to the fourth assessment report of the intergovernmental panel on climate change. Cambridge University Press, Cambridge, United Kingdom and New York, NY, USA

USAID (2016) Greenhouse gas emissions in the Philippines

Wittkamper J (2020) Investing in Amazon rainforest conservation: a foreigner's perspective (commentary). Mongabay Environ News. https://news.mongabay.com/2020/04/investing-in-amazon-rainforest-conservation/

Laura T. David Ph.D. is a Professor & Director of the University of the Philippines Marine Science Institute. Her research includes using remote sensing and numerical modelling to explore how ocean physics influences the distribution and state of ocean flora and fauna. She is currently the Chair of the Commission on Higher Education Technical Committee in Marine Science, a member of the National Panel of Technical Experts of the Climate Change Commission, and a member of the International Scientific Steering Committee of GEO Blue Planet and Future Earth Coasts.

Rhodora V. Azanza Ph.D. is a Professor Emeritus of the University of the Philippines. She is also an Academician and current President of the National Academy of Science & Technology. She has extended her expertise to government agencies and international organizations, concurrently being the vice-chairman of UNESCO-Intergovernmental Oceanographic Commission (IOC) Panel on Harmful Algal Blooms (IPHAB) and leader of IOC's Harmful Algal Blooms in Southeast Asia (HABSEA) Portal and e-learning for 7 years.

Talna Lorena P. dela Cruz is a marine scientist and a science communicator specializing in mainstreaming marine biodiversity interconnectivity, disaster resilience, and climate change issues towards public understanding and engagement.

Roselle Borja-del Rosario is a former graduate student at the University of the Philippines Marine Science Institute. Her work involved using remote sensing and numerical modelling as tools for climate research and understanding the distribution of marine resources.

Hannah Eunice L. Amihan is a researcher at the University of the Philippines Marine Science Institute specializing in the use of Geographic Information Systems (GIS) to study and help protect the country's coastal and marine habitat and resources. She also engages in various science communication efforts, one of which was the conceptualization and design development of a biodiversity museum.

Ma. Simeona M. Martinez is an Assistant Professor of the of the Department of Geography of the College of Social Sciences and Philosophy, University of the Philippines Diliman. She has a master's degree in Remote Sensing and specializes in the application of the Geographic Information Systems (GIS) in different areas of research, such as indigenous peoples' rights, disaster risk management, resource management, social vulnerability, and countermapping.

Eileen L. Peñaflor Ph.D. is a Remote Sensing Scientist and has been involved in various researches utilizing remotely sensed data in the field of coastal and watershed management, fisheries, and climate variability.

Chapter 3
Sea Level Rise and Coastal Erosion in the Philippines: Impacts and Adaptation Strategies for Coastal Communities

Fernando P. Siringan and Ma. Yvainne Y. Sta. Maria

Abstract Being archipelagic, sea level rise is a cause of climate emergency in the Philippines. Since the country's coastlines are typically narrow, rising sea level can make the coastal plains even narrower, and thus pose a threat to more than 60 percent of the population living in the coastal zone. Even without the full impact of sea level rise, many coastal areas are already eroding primarily due to storms and human activities. This chapter presents an overview of sea level rise in the Philippines and its coastal impacts with particular focus on coastal erosion, and identifies adaptation strategies that can make the Philippine coasts more resilient to coastal erosion. From 1993 to 2009, the mean rate of sea level rise (SLR) in the Philippines at 6.2 mm/y is faster than the global mean of 3.3 mm/y. Locally, relative sea level rise (RSLR) may be amplified by subsidence caused by natural or anthropogenic factors. Higher sea level would increase the landward reach of waves and tides that can trigger coastal erosion or accelerate it in areas that are already eroding. Many coastal areas, including highly developed coasts of the country, are already undergoing erosion and, in many cases, the immediate cause is human activities that directly remove and/or alter the distribution of sediments along the coast. To make the coast more resilient to sea level rise, our policymakers must address the human activities that are contributing to coastal erosion.

Keywords Sea level rise · Coastal erosion · Adaptation strategies

Key Points

- Sea level rise is a consequence of global warming and the Philippines is one of the countries at high risk where sea level is rising at rates three to four times faster than global mean sea level of 3.3 mm/y.

F. P. Siringan (✉) · Ma. Y. Y. Sta. Maria
University of the Philippines Diliman, Quezon City, Philippines
e-mail: fpsiringan@msi.upd.edu.ph

K. B. Berse et al. (eds.), *Climate Emergency in the Philippines*, Disaster Risk Reduction, https://doi.org/10.1007/978-981-99-7804-5_3

- Higher sea level would increase the landward reach of waves and tides that can trigger coastal erosion, or the narrowing or loss of beaches due to the continuous loss of sediments.
- Policymakers must address the human activities that can trigger or worsen coastal erosion through strict enforcement of existing laws such as the easement guideline and a ban on beach and offshore mining.

Summary for Policymakers

Sea level rise is a consequence of global warming. Global sea level rise results mainly from the melting of polar ice sheets and glaciers, and thermal expansion of the ocean. On a local scale, sea level rise can be amplified by subsidence caused by natural (e.g., compaction of thick soft sediments, earthquake-related) or anthropogenic factors (e.g., groundwater extraction), leading to relative sea level rise. The Philippines is one of the countries at high risk from sea level rise, for the following reasons:

- The Philippines is located in the Western Tropical Pacific (WTP) where sea level is rising at rates three to four times faster than global mean sea level of 3.3 mm/y. One of the highest rates of sea level rise in the past two decades was recorded off the eastern coast of the Philippines, at 14.6 mm/y for the period 2002–2014. Around the Philippine archipelago, the rate of global warming-related sea level rise ranges between 3 and 9 mm/y, averaging 5–6 mm/y since 1993.
- Local factors such as subsidence have accelerated relative sea level rise in some parts of the country such as in the northern delta plains of Manila Bay. Rates of relative sea level rise typically between 2 and 4 cm/y in the Pampanga and Bulacan areas have been attributed mainly to groundwater extraction. This has worsened floods in these areas in terms of frequency (i.e., floods occur even with just high tide) and duration (i.e., floods lasts for several weeks to months).
- Being located in the Pacific typhoon belt, about 18–20 tropical cyclones hit the Philippines annually. Storms can temporarily raise sea levels, allowing wave run-ups and storm surges to penetrate further inland. As such, storms can cause abrupt and severe floods and beach erosion.

Higher sea level would increase the landward reach of waves and tides that can trigger coastal erosion, or the narrowing or loss of beaches due to the continuous loss of sediments. At present, many coastal areas in the country are already undergoing erosion due to various natural and human causes, even without considering the full impact of sea level rise. The loss of coastal lands will definitely speed up with the continuing rise in sea level.

As early as now, it is imperative to formulate adaptation strategies that can enhance the resilience of coastal communities to sea level rise and coastal erosion. Policymakers must address the human activities that can trigger or worsen coastal erosion. On both national and local levels, there is a need for a strict enforcement of existing laws that can protect against coastal land loss such as the easement guideline, and a ban on beach and offshore mining.

Another strategy is to modify or craft better laws and guidelines that can improve the management of the coastal zone, as follows:

- Consider the concepts of longshore currents and geomorphic compartments for evaluating coastal development projects in Environmental Impact Assessment (EIA) guidelines. This is to ensure consideration of the connectivity of coastal processes particularly sediment dynamics, which, if disrupted, almost always leads to coastal erosion as in the case of solid-based piers. In areas where solid-based piers exist, shoreline offset occurs as the downdrift side of the pier retreats landward (erode) while its updrift segment advances seaward (accrete) due to sediment trapping.
- Rethink the design and placement of piers and ports to minimize the impact of these structures. Where appropriate, building piers and ports on stilts should take precedence over solid-based structures to ensure the continuity of longshore transport. If a solid-based pier is absolutely necessary, one must carefully consider its placement in areas where there will be minimal disruption on longshore transport.
- Improve and revise the easement guideline in consideration of rising sea levels and ongoing erosion. The current guideline is based on a fixed and uniform distance for a particular land use. We propose the current guideline be set as the minimum easement to which the projected rates of sea level rise and coastal erosion will be added. This new guideline should be made mandatory for new coastal development.
- Consider adopting a rolling easement, which is a dynamic easement that moves with the vegetation line, and thus allows for the natural migration of the beach and coastal ecosystems such as mangrove forests. This will require coastal structures to be perched on stilts.
- Claim and set aside newly accreted lands such as river deltas and spits as greenbelts or parks that can buffer against waves and sea level rise.
- Prohibit the removal of beach vegetation and forest for newly developing lands as they provide the first line of coastal defense on open sandy coasts, where mangroves do not thrive. While there are laws that prohibit the removal of mangroves, there is no such law for beach vegetation and forest.

Most importantly, there is a need to raise the awareness of Filipinos especially coastal communities and policymakers about the ongoing threat of coastal erosion and how it can be exacerbated by sea level rise. This can be done through the use of print and social media, and engagement of teachers in educating the students about coastal processes and hazards with particular focus on coastal erosion and sea level rise.

3.1 Introduction

The Philippines is located in the Western Tropical Pacific (WTP), where sea level is rising at rates two to three times faster than global mean sea level (GMSL) (Nicholls and Cazenave 2010; Nerem et al. 2010; Merrifield 2011; Cazenave and Le Cozannet 2014; Hamlington et al. 2014). For 2002–2014, the highest rate of sea level rise (SLR) of 14.7 mm/y was observed off the eastern coast of the Philippines (Rietbroek et al. 2016). On a local scale, vertical land motion can amplify or dampen the magnitude and rate of global sea level rise, leading to relative sea level rise (RSLR). Subsidence, or lowering of ground level, driven by natural (e.g., earthquakes) or anthropogenic (e.g., excessive groundwater extraction) factors can lead to accelerated rates of RSLR. In the South Harbor in Manila, the tide gauge-based RSLR rate of 16.7 mm/y is faster than the satellite altimeter-derived SLR rate of 6 mm/y, which indicates subsidence (Fig. 3.1; Resilient Seas 2012). Since tide gauges are anchored on land, these instruments can also detect the direction and magnitude of vertical land movement relative to a fixed datum. In addition to SLR/RSLR, the country faces an average of 20 tropical cyclones yearly, which can temporarily raise sea levels by causing storm surges and high-wave run-ups. With higher sea levels, waves and tides can penetrate more landward, which, in turn, can trigger erosion in presently stable or accreting shorelines or accelerate it in areas that are already eroding.

At present, coastal erosion is prevalent in many places in the Philippines, including highly urbanized coasts (e.g., northern plains of Manila Bay, Davao City) and popular beaches (e.g., Boracay Island in Aklan, Baler in Aurora). As sea level rises, coastal erosion is expected to worsen in areas that are already eroding, and ensue in presently stable or accreting coastlines. Since many Philippine coastlines are typically narrow, the continuing rise in sea level can make the coastal plains even narrower, and thus pose a threat to more than 60 percent (60%) of the population (2005 estimates) living in the coastal zone (World Bank 2005). The loss of coastal lands can displace

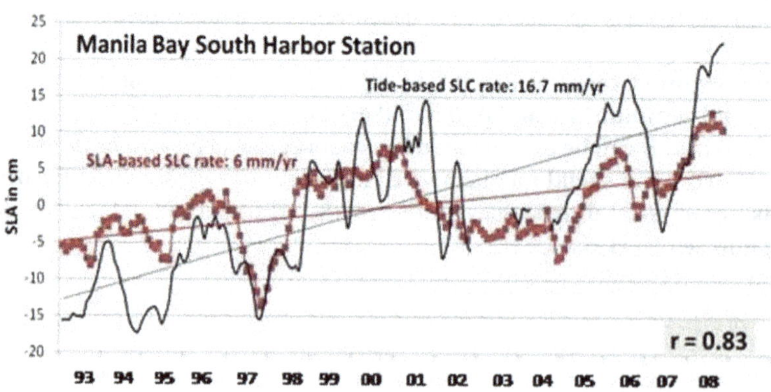

Fig. 3.1 Rates of SLR from tide gauge and satellite altimetry. Higher rate of SLR based on tide gauges may indicate subsidence in the Manila area. (*Source* Resilient SEAS 2012)

communities, endanger lives and properties, and disrupt economic activities along the coast. The loss of coastal land will also disrupt coastal ecosystems; for example, the nesting sites of sea turtles will likely vanish as beaches continue to narrow. Likewise, mangrove forests, if unable to migrate landward, will drown and eventually die off. Despite the prevalence of coastal erosion in the country, it is not yet recognized as a serious threat to coastal communities at the national level (Siringan et al. 2019). A case in point is the proliferation of activities that directly lead to coastal erosion such as beach mining (Siringan et al. 2005) despite the existence of a law prohibiting such activity (e.g., DAO 96-40 (IRR of the Philippine Mining Act of 1995 (RA 7942)). This chapter will present an overview of sea level rise and the impact of coastal erosion in the Philippines as well as identify adaptation strategies that can lessen the threat of SLR and coastal erosion.

3.2 Sea Level Rise and Its Coastal Impacts

3.2.1 Sea Level Rise in the Philippines

Global mean sea level (GMSL) has been rising since the 1900s, from rates of 1.4 mm/y for 1902–2015 to 3.6 mm/y for 2006–2015 (IPCC 2019). The same report has identified the melting of ice sheets and glaciers as the major contributor to the rapid SLR in the latter decade, accounting for ~1.8 mm/y, which exceeded the contribution of thermal expansion (~1.4 mm/y). Particularly vulnerable to SLR due to the melting of ice sheets and glaciers are regions very far away from the ice sheets, referred to as far-field region like the tropical oceans, which are expected to receive the influx of meltwater. In such regions, rates of SLR can be 30 percent (30%) faster than GMSL (Mitrovica et al. 2001, 2009 cited in Church et al. 2013). Terrestrial water storage also influences sea level rise, albeit minimally. It includes water impoundment in dams and artificial reservoirs, and groundwater depletion, which is associated, respectively, with sea level fall and sea level rise. For the period 1993–2010, total contribution of terrestrial water storage accounted for 0.38 mm/y of the total sea level budget (Church et al. 2013).

On a regional scale, sea level trends are highly variable and can be much faster than GMSL. One such region is the Western Tropical Pacific (WTP), where the Philippines is located and where rates of SLR were found to be three to four times faster than GMSL since 1993 (e.g., Nicholls and Cazenave 2010; Nerem et al. 2010; Merrifield 2011; Cazenave and Le Cozannet 2014; Hamlington et al. 2014). For the period 2002–2014, the highest rate of sea level rise of 14.7 mm/y was observed off the eastern coast of the Philippines (Rietbroek et al. 2016). In the Philippines, the rates of SLR averaged 6.2 mm/y for 1993–2009, about twice faster than GMSL of 3.3 mm/y (Nicholls and Cazenave 2010). The highest SLR rates were observed in the Pacific side (7–9 mm/y), and Celebes and Sulu seas (7–8 mm/y) followed by

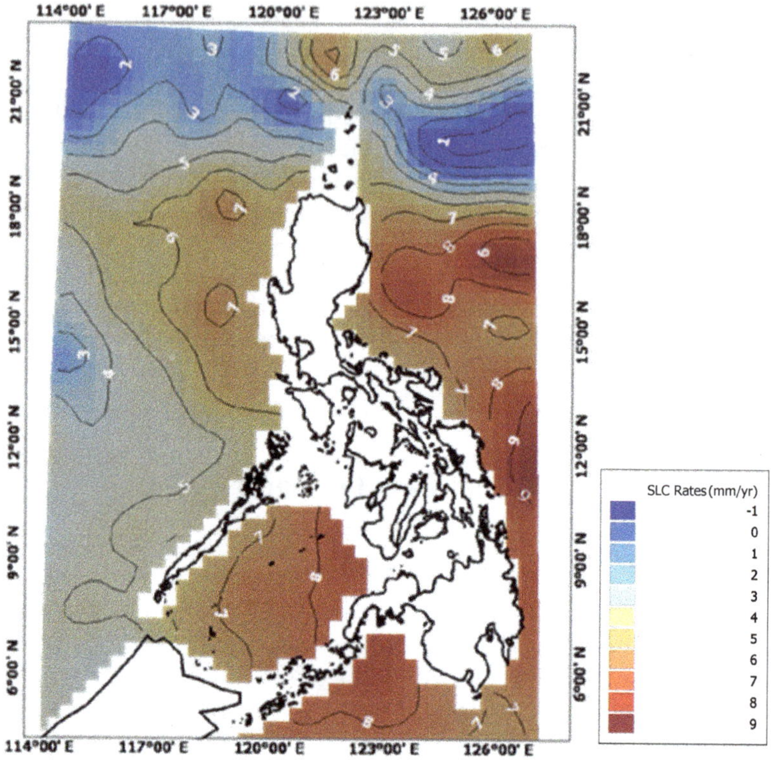

Fig. 3.2 Spatial patterns of altimeter-derived sea level rise in the Philippines: 1993–2009. (*Source* Resilient Seas 2012)

West Philippine Sea (WPS) (3–7 mm/y) (Fig. 3.2; Resilient SEAS 2012). Only the SLR rates in Luzon Strait are comparable to the GMSL.

In the Pacific, the distinct sea level pattern reveals regions of low-to-high sea levels from its eastern boundary (i.e., California) to its western boundary (i.e., Philippines). The east–west sea level pattern resembles that of the El Nino-Southern Oscillation (ENSO) (Meyssignac et al. 2012) and is largely influenced by thermosteric effects, or changes in sea level brought about by thermal expansion of the ocean (Cazenave and Le Cozannet 2014; Rietbroek et al. 2016). On an interannual scale, ENSO influences the variability of regional SLR in the WTP (including the Philippines): higher (*lower*) than normal sea levels are generally observed during La Nina (*El Nino*) (Prandi et al. 2009; Rong et al. 2009) (Fig. 3.3). Furthermore, a transition from El Nino to La Nina conditions (*La Nina to El Nino conditions*), which is associated with rising (*falling*) sea levels, may cause sea level to rise (*fall*) by as much as 45 cm in the tropical Pacific (Church et al. 2006). In the Philippines, the transition from El Nino to La Nina in 1997–1999 resulted in an abrupt rise in sea level of about 30 cm in the Pacific side (Resilient Seas 2012). Aside from more frequent La Nina

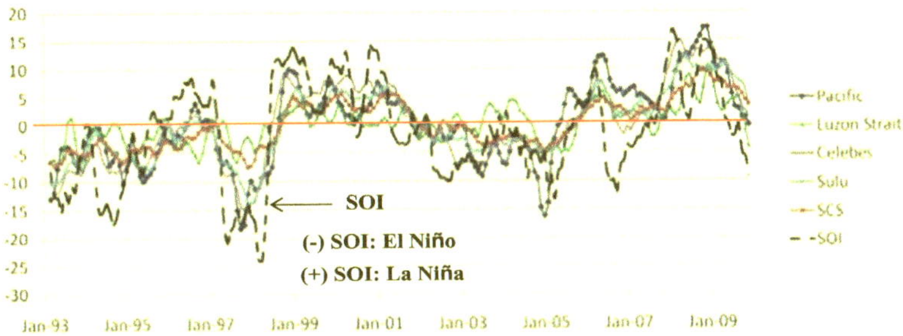

Fig. 3.3 Temporal trends of altimeter-derived sea level rise from 1993 to 2009 for the different basins in the Philippines. Superimposed is the Southern Oscillation Index (SOI), which is a proxy for ENSO: negative SOI indicates El Nino condition while positive values indicate La Nina condition. Sea levels are positively correlated with SOI: higher (lower) sea levels coincide with La Nina (El Nino) condition. (*Source* Resilient Seas 2012)

conditions, the acceleration of SLR in the WTP over the past two decades (1993–2012) has been attributed to the intensification of the Pacific trade winds, which is associated with the prevalence of the negative (cool) phase of the Interdecadal Pacific Oscillation (IPO) and Pacific Decadal Oscillation (PDO) (Merrifield 2011; Merrifield and Mathew 2011; Hamlington et al. 2014; England et al. 2014). Both PDO and IPO are Pacific climate cycles characterized by a shift from warm to cool mode every 20–30 years but with different geographic extent: IPO is a basin-wide climate mode while PDO occurs in North Pacific (Mantua and Hare 2002; Han et al. 2014; England et al. 2014; Easterbrook 2016). These climate modes exhibit similar behavior: during the positive phase of IPO/PDO, a warm tropical Pacific and weakened trade winds prevail, whereas its negative phase is characterized by a cool tropical Pacific and intensifying trade winds. The spatial pattern of PDO resembles that of ENSO but they have different time scales: PDO lasts for 20–30 years while ENSO occurs every 2–7 years (Mantua and Hare 2002; Deser and Kevin 2016).

On a local scale, SLR may be further amplified in areas where subsidence is occurring, leading to relative sea level rise (RSLR), such as in the northern delta plains of Manila Bay (Siringan et al. 2000; Rodolfo and Siringan 2006) (Fig. 3.1). In these sites, rates of RSLR due to subsidence averages a few centimeters per year (2–4 cm/y) (Siringan and Rodolfo 2003; Rodolfo and Siringan 2006; Resilient Seas 2012). Given an average RSLR rate of 4 cm/y, the full impact of a 1-m SLR will be felt in just 25 years, which is way faster than the IPCC projection of 0.84 m in 2100 under the worst-case scenario of continuing rise in emission (Representative Concentration Pathway (RCP) 8.5) (IPCC 2019). At present, the subsiding inland towns are already experiencing episodic flooding during normal high tide levels, even without rains, and flooding triggered by monsoon rains and storms takes months to recede (Rodolfo and Siringan 2006).

Owing to its geographic location, the Philippines can be considered as one of the hotspots of SLR for the following reasons:

1. The effects of steric changes, which account for 75 percent (75%) of SLR for the period 2002–2014, will continue to be felt for centuries even after GHG concentrations have stabilized owing to the large heat capacity of the ocean (Church and White 2011). It is likely that this scenario will also persist in the Philippines since WTP is largely affected by thermal expansion of the ocean.
2. The contribution of land-ice melt to global sea level has steadily increased and is about 1.8 mm/y for 2006–2015 due to the combined mass loss from Greenland and Antarctic ice sheets (IPCC 2019). Considering land-ice melt, projections of SLR by 2100 could reach 1–2 m (IPCC 2019; Dangendorf et al. 2019). Being located in the far-field region, the Philippines is expected to experience acceleration in sea level rise, about 30 percent (30%) higher than the global mean (Mitrovica et al. 2001, 2009 cited in Church et al. 2013), due to the influx of meltwater into the tropical oceans.
3. At a local scale, sea level rise has been amplified by subsidence caused by excessive groundwater extraction such as in the northern delta plains of Manila Bay (Siringan et al., 2001). It is likely that subsidence is also occurring on densely populated coastal plains that depend on groundwater resources. However, little is known about vertical land movement in other places due to limited number of tide gauges.
4. Superimposed on the above factors is the influence of climate variability on sea level patterns in the Philippines. During the negative phase of IPO/PDO and La Nina conditions, accelerated rates of SLR are expected due to strengthening trade winds. In contrast, slower rates of sea level rise or even sea level fall can be expected during IPO/PDO positive and El Nino conditions. The PDO has started shifting to a positive phase in 2014 (https://sealevel.jpl.nasa.gov/science/elninopdo/pdo/), and thus the Philippines is experiencing a slight lowering of sea level. Projections suggest that multidecadal climate variability would intensify the Pacific easterlies in the coming decades and from mid-2060s to mid-2080s that can accelerate SLR while weakening trade winds projected to occur in 2030–2050s may slow down SLR (Luo et al. 2012).

3.2.2 *Sea Level Rise and Coastal Erosion*

Coastal erosion is defined as the narrowing or disappearance of beaches due to the continuous loss of beach sediments (Siringan et al. 2019). Erosion ensues where there is sediment deficit to the coast caused by either diminishing supply from land and/or offshore sources, or enhanced removal of sediments by natural processes and human activities. Coastal erosion may be slow (i.e., less than 0.5 m/y) or rapid (i.e., >2 m/y), and mild (i.e., few tens of meters over several decades) or severe (i.e., several tens to hundred meters in a single storm event). It can happen instantly as in the case of storm-induced erosion. Erosion may also occur slowly over time as in SLR-driven erosion. Long-term erosion takes place when alternating episodes of erosion and accretion (land gain) result in net erosion through time, with the magnitude of beach loss ranging from several tens to hundreds of meters over several years or decades.

Sea level rise (SLR) is considered a primary driver of coastal erosion due to the worldwide prevalence of erosion even in sparsely populated and relatively undeveloped sandy coasts (e.g., Bruun 1962; IPCC 1990; Leatherman 1990; Leatherman et al. 2000; Zhang et al. 2004). SLR can trigger erosion by allowing waves and tides to break closer to shore, which enhances the removal of sediments farther up the beach (Leatherman 1990). With a rise in sea level, the beach is expected to move landward and upward proportional to the rate of SLR (Bruun 1962). In the eastern coast of USA, the magnitude of erosion was found to be 91–180 times the rate of SLR (Zhang et al. 2004), which is within the 50–200 times predicted by the Bruun rule (Bruun 1962). However, SLR will not always lead to coastal erosion since coastal response is also controlled by other factors such as the antecedent geology and geomorphology, availability of sediment supply, nearshore processes, and human interference (Morton 1979; SCOR WG 89 1991; Brunel and Sabatier 2009; Romine et al. 2013; Romine et al. 2016). In Hawaii, headland beaches located on the northern end are found to be more erosional than the embayed beaches since the northerly swells drive longshore transport to the south favoring sediment deposition on the adjacent embayed beaches (Romine et al. 2016). They concluded that accretion can be expected on the portion of the beach that is at the receiving end of the longshore drift. Along the French Mediterranean coast, the pocket beaches of Provence are found to be more vulnerable to SLR due to its relatively narrow beaches and the greater influence of SLR, which accounts for 60 percent (60%) of the shoreline retreat in Provence compared to eight percent (8%) in the open beaches of Camargue, which have gentler slopes and higher rates of SLR (Brunel and Sabatier 2009). Furthermore, they observed that rising sea levels worsened shoreline retreat where the beaches were already being eroded and reduced the advance of the shoreline on beaches that were accreting (e.g., mouth, spit, gulf). Similarly, along the eastern coast of the USA, prediction using Bayesian networks indicates that under a scenario of higher rates of SLR (>3.3 mm/y), there is a remarkable increase in the proportion of eroding areas and a subsequent decrease in the proportion of areas that are presently stable (Gutierrez et al. 2011).

3.3 Coastal Erosion in the Philippines

Coastal erosion is prevalent in the Philippines. Most of the coastal areas that were studied and visited by the Geological Oceanography Laboratory of the Marine Science Institute in UP Diliman over the past two decades are undergoing erosion (Siringan et al. 2013; Siringan et al. 2019). This includes some highly developed coasts (e.g., northern delta plains of Manila Bay, Davao City) and popular beaches (e.g., Boracay Island in Aklan, Baler in Aurora). The immediate causes of erosion vary from one area to another and consist of natural and/or human factors. Among the natural factors, the impact of storms was the most documented. Several storm events triggered abrupt shoreline retreat in the order of several tens of meters (e.g.,

20 m in Caba, La Union in Siringan and Jaraula 2001; 40 m in Botolan, Zambales) to a little over a hundred of meters in Tanauan, Leyte (Doctor 2018). Coastal erosion worsens during periods of successive storms by preventing beach recovery as in La Union and the Verde Island Passage (VIP) (e.g., Siringan et al., 2005; Siringan et al. 2008). In the VIP, where most coastlines were stable to accreting, erosion dominated during a period of more frequent storms (Siringan et al. 2008). Similarly, erosion observed in San Fernando Bay in La Union was attributed to successive storms in the 1990s that hindered beach recovery (Siringan et al. 2005). However, beach recovery can be expected in areas where sediment supply is sufficiently high and have low exposure to typhoon. In the case of eastern Leyte, coastal segments that were severely eroded during typhoon Yolanda underwent natural recovery due to the availability of sediments and non-occurrence of major storms thereafter (Doctor 2018). Another documented case is erosion due the switching of river mouth, which can be due to natural or man-made diversion. Shoreline retreat occurs in the vicinity of the formerly active river mouth as sediment supply is reduced in this area. In contrast, shoreline advance takes place in the vicinity of the presently active channel due to higher sediment discharge. Such is the case in the bayhead of Lingayen Gulf where the delta associated with the Agno-Labrador River retreated by as much as 1 km several years after the diversion of the active channel to the Agno-Lingayen River where a delta continues to form to this day (Mateo and Siringan 2007).

In many areas, the immediate causes of coastal erosion are human activities that directly alter the supply and distribution of sediments along the coast. By directly removing sediments from the beach and offshore waters, beach and offshore mining results in coastal erosion. Along the 77-km-long siliciclastic coasts of La Union, the volume of black (magnetite) sand extracted in the 1960–70 s, equivalent to two million cubic meters, can be translated to a shoreline retreat of 30 m with 1 m lowering of beach elevation (Siringan et al. 2005). The removal of black sand, which is rich in iron and other metals, and thus denser than gray or brown sands, also strips the beach of an armor, thereby exposing the remaining lighter sediments to wave erosion. Moreover, the process of excavating black sand loosens the naturally compacted and partially cemented sand layer underneath, making them easily erodible by waves. This explains why the beach remains prone to erosion long after mining activity has stopped. In Luna, La Union, picking of pebbles and gravel on the beach, a form of local livelihood, is responsible for 87 percent (87%) of the 1.5 m/y of shoreline retreat (Siringan et al. 2006). In the predominantly stable and accreting coasts in the Verde Island Passage (VIP), localized erosion was observed in the vicinity of hard coastal structures such as seawalls and piers (Siringan et al. 2009). Generally, chronic erosion is prevalent along coastlines fortified with seawalls and groins, and in the vicinity of solid-based piers. Seawalls promote scouring and offshore transport of sediments that eventually lead to the narrowing and loss of beaches in front of these structures (e.g., Fletcher et al. 1997; French 2001; Siringan et al. 2005). Seawalls constructed along Bulabog Beach in Boracay Island after the onslaught of typhoon Frank might have contributed to pervasive erosion in the area by preventing natural beach recovery (Bathy and Erode 2015). Ironically, groins, or perpendicular structures that promote sedimentation, can induce erosion elsewhere by trapping sediments efficiently. In

Agoo, La Union, a groin field consisting of more than 50 groins, while somewhat effective in promoting accretion, has triggered erosion on the coastal segment immediately south of the groin field by reducing sediment supply to the area (Siringan et al. 2005). Similar to groins, solid-based piers are also effective in trapping sediments, and thus can disrupt the longshore currents, which are shore-parallel wave-generated currents responsible for redistributing sediments continuously along a linear coastline (Fig. 3.4). Since solid-based piers trap sediment updrift of it, sediments reaching the downdrift segments are reduced, leading to erosion. Over time, there will be a visible shoreline offset at the opposite sides of the solid-based pier, manifested as the discrepancy in shoreline position between the accreting updrift coastal segments and the eroding downdrift segment (Siringan et al. 2019). In La Union, a 130-m-long solid-based pier built just south of Bauang River resulted in as much as 100-m shoreline offset between the coastal segment north and south of it (Fig. 3.5). Severe erosion in the area has caused the land area of barangay Pilar, located immediately south of the pier, to significantly decrease. Shoreline offset in the order of several tens to a little over one hundred meters has also been observed in Bantayan Island in Cebu, Kiamba in Sarangani and even in far-flung coastal areas where solid-based fish ports are common.

On carbonate or reef-fringed coasts, the magnitude of erosion is within several meters to a few tens of meters. In La Union, shoreline retreat along its carbonate coasts is an order of magnitude lower than that of siliciclastic shorelines (Siringan et al. 2019). Along Patar Beach in Bolinao, Pangasinan, coastlines fronted by a wide reef flat appear to be stable to accreting (Bathy and Erode 2015). Since carbonate coasts are typically narrow, erosion of a few meters particularly for tourist destinations like Boracay Island in Aklan and El Nido in Palawan has become a major concern (Siringan et al. 2019). Since purely carbonate coasts and islands are often supply limited with nearby coral reefs and associated habitats as the sole source of beach materials, maintaining and improving the health of these ecosystems will ensure continuous sediment supply to the adjacent beaches. Just as important, in a healthy reef, new growths where the waves break are able to maintain or even increase its

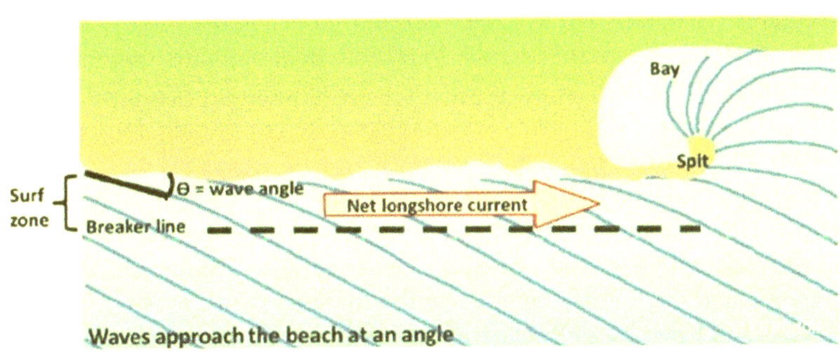

Fig. 3.4 Longshore currents. These are wave-generated currents that run parallel to the shoreline. They are responsible for transporting and redistributing voluminous amount of sediments along the coast. (Diagram courtesy of KS Rodolfo; *Source* Siringan et al. 2019)

Fig. 3.5 Shoreline offset at the vicinity of a solid-based pier south of Bauang River in Bauang, La Union. Interruption of longshore currents led to a 100-m wide shoreline offset. (Base map: Google Earth dated July 13, 2009)

capacity to dampen the waves before they approach the beach. In the Philippines, a lot of structures have been built on the reef and its beach. Erosion is enhanced by solid-based structures by modifying the longshore drift and directing the sediments beyond the edge of the reef. In areas where sedimentation is enhanced on the reef, the corals and associated organisms are smothered and eventually buried.

Although sea level rise is a major driver of coastal erosion (Leatherman 1990), only a few studies have looked at the effects of SLR on coastal erosion in the country. In Boracay Island, the higher magnitude of erosion for the period 2002–2010 was found to coincide with a period of faster rise in sea level (Bathy and Erode 2015). In Guimaras Island, SLR was found to be the main driver of erosion in the highly embayed coasts of Ravina and Poblacion; however, for the whole island, the destruction of mangroves emerged as the main contributor to coastal erosion (Aggangan 2015). Despite the presence of actual measurement of sea level rise in Sibutu Island in Tawi-Tawi, the contribution of extraction of beach materials to beach erosion cannot be discounted (Bathy and Erode 2015).

3.4 Coastal Adaptation Strategies

3.4.1 Address Human Activities that Cause Coastal Erosion

The previous section has identified several human activities that contributed to coastal erosion. A logical step is to regulate human activities that directly induce coastal erosion such as beach mining. This can be done by a strict enforcement of existing laws regulating such activities, or crafting new laws or guidelines that would be more responsive to the present condition (Siringan et al. 2019).

A. Enforce the existing laws

A number of existing laws, if strictly enforced, can already lessen the impact of anthropogenic drivers of coastal erosion by regulating or prohibiting activities that are detrimental to the coastal zone. Some of these laws are presented in Table 3.1.

Despite the existence of such laws, there are certain loopholes in its implementation. A case in point is beach mining. While there is a ban on the extraction of beach materials 200 m from the mean low tide along the beach (RA 7942), the same law provides the Provincial Governor the authority to issue permits for commercial and industrial sand and gravel permit (Section 46 and 47), exclusive sand and gravel permit (Section 48), and Government Gratuitous Permit (Section 49). It may explain the proliferation of extraction of black sand deposits along the coast of Cagayan despite the existence of a law prohibiting coastal mining. In the case of the easement, there is ambiguity in the high tide datum, which delineates the seaward limit of the easement, as it is not specified in the current guideline whether it refers to any high tide level, mean high tide level, or highest spring tide level or king tide. Within a year, the king tide represents the extreme high tide level, and thus should be the basis of the seaward limit of the easement since it can provide better buffering against extreme water levels. The delineation of the easement should be done prior to any development or construction to avoid the need for its ad hoc implementation. The Department of Environment and Natural Resources (DENR) took the lead in clearing the easement of Boracay and El Nido of establishments and properties encroaching on it. What remains unclear is how the easement guideline is being implemented in areas that are newly developing, and specifically which agency is tasked to physically delineate or mark the easement prior to development and declare it a "no-build" zone.

Table 3.1 Pertinent laws related to coastal zone management (*Source* Siringan et al. 2019)

Activity	Relevant law
Easement as a "no-build zone" for coastal zoning and development	"Structures of any kind cannot be built on the banks of rivers and streams and the shores of the seas and lakes throughout their entire length within a zone of three (3) meters in urban areas, twenty (20) meters in agricultural areas and forty (40) meters in forest areas. These zones are subject to the easement of public use only for recreation, navigation, floatation, fishing and salvage." (PD 1067 (Water Code), Article 51)
Beach mining	"No extraction, removal and/or disposition of materials shall likewise be allowed in offshore areas within five hundred (500) meters distance from the coast and two hundred (200) meters from the mean low tide level along the beach." (DAO 96-40 (Revised IRR of RA 7942)) "It shall likewise be unlawful for any person or corporation to gather, sell or export white sand, silica, pebbles and any other substances which make up any marine habitat." (RA 8550 (Fisheries Code), Section 92) Under DENR Administrative Order (DAO) 2010-21, the revised rules and regulations of RA 7942, the following were identified as closed areas for mining: (a) "Areas proclaimed as marine reserves/marine parks and tourist zones as defined by law and identified initial components of the National Integrated Protected Areas System (NIPAS) pursuant to R.A. No. 7586 and such areas expressly prohibited thereunder, as well as under Department Administrative Order No. 25, Series of 1992, and other laws"; (b) "In case of seabed/marine aggregate quarrying, offshore areas less than 1,500 m from the mean low tide level of land or island(s) and where the seabed depth is less than 30 m measured at mean sea level."
Mangrove protection	Presidential Decree 705 (Revised Forestry Code of the Philippines), that prescribes the "retention of a 20-m wide mangrove strip to protect coastal areas from strong winds and typhoons." At least twenty-five (25%) but not more than forty (40%) per cent of bays, foreshore lands, continental shelf of any fishing ground shall be set aside for the cultivation of mangroves to strengthen the habitat and spawning grounds of fish. (Section 81)" (RA 8550 (Fisheries Code), Section 81)
Protection of coral reef and seagrass beds	It shall be unlawful for any person, natural or juridical, to fish with gear method that destroys coral reefs, seagrass beds, and other fishery marine life habitat as may be determined by the Department. (RA 8550 (Fisheries Code), Section 92)

B. Revise and craft new laws or guidelines

Another strategy is to modify or craft better laws and guidelines to curb unregulated human activities in the coastal zone. Based on our two decades of work on coastal erosion, we have identified the following gaps in the current guidelines and laws.

Include the concept of longshore currents and geomorphic compartments in the EIA guideline for assessing coastal development projects. Assessing the impact of coastal infrastructures and other development projects typically requires information on water circulation and water quality. However, its impact on coastal sediment dynamics is often ignored. A crucial element of coastal sediment dynamics is longshore currents, which transport and redistribute voluminous amounts of sediments to different coastal segments that may run uninterrupted within a geomorphological compartment (Fig. 3.4). A geomorphological compartment may consist of either an embayment or linear coasts bounded by cliffs and/or rivers (Fig. 3.6). The interruption of the longshore transport of sediments or longshore drift within geomorphological compartments is the main cause of coastal erosion in many places as in the case of solid-based pier (see Fig. 3.5). It is therefore necessary to consider the impact of coastal projects on the continuity of longshore transport to assess the degree of its potential impact on coastal sediment dynamics as well as determine its area of influence within a geomorphological compartment. More importantly, the use of geomorphological compartments instead of political boundaries as spatial scale when conducting assessment and/or planning for coastal protection is strongly recommended in order to ensure the continuity of coastal processes such as longshore currents (SagipBaybayPH 2018).

Improve the placement and design of piers and ports. One of the coastal infrastructures that interrupt the longshore currents are solid-based piers and ports. In our archipelagic country, piers and ports are necessary for moving people and products across the different islands and as such are widely distributed all over the country (Fig. 3.7). There are various designs of ports and piers such as port facilities that are embayed (Fig. 3.7b) or reclaimed, and solitary piers and series of piers (Fig. 3.7c and d). Greatest shoreline offset has been observed in the vicinity of solitary piers with solid bases especially if they are placed close to the sediment source as in the case of the BPPC pier Bauang, La Union (see Fig. 3.5). To arrest erosion, the downdrift side of solitary solid-based piers is fortified with seawalls as in Maconacon, Isabela (Fig. 3.7c). Port facilities may also disrupt longshore transport if they are built on reclaimed lands and encroach on the foreshore or nearshore waters. Even embayed port facilities may interrupt coastal sediment dynamics if placed along the direct path of longshore transport (Fig. 3.7b). A series of solid-based piers line the coast of other areas. In Samal Island, the series of groins also serves as piers, which extend all the way to the edge of the reef flat (Fig. 3.7d). As such, this series of groins/piers disrupt sediment transport along the coast and even promote transport of sediments to deeper waters, where these sediments remain and become lost to the coastal sediment budget (Siringan et al. 2019). The above examples clearly demonstrate that solid-based piers and ports are major contributors to erosion.

To minimize the impacts of solid-based piers and ports on the adjacent coastlines, there is a need to rethink the design and placement of ports and piers. Where appropriate, building piers and ports on stilts should take precedence over solid-based structures to ensure continuity of longshore transport especially in places close to the sediment source or along linear coastline. However, if a solid-based design is absolutely necessary, one must carefully consider the placement of the pier or port where there will be minimal disruption on the longshore drift. An example would be close to the cliff end of an embayment. However, care should be taken not to place these structures on or near living reefs and other coastal habitats so as to prevent heavy sedimentation that can smother and eventually decimate them. A hybrid of structure on stilt close to land and solid-based beyond the surf zone can be considered.

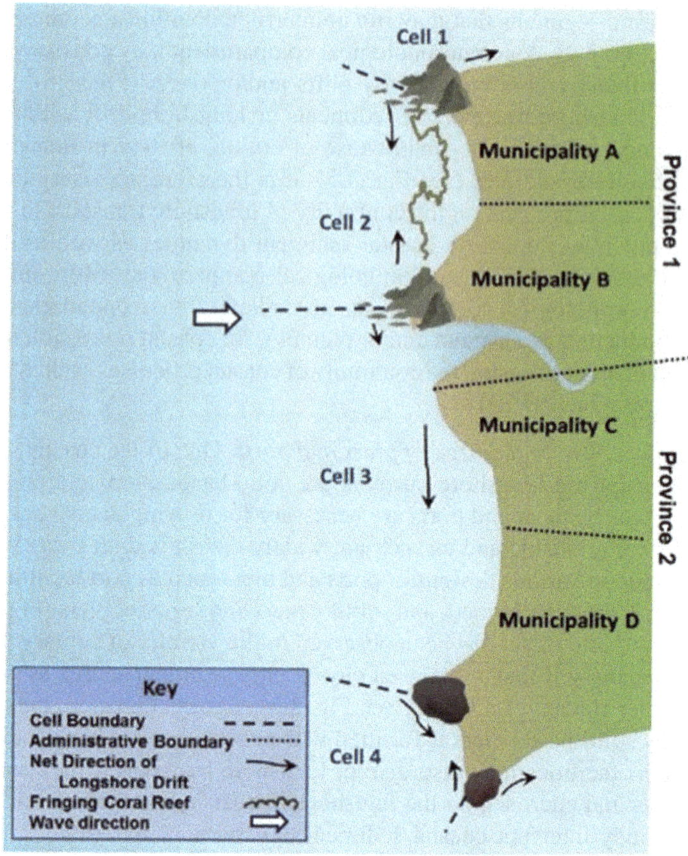

Fig. 3.6 Geomorphological compartment as spatial scale of impact assessment on coastal sediment dynamics. A geomorphological compartment may consist of an embayment bounded by cliffs or a linear coastline bounded by rivers and/ or cliffs on either end. Within a geomorphological boundary, longshore currents are responsible for the redistribution of sediments within the same littoral cell, resulting in longshore drift. Notice that the longshore drift crosses political boundaries and adjacent municipalities may share the same littoral cell. Such cases warrant cooperation and coordination among contiguous political units to effectively manage a shared coastline. (*Source* Siringan et al. 2019)

These recommendations should be considered by the agencies mandated to build and operate piers and port such as the Philippine Ports Authority (PPA) for commercial, industrial, and tourism purposes, and Philippine Fishery Development Authority (PFDA) for fisheries. Most of the piers mapped on Google Earth (Fig. 3.7) appear to have solid foundation as shoreline offset is visible even in remote areas. Many of these piers are short and cater to municipalities, and yet still capable of interrupting longshore transport, and hence can induce erosion over time. The Local Government Units (LGUs) should also be informed of the adverse impact of even short solid-based piers to minimize erosion on the adjacent coastal segment.

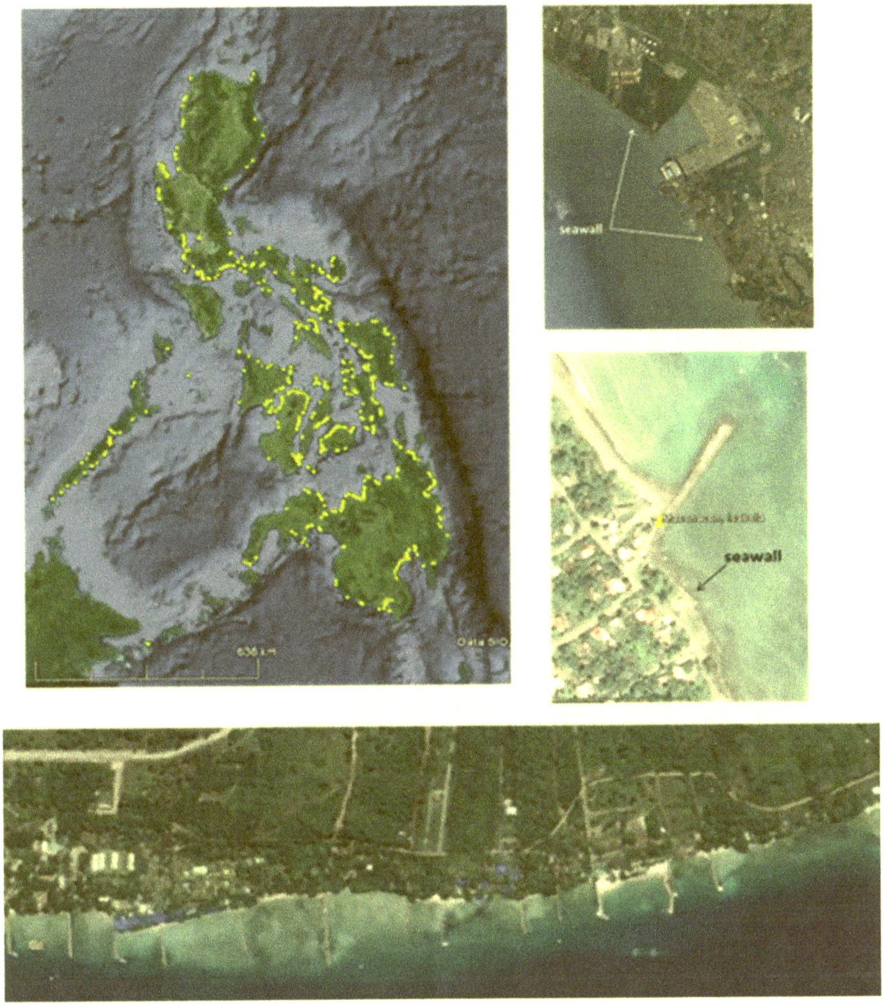

Fig. 3.7 Location of piers and ports in 2015 using Google Earth (**a**). Examples are embayed port in Batangas (**b**), solitary pier with downdrift side fortified with seawalls in Maconacon, Isabela (**c**) and series of piers/groins in Samal Island built all the way to the edge of the reef flat (**d**). *Source of images* Google Earth

Revise the current fixed easement guideline. The current easement guideline in the Water Code of the Philippines (PD 1067) is based on a fixed and uniform distance depending on land classification (i.e., 3 m in urban, 20 m in agricultural, and 40 m in forested areas). With the growing threat of SLR, there is a need to revise the current easement guideline to better buffer against SLR and coastal erosion. There is a guideline in the latest Comprehensive Land Use Plan (CLUP) Guidebook (HLURB 2013) in which the LGUs may declare an expanded easement for high-risk area based on historical trends and projections of SLR and coastal erosion. Since this is just a guideline, the inclusion of expanded easement in the CLUP is optional; a law is needed to make it mandatory. In the revision, the current easement in PD 1067 should be set as the minimum easement to which the projected rates of SLR and coastal erosion will be added (SagipBaybayPH 2018).

A rolling easement should also be considered instead of the current fixed-distance easement. A rolling easement is a dynamic public easement that moves with the vegetation line and prohibits hard structures within its limits (Titus 1998; 2011). Structures must be perched on stilts to allow the natural migration of the beach and associated coastal ecosystems. Beaches are expected to naturally migrate landward and upward with rising sea levels.

Lastly, there is also a need to revise the provision on Article 4 of the Spanish Law of Waters regarding shoreline accretion, which states that "lands added to the shores by accretion and alluvial deposits caused by the action of the sea form part of the public domain." However, the same law also allows the government to declare such new lands to be the property of the owners of the estates adjacent to it once these new lands have stabilized. With rising sea levels, the newly accreted lands can serve as buffer against extreme water levels. Instead of declaring these new lands private property, the government should assert the public domain nature of these lands and convert them into greenbelts and parks that can further buffer against waves and sea level rise.

Ban the removal of natural vegetation on the beach and within the easement. On open sandy coasts, the first line of coastal defense is the natural beach forest and vegetation. Beach vegetation, particularly the creeping variety, such as *Ipomoea* sp., is effective in stabilizing beach sediments, making them less prone to wave erosion. Unfortunately, beach forest and vegetation are considered nuisance, and thus are removed to clear the beach for easy access and to improve its aesthetics.

While there is a law that promotes "the retention of a 20-m wide mangrove strip to protect coastal areas from strong winds and typhoon" (PD 705), there is no such law pertaining to the conservation of the natural beach forest and vegetation. Owing to its role for the protection of open sandy coasts, where mangroves do not thrive, there is a need for a law or ordinance that would regulate or ban the removal of beach vegetation. Along with this, coastal developers should be encouraged to consider an elevated boardwalk, instead of removing the beach vegetation, in providing access to the beach.

3.4.2 Hard Engineering Solutions Versus Ecosystem-based Adaptation (EbA)

Another popular coastal adaptation strategy is the protection of coastal communities with hard engineering structures such as seawalls and groins. While effective in protecting properties, seawalls are known to enhance erosional processes, causing beaches fronting it to narrow and eventually disappear (Fletcher et al. 1997; French 2001; Siringan et al. 2005). Perpendicular structures such as groins can successfully trap coastal sediments, but induce erosion elsewhere, similar to the effects of solid-based piers.

In the Philippines, where a large portion of the coastline is lined with coral reefs and associated ecosystems, these coastal habitats can dampen the impact of incoming waves, and thus provide protection for the nearby coastal communities. Coral reefs, seagrasses, and mangroves also supply sediments to the adjacent coast, thereby contributing to beach stabilization. Along sandy open coasts, beach vegetation and forest can help stabilize beach sediments against wave action. Where these coastal habitats are present, protecting and rehabilitating them is a "win–win" situation not only for coastal protection but also for the numerous ecosystem services derived from them. For this reason, the conservation and rehabilitation of the coastal habitats should take precedence over hard protection strategies. Hard coastal structures should only be tapped as a last resort and in conjunction with other strategies that can potentially lessen their environmental impacts.

3.4.3 Engage Stakeholders and Educators in Citizen Science

Despite the growing threat of coastal erosion with the continuing rise in sea level, this problem either remains unknown or largely ignored. Since many Filipinos are living in the coastal zone, they must be informed about the threat of sea level rise and coastal erosion and how to deal with this problem. Specifically, they must be made aware of how certain human activities can trigger or worsen coastal erosion. More importantly, this issue must reach our government and policymakers, which has continued to promote reclamation projects despite the inevitability of sea level rise.

With DOST funding, our previous project embarked on a communication, education and public awareness (CEPA) campaign through the production of print materials (e.g., guidebook, brochures, posters, policy brief, modular primers), creation of a webpage (www.sagipbaybayph.com) and social media presence, and engagement of tertiary educators in promoting coastal erosion awareness. The first two activities were meant to raise the awareness of the general public and those in the national and local government about coastal erosion. It is hoped that the general public, especially the stakeholders can participate in citizen science, specifically in reporting where erosion is occurring. The third activity sought to educate the students about coastal processes and hazards with particular focus on coastal erosion by providing teaching materials (i.e., lectures in PowerPoint, field exercises, field and data processing manuals) related to coastal processes and hazards. These teaching materials can be integrated in their oceanography, geography, and earth science classes. These materials were rolled out through training workshops of tertiary teachers hailing from selected SUCs all over the country. To ensure the continuation of this CEPA campaign beyond project duration, electronic files of the print (posters, brochures, and modular primers) and teaching materials were made available for download via the sagipbaybay website. Since there is no more fund for maintaining the webpage, we are exploring other avenues to make these materials available for everyone. For the past 3 years, training workshops for tertiary teachers on coastal erosion management were conducted with funding support from DOST. Such training workshops can still be continued in order to reach out to more SUCs and transfer these teaching materials to them, if and when funding becomes available.

Another important consideration for coastal adaptation is the identification of eroding areas, the rate/magnitude, and the immediate causes of erosion, which can inform strategies for intervention and/or adaptation. There is a need for a Philippine-wide delineation of shoreline position changes to be able to have an inventory of eroding coastal segments using uniform methods and appropriate scale. In this respect, the Mines and Geoscience Bureau (MGB) of the DENR is the agency tasked with coastal erosion mapping in the country, and thus can take the lead in this activity. Another data gap concerns relative sea level rise, which can be an order of magnitude greater than global or regional sea level rise. Little is known about vertical land movement in other coastal areas due to limited number of tide gauges. Increasing the number of tide gauges can provide better estimates of the contribution of vertical land movement to sea level rise. Alternatively, a nationwide assessment of ground movement specifically subsidence using interferometry is also recommended. With this information, estimates of sea level rise can be further refined. To keep up with sea level rise, more people are needed to document and investigate coastal changes in different areas in the country.

3.5 Conclusion

Sea level rise is expected to accelerate as ice sheets and glaciers continue to melt due to global warming, which will worsen marine flooding, saltwater intrusion, loss of coastal habitats and coastal erosion on low-lying coastal areas, and island systems. In the Philippines, coastal erosion is already taking place in many areas. It is expected to worsen with the continuing rise in sea level—the typically narrow coastlines will become even narrower or may even disappear—affecting millions of Filipinos living on the coastal zone. Problems arising from this will only intensify due to the growing number of Filipinos occupying the coastal zone. It is thus imperative that adaptation measures be put in place to increase the resilience of coastal communities to sea level rise. Above all, shifting the focus of development from coastal to inland areas should accompany coastal adaptation efforts.

References

Aggangan JS (2015) Historical shoreline changes and possible causes of erosion in the island system of Guimaras, Philippines. A thesis submitted for the Degree of Master of Science in Marine Science

Bathy and Erode (2015) Acquisition of detailed bathymetry for coastal erosion management. A terminal report submitted to the Department of Science and Technology, Philippine Council for Agriculture, Aquatic and Natural Resources Research and Development

Brunel C, Sabatier F (2009) Potential influence of sea-level rise in controlling shoreline position on the French Mediterranean Coast. Geomorphology 107(1–2):47–57

Bruun P (1962) Sea-level rise as a cause of shore erosion. J Waterw Harb Div 88(1):117–132

Cazenave A, Cozannet GL (2014) Sea level rise and its coastal impacts. Earth's Future 2(2):15–34

Church JA, White NJ, Hunter JR (2006) Sea-level rise at tropical Pacific and Indian Ocean Islands. Glob Planet Change 53(3):155–168

Church JA, White NJ (2011) Sea-level rise from the late 19th to the early 21st century. Surv Geophys 32(4–5):585–602

Church JA, Clark PU, Cazenave A, Gregory JM, Jevjereva S, Leverman A, Merrifield MA et al (2013) Sea level change. PM Cambridge University Press

Dangendorf S, Hay C, Calafat FM, Marcos M, Piecuch CG, Berk K, Jensen J (2019) Persistent acceleration in global sea-level rise since the 1960s. Nat Clim Chang 9(9):705–710

DENR Administrative Order No. 2010-21 (2010) Providing for a consolidated department of environment and natural resources administrative order for the implementing rules and regulations of Republic Act No. 7942, otherwise known as the Philippine Mining Act of 1995

DENR Administrative Order 96-40 (1996) Implementing rules and regulations of RA 7942 (The Philippine Mining Act of 1995)

Deser C, Trenberth K, National Center for Atmospheric Research Staff (eds) (2016) The climate data guide: Pacific Decadal Osciallation (PDO): Definition and Indices. http://climatedataguide.ucar.edu/climate-data/pacific-decadal-oscillation-pdo-definition-and-indices. Last Accessed 06 Jan 2016

Doctor MA (2018) Shoreline response of Eastern Leyte to single extreme Super Typhoon Yolanda and to long term forcing mechanism. MS thesis. University of the Philippines, Diliman, Quezon City

Easterbrook D (ed) (2016) Evidence-based climate science: data opposing CO_2 emissions as the primary source of global warming. Elsevier

England MH, McGregor S, Spence P, Meehl GA, Timmerman A, Cai W, Gupta AS, McPhaden MJ, Purich A, Santoso A (2014) Recent intensification of wind-driven circulation in the Pacific and the ongoing warming hiatus. Nat Clim Chang 4(3):222

Fletcher CH, Mullane RA, Richmond BM (1997) Beach loss along armored shorelines on Oahu, Hawaiian Islands. J Coast Res 209–215

French P (2001) Coastal defences: processes, problems and solutions. Routledge, London, p 366

Gutierrez BT, Plant NG, Thieler ER (2011) A Bayesian network to predict coastal vulnerability to sea level rise. J Geophys Res Earth Surf 116(F2)

Hamlington BD, Strassburg MW, Leben RR, Han W, Nerem RS, Kim KY (2014) Uncovering an anthropogenic sea-level rise signal in the Pacific Ocean. Nat Clim Chang 4(9):782–785

Han W, Meehl GA, Hu A, Alexander MA, Yamagata T, Yuan D, Ishii M et al (2014) Intensification of decadal and multi-decadal sea level variability in the western tropical Pacific during recent decades. Clim Dyn 43(5–6):1357–1379

Housing and Land Use Regulatory Board (2013) CLUP guidebook: a guide to comprehensive land use plan preparation volume I: the planning process, p 247

IPCC (Intergovernmental Panel on Climate Change) (1990) Climate change: the IPCC scientific assessment. Report prepared by Working Group II

IPCC (2019) Summary for policymakers. In: Pörtner H-O, Roberts DC, Masson-Delmotte V, Zhai P, Tignor M, Poloczanska E, Mintenbeck K, Nicolai M, Okem A, Petzold J, Rama B, Weyer N (eds) IPCC special report on the ocean and cryosphere in a changing climate. (In press)
Leatherman SP (1990) Modelling shore response to sea-level rise on sedimentary coasts. Progress in physical geography, 14(4):447–464
Leatherman SP, Zhang K, Douglas BC (2000) Sea level rise shown to drive coastal erosion. EOS Trans Am Geophys Union 81(6):55–57
Luo J-J, Sasaki W, Masumoto Y (2012) Indian Ocean warming modulates Pacific climate change. Proc Natl Acad Sci 109(46):18701–18706
Mantua NJ, Hare SR (2002) The Pacific decadal oscillation. J Oceanogr 58(1):35–44
Mateo ZRP, Siringan FP (2007) Tectonic control of high-frequency Holocene delta switching and fluvial migration in Lingayen Gulf Bayhead, northwestern Philippines. J Coast Res 182–194
Merrifield MA, Mathew EM (2011) Regional sea level trends due to a Pacific trade wind intensification. Geophys Res Lett 38(21)
Merrifield MA (2011) A shift in western tropical Pacific sea level trends during the 1990s. J Clim 24(15):4126–4138
Meyssignac B, Salas y Melia D, Becker M, Llovel W, Cazenave A (2012) Tropical Pacific spatial trend patterns in observed sea level: internal variability and/or anthropogenic signature? Clim Past 8(2):787–802
Morton RA (1979) Temporal and spatial variations in shoreline changes and their implications, examples from the Texas Gulf Coast. J Sediment Res 49(4):1101–1111
Nerem RS, Chambers DP, Choe C, Mitchum GT (2010) Estimating mean sea level change from the TOPEX and Jason altimeter missions. Mar Geod 33(S1):435–446
Nicholls RJ, Cazenave A (2010) Sea-level rise and its impact on coastal zones. Science 328(5985):1517–1520
Prandi P, Cazenave A, Becker M (2009) Is coastal mean sea level rising faster than the global mean? A comparison between tide gauges and satellite altimetry over 1993–2007. Geophys Res Lett 36(5)
Presidential Decree No. 1067 (1976) The water code of the Philippines
Presidential Decree No. 705 (1975) Revised forestry code of the Philippines.
Republic Act 8550 (1998) The Philippine fisheries code
Resilient Seas (2012) Philippine-wide assessment of vulnerability to sea-level rise. In: Year 3 terminal report for project 2: retrospective analysis of climate change from coastal erosion trends and uplifted coral reef assemblages. Remote Sensing Information for Living Environments and Nationwide Tools for Sentinel Ecosystems in our Archipelagic Seas Program, Department of Science and Technology, Philippine Council for Agriculture, Aquatic and Natural Resources Research and Development
Rietbroek R, Brunnabend S-E, Kusche J, Schröter J, Dahle C (2016) Revisiting the contemporary sea-level budget on global and regional scales. Proc Natl Acad Sci 113(6):1504–1509
Rodolfo KS, Siringan FP (2005) Global sea-level rise is recognised, but flooding from anthropogenic land subsidence is ignored around northern Manila Bay Philippines. Disasters 30(1):118–139
Rodolfo KS, Siringan FP (2006) Global sea-level rise is recognised, but flooding from anthropogenic land subsidence is ignored around northern Manila Bay, Philippines. Disasters 30(1):118–139
Romine BM, Fletcher CH, Barbee MM, Anderson TR, Frazer LN (2013) Are beach erosion rates and sea- level rise related in Hawaii? Glob Planet Change 108:149–157
Romine BM, Fletcher CH, Frazer LN, Anderson TR (2016) Beach erosion under rising sea level modulated by coastal geomorphology and sediment availability on carbonate reef fringed island coasts. Sedimentology 63(5):1321–1332
Rong Z, Liu Y, Zong H, Xiu P (2009) Long term sea level change and water mass balance in the South China Sea. J Ocean Univ China 8(4):327
SagipBaybay PH (2018) Policy paths for addressing coastal erosion in the Philippines: a policy brief. Prepared for the DOST-PCAARRD project. CEPA (Communications, Education and

Policy Awareness) and Policy Review Towards Improving Coastal Erosion Management in the Philippines, p 6

Scientific Committee on Ocean Research (SCOR) Working Group 89 (1991) The response of beaches to sea-level changes: a review of predictive models. JCR 7(3):895–921

Siringan FP, Jaraula CM (2001) Land configuration changes along the Aringay-Bauang coastal zone: possible causes and mitigation measures. A terminal report submitted to a concerned citizen

Siringan FP, Sta Maria MYY, Pang-ot GP (2006) Coastal erosion in Northern La Union: trends and possible mitigating measures. A terminal report submitted to the Provincial Government of La Union, p 102

Siringan FP, Sta Maria MYY, Jordan JCJ (2019) Coastal erosion management in the Philippines: a guidebook, p 155. The Marine Science Institute, University of the Philippines Diliman, Quezon City, Philippines

Siringan FP, Sta Maria MYY, Follosco NM (2013) Coastal erosion in the Philippines: a national assessment report. Submitted to the COBSEA YEOSU project: addressing the challenge of sea-level rise and coastal erosion in the East Asian Seas initial implementation of the COBSEA regional programme on coastal erosion

Siringan FP, Maeda Y, Rodolfo KS, Omura A (2000) Short-term and long-term changes of sea level in the Philippine Islands. In: Mimura M, Yokoki H (eds) Global change and Asia Pacific coasts: proceedings of APN/SURVAS/LOICZ joint conference on coastal impacts of climate change and adaptation in the Asia-Pacific region, pp 143–149

Siringan FP, Rodolfo KS (2003) Relative sea level changes and worsening floods in the Western Pampanga Delta: causes and some possible mitigating measures. Sci Diliman 15(2):1–12

Siringan FP, Berdin RD, Sta Maria MYY (2005) Coastal erosion in San Fernando, La Union: trends and possible mitigating measures. A terminal report submitted to the City Government of San Fernando, La Union

Siringan FP, Berdin RD, Jaraula CMB, Remotigue CT, Sta Maria MYY, Zamora PB (2005) A challenge for coastal management: large and rapid shoreline movements in the Philippines in know risk, pp 218–219. United Nations

Siringan FP, Soria JLA, Ramos RD (2008) Changes in shoreline positions and vulnerability to marine flooding of the Verde Island Passage Corridor. A terminal report submitted to Conservation International—Philippines

Titus JG (1998) Rising seas, coastal erosion, and the takings clause: how to save wetlands and beaches without hurting property owners. Md L Rev 57:1279

Titus J (2011) Rolling easements. Climate ready estuaries EPA program, p 169. United States Environmental Protection Agency

World Bank (2005) The Philippine environment monitor 2005: coastal and marine resource management

Zhang K, Douglas BC, Leatherman SP (2004) Global warming and coastal erosion. Clim Change 64(1–2):41

Fernando P. Siringan Ph.D., is a Professor at the Marine Science Institute, University of the Philippines Diliman where he served as its Director from 2016 to 2019. Dr. Siringan has been involved in studies that include paleo-climate reconstructions, harmful algal blooms, submarine groundwater discharge, geophysical mapping of mesophotic corals, mapping of offshore active faults, subsidence and sea level rise, and coastal erosion. He recently published a guidebook on coastal erosion management of the Philippines.—E-mail: fpsiringan@up.edu.ph.

Ma. Yvainne Y. Sta. Maria is a Ph.D. candidate at the Institute of Environmental Science and Meteorology, University of the Philippines Diliman, and is investigating the contribution of relative sea-level rise to coastal erosion in selected Philippine coastlines. She is one of the co-authors of the guidebook on coastal erosion management in the Philippines.—E-mail: ystamaria@msi.upd.edu.ph.

Chapter 4
Epidemics, Diseases, Health and Safety Emergencies Arising from Climate Change and Disasters in the Philippines

Jinky Leilanie Del Prado Lu

Abstract The study aims to present the data on the frequency and severity of disasters brought about by climate change in the Philippines, the common diseases in affected communities, the immediate health effects after a natural disaster and the preceding environmental risk factors, as well as the epidemics that have plagued the country for the past years. Data were gathered from records of local and international agencies. The Philippines is highly vulnerable and has actually hosted floods, landslides, typhoons resulting from climate change. The Philippines is fourth among countries that are most exposed to multiple hazards due to natural disasters with a risk exposure covering 22.3% of total area, and 36.4% of population being affected. Climate change is an aggravating factor to the occurrence of natural disasters such as typhoons and droughts as these have the potential to adversely affect human societies. Additionally, the Philippines is the third country with the highest risk for natural disasters, after Vanuatu and Tonga, based on the World Risk Index 2018. The study showed that there are immediate adverse health effects of natural disasters. Drought brings about protein malnutrition; destruction of water and sewerage systems bring about unsanitary conditions that harbor communicable diseases; and variations in climate implicate changes in patterns of occurrences of vector-borne diseases. The study also showed that most common communicable diseases arising from disasters and emergencies include diarrhea, acute respiratory infection, measles and malaria. In the international scene, cholera, diarrhea, hepatitis A and E, measles, meningitis, tetanus, diarrhea, leptospirosis, acute respiratory syndrome, coccidiomycosis, and malaria were reported to be the major epidemics after a type of natural disaster. Natural disaster can inhibit economic development of the nation, and therefore, the Philippines is challenged to come up with and implement a comprehensive disaster preparedness and mitigation measures for disasters and health epidemics.

Keywords Climate change · Natural disasters · Epidemics · Health and safety emergencies · Diseases brought about by climate change

---

J. L. D. P. Lu (✉)
University of the Philippines Manila, Manila, Philippines
e-mail: jdlu@up.edu.ph

K. B. Berse et al. (eds.), *Climate Emergency in the Philippines*, Disaster Risk Reduction, https://doi.org/10.1007/978-981-99-7804-5_4

Key Points

- The climate emergency exacerbates the risk of epidemics, diseases, and health emergencies in the Philippines.
- If nothing is done, then millions of lives will be at risk from the direct impact of disasters and the health effects that arise as communities cope and recover.
- Policymakers must advocate for sustainable and climate-protective practices to avoid the negative effect of human activities on the environment. They must also advocate for measures that protect and support communities during disaster events.

Summary for Policymakers

The problem of the occurrence of regular environmental disasters (e.g., cyclones, floods, droughts, landslides) and the subsequent health concerns (e.g., epidemics, fatalities) that accompany it is as direct result of climate change impacts and the Filipino people remain the most vulnerable to these.

- In general, climate-related health impacts can be categorized as either direct or indirect. Direct impacts include that of heat stress, which can directly cause cardiovascular and respiratory problems and even deaths. Indirect health impacts include ecological mediations, vector-borne diseases, marine-borne diseases, and air pollution (Lu 2016). Examples are in the case of increased exposure to extreme thermal conditions where a direct effect is an increase in heat- and cold-related deaths and illnesses, while a direct effect is an increased incidence of transmission of infectious diseases. The occurrence of extreme weather events has a direct health impact through the increase in the frequency and severity of injuries, deaths, and psychosocial disorders. The Philippines is fourth among countries that are most exposed to multiple hazards due to natural disasters. The risk exposure to disaster covers 22.3% of the total area and 36.4% of the population of the country. According to the Institute for International Law of Peace and Armed Conflict—IFHV (2018) World Risk Report, the Philippines is ranked third in the highest disaster risk, with Vanuatu being first and Tonga second. They also report that the country lacks coping and adaptive mechanisms to mitigate the impact of climate change. There have been several natural disasters that plagued the Philippines. In terms of fatalities and health emergencies, storm, flood, and earthquakes caused lives and damaged properties. Ground shaking was more frequent than tsunami in the country. Viral infectious diseases caused more deaths than bacterial and parasitic infectious diseases.
- There are observed health implications due to the effect of climate change. Dengue fever cases in the Philippines have been noted to be associated with temperature increase in the last decade (Perez 2009). As the temperature rises, so do the cases of dengue across the country. Among the top five natural disasters causing massive number of health emergencies and fatalities, earthquake caused the most number of deaths, while storm adversely affected the most number of people. Flooding was the most devastating economically. In the Philippines, the top natural

disasters that have occurred from 1900 to 2011 were storms and earthquakes; however, storms occurred more frequently and caused more harm to people among the two disasters. These damages not only translate to losses within the health dimension but include the economic dimension as well. The damages of climate-change-related disasters (e.g., storms, floods) are further aggravated by the lack of economic resources. In terms of vulnerability based on the economic risk to gross domestic product (GDP), the Philippines ranked tenth place. The country is estimated to have 50.3% of total area at risk, 81.3% of population in areas at risk, and 85.2% of GDP in areas at risk.

The most common communicable diseases arising from disasters and emergencies include diarrhea, acute respiratory infection, measles, and malaria (WHO 2006). Communities and populations who have survived through disasters are forced to seek temporary shelter until it is safe to return to their original homes. However, makeshift shelters are not equipped with proper sanitation faculties to accommodate the families displaced, increasing the risk of communicable diseases post-disaster. The most common communicable diseases arising from disasters and emergencies include diarrhea, acute respiratory infection, measles, and malaria.

- Even after the occurrence of a natural disaster, the threat to health from communicable diseases remains and even worsens. In reality following natural disasters, the disaster effects on size, health status, and living conditions of the displaced populations increase the risk of outbreak development (Toole 1997 as cited in WHO 2006). In general, the overall risk for development of communicable disease outbreaks is lower than perceived. However, the risk of transmission of certain endemic and epidemic-prone diseases can increase following natural disasters

Invest in research and development for projecting the impacts of climate change to health, and its relationship to epidemics and health emergencies.

To address this, policymakers at the national level must encourage sustainable practices nationwide to reduce the effects of climate change and support the local government in identifying vulnerable communities and provide ways to prevent and mitigate the risks identified.

- Conduct risk and vulnerability assessments of all communities to identify those at significant risk for the health-related effects of climate change.
- Advocate for the construction of well-equipped facilities that can serve as temporary safe shelters for communities displaced by environmental disasters among other mitigation and prevention interventions for specific disasters.
- Provide pathways that allow affected communities' access to government resources for the rapid rehabilitation and protection of vulnerable communities.

To address this, policymakers at the local level must work with the national government to identify vulnerable communities within their jurisdiction and initiate protective and preventive measures.

- Local policymakers should work side by side with communities to increase their awareness on the risks present and ways to address this such as through training workshops and capability building sessions.
- Local policymakers should push for the creation and empowerment of a disaster risk reduction unit to carry out preventive measures and disaster responses.
- Local policymakers should work with the national government to ensure that business, institutions, and individual citizens registered under their area comply with sustainable and climate-protective practices.

4.1 Introduction

Climate change is considered the biggest global health threat of the twenty-first century claiming approximately thousands of lives yearly and will continue to affect the lives of people through disasters brought about climate change, climate-sensitive diseases, and other health conditions that are exacerbated by climate change (PAHO, n.d.). It is projected that climate change will lead to an unacceptable and catastrophic risk to the human health directly and indirectly (Watts et al. 2015).

Climate change has direct and indirect impact on health. The direct impacts include the changing climate, particularly heat waves, droughts, heavy storms, floods, and sea level rise (PAHO n.d.; Watts et al. 2015). The indirect impacts of climate change include the vector-borne and airway diseases, food and water insecurity, undernutrition, forced displacement, and mental health illnesses, which result from the modifications brought about climate change to the environment (PAHO n.d; Watts et al. 2015). The detrimental effects of climate change to the social and environmental determinants of health, specifically air, water, food, and shelter, were estimated to cost an additional 250,000 fatalities annually by 2030 to 2050 due to malnutrition, malaria, diarrhea, and heat stress, and a direct cost of approximately USD $2–4 billion per year by 2030 for health (WHO 2021). Environmental factors that are harmful to health, particularly air pollution, extreme weather events, forced displacement, food insecurity, pressures on mental health, and diseases, were responsible for almost 13 million fatalities globally, which represented almost 22% of global fatalities and 26% of fatalities for children below 5 years old in 2012 (Prüss-Üstün et al. 2016).

Climate change aggravates natural disasters. Studies showed that global climate change accounted for the increasing magnitude and frequency of natural disasters. Kahn (2005) noted that the frequency and severity of natural disaster can be caused, augmented, and amplified by climate change. The Economics of Climate Adaptation (ECA) (2009) also showed that climate change is directly associated with the intensity of natural hazards.

The increasing surface temperature globally increases the risk for more frequent droughts and increased intensity of powerful storms through the increased heat in the atmosphere, warmer ocean surface temperature, and rising sea levels, which increases wind speeds in tropical storms (US Geological Survey n.d.). The devastating effects of natural disasters include changes in the physical landscape, sociophysical, biochemical, and increasing the exposure and vulnerability of the community to environmental stressors, which increases the vulnerability of the impacted communities and displaced individuals to the negative economic and health impact of natural disasters.

Climate change negatively impacts the environment by decreasing the sea ice and water resources, and increasing permafrost thawing, heat waves, and precipitation. In Asia, it is projected that fatalities related to flood- and drought-related diseases will increase, freshwater availability will be reduced by 2050, and coastal lines will be at risk due to increased flooding (US Geological Survey n.d.). Tropical-cyclone-related hazards and heavy precipitation were found to increase with increasing and worsening climate change in Japan and East Asia (Mori et al. 2021). Cappelli et al. (2021) analyzed natural disasters between 1992 and 2018 in 149 countries and determined the connection between climate-change-induced disasters, inequality, and vulnerability. Inequality to access and availability of technologies, infrastructure, and preventive measures was found to be inversely proportional with extreme climatic events, showing that countries with higher levels of inequalities, including gender differences in vulnerability, reported increasing frequency of extreme climatic events (Cappelli et al. 2021).

Asia is considered as one of the regions considered the most prone to natural disasters or one-third of disasters that occurred globally due to its geographic location and diverse topography (Jha et al. 2018). Asian countries, particularly the Philippines, Bangladesh, Cambodia, India, Pakistan, and Viet Nam, are the most vulnerable and susceptible to natural disasters (Jha et al. 2018). It was estimated that natural disasters costed Asia approximately 350,000 fatalities, almost USD $500 billion in total damages, and more than billion affected lives (Jha et al. 2018).

A recipient of some of the worst natural disasters arising from climate change, the Philippines suffered almost 75% of the world's natural disasters that happened in the years 1970–1977 and occurred in Asia and the Pacific Region, including other Pacific Island countries. Located along the typhoon belt in the Pacific, the Philippines is highly prone to hydrometeorological events, which is 80% of the country's natural disasters, and experiences an average of 20 typhoons annually (Jha et al. 2018). The country experienced changes in paths of typhoon and increasing intensity, and frequency of hot days and warm nights. It is expected that the average precipitation deviation will increase by 8 mm per month and could be associated with an additional disaster in the country every 3 years (Thomas 2013, as cited by Jha et al. 2018). Also, the Philippines is located in the middle of the most active earthquake zones in the world, The Pacific Ring of Fire, which makes the country more vulnerable to frequent volcanic eruptions and earthquakes.

Climate-related disasters are the manifestations of global warming. The impact of climate change and disasters is detrimental especially to developing regions with high levels of poverty, and poor and inadequate health infrastructure and healthcare system that is unable to adopt, prepare, and respond to health risks posed by climate change. Building resilience against climate change and climate-sensitive natural disasters in the Philippines requires understanding on how these threats impacted public health in the country.

This study aims to present the frequency and severity of natural disasters in the Philippines and the attendant epidemics after a type of natural disaster brought about by climate change using information from previously published studies related to this topic.

The study tries to answer the following questions:

1. What are the determinants affecting the frequency and magnitude of disasters brought about by climate change in the Philippines?
2. What are the critical health consequences of a disaster and its associated environmental risk factors?
3. What are the prevalent diseases observed in the affected communities as an aftermath of a disaster?
4. What are the recurring epidemics afflicting the Philippines in the past years?

4.2 Methodology

The study looked into trends of natural disasters related to climate change in the Philippines. Data also included profiling of types of diseases, health and safety emergencies in the aftermath of natural disasters. Both local and international data were looked into. The data used in this study were gathered from the records of Emergency Events Database (EM-DAT) of the Office of U.S. Foreign Disaster Assistance/ Centre for Research on the Epidemiology of Disasters (OFDA/CRED) International Disasters Database, the World Economic Forum's Global Risks Report World Risk Index, Independent Evaluation Group (IEG) of the World Bank, United Nations International Strategy for Disaster Reduction Secretariat (UNISDR), and the National Disaster Risk Reduction and Management Council (NDRRMC) formerly known as National Disaster Coordinating Council (NDCC). Then discursive and analytic methods were employed in building the thematic arguments and frameworks in the study on the relations of natural disasters, climate change, and health/ wellbeing.

4.3 Results and Discussion

Highly vulnerable to hydrometeorological disasters, particularly floods, landslides, and typhoons, the Philippines is included by the International Red Cross and Crescent Societies as one of the disaster-prone countries in the world. Philippines is an archipelago that is susceptible to tsunamis and typhoons (Esteban and Fabian, 2004). Situated in the typhoon belt of the North Pacific Basin, the Philippines is at increased risk to typhoons and El Nino phenomenon (Co, 2010).

4.3.1 Climate Change, Natural Disasters, and Epidemics

Climate is a long-term change in the average weather patterns that define the Earth's local, regional, and global climates. These changes accompany various observed effects to the planet (NASA, 2019). In relation to the occurrence of natural disasters, changes in the global climate worsen climate hazards and increase the risk of extreme weather conditions (Oxfam International). With increasing global surface temperatures, more droughts are produced, and water vapor is evaporated that augments the development of powerful storms. Furthermore, rising sea levels in high areas not usually reached by sea are then subjected to erosive effects of the waves and currents.

Both Global Climate Risk Index reports in 2015 (Kreft et al., 2015) and 2019 (Institute for Economics and Peace, 2019) listed the Philippines as the number one most affected country by climate change. The country has had ten of its deadliest typhoons from 1947 to 2014 although five of these have only occurred in 2006 (The Climate Reality Project, 2016). This may possibly be attributed to the worsening climate conditions over the past years. Damages brought about by typhoon cause displacement of populations that contribute to increasing cases of communicable diseases, most notable are those brought about by disrupted sources of clean water (Watson et al. 2007).

Apart from typhoons, droughts or El Niño have been occurring as a result of climate change and these have had grave health consequences. Changes in temperature are seen to be positively correlated to increase in dengue and malaria cases. The 2015 El Niño in the Philippines saw a 60 percent (60%) increase in the number of dengue cases (USAID, 2017). The Philippines Atmospheric, Geophysical, and Astronomical Services Administration (PAGASA, n.d.) expects a projected 1°C increase in 2020 and a 2°C increase in 2050. Given these projected increases, a possible worsening of climate-sensitive health conditions can be expected.

In general, Climate-related health impacts can be categorized as either direct or indirect. Direct impacts include that of heat stress, which can directly cause cardiovascular and respiratory problems and even deaths. Indirect health impacts include ecological mediations, vector-borne diseases, marine-borne diseases, and air pollution (Lu 2016). Climate and natural disaster-related health conditions will be discussed in the succeeding sections. The table below summarizes the possible climate-related health impacts.

4.3.2 Disaster Statistics: Philippine Setting

The Philippines is fourth among countries that are most exposed to multiple hazards due to natural disasters. The risk exposure to disaster covers 22.3% of the total area and 36.4% of the population of the country. See Table 4.2.

According to the Institute for International Law of Peace and Armed Conflict—IFHV (2018) World Risk Report, the Philippines is ranked third in the highest disaster

Table 4.1 Summary of climate-related health impacts

Implications of climate change	Potential health impacts	
	Direct	**Indirect**
↑ exposure to thermal extremes	↑ rates of heat- and cold-related deaths and illnesses	↑ incidence and transmission of infectious diseases
↑ extreme weather events (intensity and frequency)	↑ frequency and severity of injuries, deaths, and psychosocial disorders	–
↑ exposure to UV-B	↑ risks of skin cancer, burns, infectious diseases, eye damage (cataracts), immunosuppression	–
↑ range and activity of existing and new infective agents (i.e., as food-borne, water-borne, vector- and rodent-borne agents)	–	↑ exposure to new and existing vector-borne diseases; ↑ incidence of diarrheal and other infectious diseases; emergence of new diseases
Sea level rise	–	Psychosocial disruption
Worsened air pollution (contaminants, pollens, and spores)	–	↑ incidence of respiratory diseases (i.e., asthma and allergies) and cardiovascular diseases; ↑ exposure to environmental contaminants and subsequent impacts to health development
Agricultural disruption	–	Malnutrition

risk, with Vanuatu being first and Tonga second. The report attributes these high disaster risk indices to the countries' very high exposure to occurrences of cyclones and earthquakes. The calculation for the World Risk Index is based on four components: (1) exposure to earthquakes, cyclones, floods, droughts, and sea level rise; (2) susceptibility depending on infrastructure, food supply, and economic framework conditions; (3) coping capabilities depending on governance, health care, social and material security; and (4) adaptive capacities related to coming natural events, climate change, and other challenges (Bündnis Entwicklung Hilft 2018). The top 10 countries with the highest World Risk Index are presented in the table below.

Highly susceptible and prone to natural disasters, the Philippines has been identified as one of the top countries at risk for natural disasters worldwide. According to the Modeled Number of People Present in Hazard Zones by the United Nations International Strategy for Disaster Reduction Secretariat (UNISDR) (2009), the Philippines ranked 2nd out of 89 countries at risk of cyclone, 2nd out of 153 for risk of earthquake, 4th out of 162 for risk of landslide, 5th out of 265 for risk of tsunami, 8th out of 162 for risk of flood, and 33rd out of 184 for risk of drought.

Table 4.2 Countries most exposed to multiple hazards

Country	Percent of total area exposed*	Percent of population exposed*	Maximum number of hazards*	Index	Exposure	Vulnerability	Susceptibility	Lack of coping capacities	Lack of adaptive capacities	Percent of total area at risk***	Percent of population in areas at risk***	Percent of GDP in areas at risk***
1. Philippines	22.3	36.4	5	25.14	49.94	50.33	29.58	81.57	39.83	50.3	81.3	85.2
2. Guatemala	21.3	40.8	5	20.60	38.50	53.50	32.98	81.35	46.17	52.7	92.1	92.2
3. Vanuatu	28.8	20.5	3	50.28	86.46	58.15	36.07	86.37	52.00	-	-	-
4.Solomon Islands	7.0	4.9	3	23.39	49.94	50.33	29.58	81.57	39.83	–	–	–
5. Taiwan, China	73.1	73.1	4	–	–	–	–	–	–	–	–	–
6. Costa Rica	36.8	41.1	4	–	–	–	–	–	–	51.9	84.8	86.6
7. Vietnam	8.2	5.1	3	–	–	–	–	–	–	33.2	75.7	89.4
8. Guatemala	–	–	–	20.60	38.50	53.50	32.98	81.35	46.17	52.7	92.1	92.2
9. Bangladesh	–	–	–	17.38	29.95	58.03	33.72	84.96	55.42	71.4	83.6	86.5
10. Ecuador	13.9	23.9	5	–	–	–	–	–	–	–	–	–
11. Chile	12.9	54.0	4	–	–	–	–	–	–	–	–	–
12. Japan	10.5	15.3	4	–	–	–	–	–	–	–	–	–
13. Tonga	–	–	–	29.42	55.92	52.61	28.93	80.06	48.82	–	–	–
14. Papua New Guinea	–	–	–	20.88	31.05	67.24	55.58	83.80	62.35	–	–	–
15.Brunei Darussalam	–	–	–	18.82	52.71	35.71	14.80	61.50	30.83	–	–	–

(continued)

Table 4.2 (continued)

Country	Percent of total area exposed*	Percent of population exposed*	Maximum number of hazards*	Index	Exposure	Vulnerability	Susceptibility	Lack of coping capacities	Lack of adaptive capacities	Percent of total area at risk***	Percent of population in areas at risk***	Percent of GDP in areas at risk***
16. Fiji	–	–	–	16.58	35.51	46.68	22.33	75.48	42.24	–	–	–
17. El Salvador	–	–	–	–	–	–	–	–	–	88.7	95.4	96.4
18. Jamaica	–	–	–	–	–	–	–	–	–	94.9	96.3	96.3
19. Dominican Republic	–	–	–	–	–	–	–	–	–	87.2	94.7	95.6
20. Albania	–	–	–	–	–	–	–	–	–	86.4	88.6	88.5
21. Colombia	–	–	–	–	–	–	–	–	–	21.2	84.7	86.6

References:
*IEG stands for Independent Evaluation Group of the World Bank, 2005.
**World Disaster Risk Index. Adapted from Institute for International Law of Peace and Armed Conflict, 2018.
***Countries with Highest Vulnerability based on Economic Risk to GDP from Two or More Hazards (based on *IEG). IEG stands for Independent Evaluation Group of the World Bank. World Bank, 2005.

There have been several natural disasters that plagued the Philippines. Table 4.5 summarizes the natural disasters that have occurred in the Philippines from 1900 to 2019. In terms of fatalities and health emergencies, storm, flood and earthquakes caused lives and damaged properties. Ground shaking was more frequent than tsunami in the country. Viral infectious diseases caused more deaths than bacterial and parasitic infectious diseases.

4.3.3 Epidemics as Type of Natural Disaster

There are three common categories of disasters: (1) biological, (2) geophysical, and (3) hydrometeorological. Biological encompasses epidemics and vector-borne diseases, manifesting as bacterial (e.g., cholera and meningococcal), viral (e.g., acute, acute diarrheal syndrome, acute respiratory syndrome (SARS), dengue/dengue hemorrhagic fever, and influenza), and parasitic infectious diseases. Meanwhile, geophysical disasters pertain to earthquakes, tsunamis, and volcanic eruptions as a result of geological forces. Lastly, hydrometeorological disasters cover floods, tidal waves, storms, typhoons, landslides, avalanches, droughts, forest fires, and instances of extreme rainfall brought about by extreme temperatures and climatic conditions.

In relation to dengue being the number one cause of death in the biological category of disasters, dengue fever cases in the Philippines have been noted to be associated with temperature increase in the last decade (Perez 2009). Perez (2009) observed an increasing trend of cases as temperature increases across the country between 1992 and 2005. In fact, the most number of cases (35,600 cases) and the highest temperature (0.9 Celsius increase) were reported in 1998 (Perez 2009).

Among the top five natural disasters causing massive number of health emergencies and fatalities, earthquake caused the most number of deaths, while storm adversely affected the most number of people affected. Flooding was the most devastating economically. The top natural disasters that have frequently occurred from 1900 to 2011 in the Philippines include earthquake, flood, and storm.

In terms of vulnerability based on the economic risk to gross domestic product (GDP), the Philippines ranked tenth place. The country is estimated to have 50.3% of total area at risk, 81.3% of population in areas at risk, and 85.2% of GDP in areas at risk.

4.3.4 Types of Communicable Diseases in Disasters and Emergencies

The presence of a large number of dead bodies is commonly expected to result in higher chances for an outbreak to occur (de Ville de Goyet, 2004). In reality following natural disasters, the disaster effects on size, health status, and living conditions of the displaced populations increase the risk of outbreak development (Toole 1997 as cited in WHO 2006).

In general, the overall risk for development of communicable disease outbreaks is lower than perceived. However, the risk of transmission of certain endemic- and epidemic-prone diseases can increase following natural disasters (WHO 2006).

Natural disasters may result in infectious diseases during outbreaks due to population displacement and environmental stressors that exacerbate vulnerability of the impacted communities to various diseases. Infectious diseases include diarrheal diseases, acute respiratory infections, malaria, leptospirosis, measles, dengue fever, viral hepatitis, typhoid fever, meningitis, tetanus, and cutaneous mucormycosis (Isidore et al. 2012). See Table 4.3.

4.3.5 Outbreak of Diseases by Type of Disaster

Immediate adverse health effects occur after natural disasters. Some of these immediate health effects are presented in Table 4.4. Malnutrition, respiratory and gastrointestinal illnesses are found to be associated with drought. Trauma-related injuries and fatalities, including crush injuries, are the immediate health effects for earthquakes. Burn injuries, respiratory illness, and fatalities occur mostly for fires. Drowning, diarrheal diseases, respiratory infections, and bites are health effects of floods. Disease outbreaks after flooding and in a large-scale population displacement are more commonly reported compared to earthquake, volcanic eruptions, and tsunamis, and small-scale population displacement (Watson et al., 2009). Hyperthermic illness, respiratory problems, and permanent disabilities resulting from stroke are the immediate health effects of heat waves. Water-borne and vector-borne diseases and trauma-related injuries are commonly reported for hydrometeorological disasters, specifically hurricanes, typhoons, and cyclones. Crush injuries, fractures, and soft-tissue injuries are the immediate health effects for tornadoes. Respiratory illnesses, burn injuries, and infections are some of the immediate health effects of volcanic eruptions. Frostbite, cardiovascular and respiratory illnesses, and food- and water-borne diseases are the immediate health effects of winter storms.

In the documented list of epidemic occurrences (from 2000 to 2019), dengue accounted for the highest fatality count, with 737 recorded deaths and 123,939 reported cases. These findings underscore the significance of addressing the health risks in the country, more particularly in the developing disaster management and public health strategies.

Among the natural disasters, flooding emerged as the predominant catalyst for the onset of epidemic outbreaks. Meanwhile, at the global level, it was found that the common epidemics that arise as an aftermath of various disasters include, but are not limited to, cholera, diarrhea, hepatitis A and E, measles, meningitis, tetanus, leptospirosis, acute respiratory syndrome, coccidiomycosis, and malaria.

Table 4.3 List of the common diseases in affected communities after a natural disaster

Disease	Environmental risk factors	Health hazards
Acute upper respiratory tract infections	Overcrowding, Poor Hygiene (WHO 2002); destruction of healthcare infrastructures[2]	Influenza and pneumonia may cause severe complications, especially in groups at risk (WHO 2002); polymicrobial respiratory infections including measles and tuberculosis (Mayrouli et al., 2021)
Diarrhea	Contaminated drinking water or food or poor sanitation (WHO 2002); temperature of air and water, precipitation pattern, extreme rainfall, and seasonal variation (CDC, 2020)	Dehydration, especially in children and elderly, shown by dark coloration of the urine, dry tongue, or leathery skin (WHO 2002); salmonellosis and campylobacteriosis (CDC, 2020)
Measles	Crowding, poor hygiene (WHO, 2002); low baseline immunization coverage, population displacement, poor access to safe water and sanitation facilities (WHO, n.d.)	Severe constitutional symptoms, high case fatality rate (WHO 2002); severe complications leading to blindness, severe diarrhea, ear infection, pneumonia (WHO, n.d.)
Malaria	Breeding of *Anopheles* mosquitoes in stagnant water bodies (WHO 2002); temperature, humidity, rainfall, soil quality, elevation/slope, land cover, presence of natural enemies of mosquitoes and larvae, land use, land change, deforestation, housing type, infrastructure (water, sanitation, waste collection), urbanization, development projects, disasters by human-made changes (Castro 2017)	Disease may rapidly become fatal unless medical care is provided within the last 48 h (WHO 2002); fever and flu-like symptoms, nausea, vomiting, diarrhea, anemia and jaundice, kidney failure, seizures, mental confusion, coma, and death (CDC, n.d.)
Meningococcal meningitis	Crowding (WHO 2002); enclosed environments with known TB patients (CDC, n.d.)	Often fatal if untreated at an early stage; neurological problems in survivors (WHO 2002); nausea, vomiting, photophobia, altered mental status (CDC, n.d.)
Shigella dysentery	Contaminated drinking water or food, or poor sanitation, poor hygiene (WHO 2002)	Case fatality rate may be high[1]; bloody diarrhea, fever, stomach pain, severe stomach cramping, and dehydration (CDC, n.d.)

(continued)

Table 4.3 (continued)

Disease	Environmental risk factors	Health hazards
Viral hepatitis A	Poor hygiene and sanitation (WHO 2002); contaminated food and water, direct contact with an infected person (WHO 2022)	Long term disabling effects (WHO 2002); fever, malaise, loss of appetite, diarrhea, nausea, abdominal discomfort, dark-colored urine, jaundice (WHO 2022)
Louse-borne typhus	Unhygienic conditions leading to lice infestations (WHO 2002); extreme overcrowding (CDC, n.d.)	May be fatal without treatment (WHO 2002); fever and chills, headache, rapid breathing, body and muscle aches, rash, cough, nausea, vomiting, confusion (CDC, n.d.)
Typhoid fever	As for diarrhea (WHO 2002); contaminated food and water, poor sanitation, infected human waste (NHS 2021)	Without appropriate medical care, may lead to fatal complications in a few weeks (WHO 2002); fever, stomach pain, constipation, diarrhea, internal bleeding (NHS 2021)
Cholera	As for diarrhea (WHO 2002); contaminated food and water (CDC, n.d.)	As for diarrhea (WHO 2002); profuse watery diarrhea, vomiting, thirst, leg cramps, restlessness or irritability, severe dehydration that can lead to shock, coma, and death (CDC, n.d.)
Dengue and Dengue hemorrhagic fever (DHF)	Breeding of *Aedes* mosquitoes in natural or artificial containers, filled with water (WHO 2002)	Dengue usually runs a mild curve. DHF, however, is often accompanied by heavy hemorrhages, which may be fatal (WHO 2002); abdominal pain, persistent vomiting, fever, hemorrhagic manifestations, and altered mental state, signs of shock (CDC, n.d.)
Diphtheria	Crowding, poor hygiene (WHO 2002)	A secretion is deposited in the respiratory tract, which can lead to asphyxiation (WHO 2002); difficulty in breathing, heart rhythm problems, and fatality (CDC, n.d.)

(continued)

Table 4.3 (continued)

Disease	Environmental risk factors	Health hazards
Tetanus	Poor hygiene, injury (WHO 2002); soil, dust, manure contaminated with spores of tetanus bacteria (CDC, n.d.)	Fatal (WHO 2002); tightening of the jaw muscles, difficulty of swallowing and breathing, painful muscle stiffness, seizures, involuntary muscle spasms (CDC, n.d.)
Rabies	Bite from an infected animal host (WHO 2002); having a non-rabid pet known to the victim (Warner, 2010)	Fatal if untreated (WHO 2002); severe injuries needing debridement, cellulitis (Warner 2010)
Relapsing-fever (louse borne or tick borne)	Unhygienic conditions leading to lice or tick infestations (WHO 2002); cave exposures (CDC, n.d.)	Often fatal in untreated persons, depending on immunity levels (WHO 2002); recurring bouts of fever, headache, muscle and joints aches, nausea (CDC, n.d.)
Heat stress	Excessive temperatures (WHO 2002)	Risk of coma (WHO 2002); heat stroke, heat exhaustion, heat cramps, heat rashes (CDC, n.d.)

Adapted from WHO (2002).

4.3.6 International Data on Epidemics

Bacterial infectious diseases accounted the greatest number of fatalities, followed by parasitic and viral infectious diseases. Although Africa reported the greatest mean number of events related to bacterial infectious diseases, Asia recorded the most number of people killed per event. Generally, Asia had the highest mean number of fatalities in all three groups of epidemics.

4.3.7 Cases of Natural Disasters and Epidemics in the Philippines

Cases of disease epidemics were noted in the occurrence of natural disasters in the Philippines. Typhoons account for most of the natural disaster occurrences in the country. In 2011, the Health Emergency Management Bureau (HEMB) created the Surveillance for Post Extreme Emergencies and Disasters (SPEED), which is a real-time syndromic surveillance system that allows the early detection and monitoring of post-disaster disease trends (Salazar et al. 2016). In this surveillance system,

Table 4.4 Immediate health effects of natural disaster

Type of natural disaster	Immediate health effects
Drought	Protein malnutrition (kwashiorkor) or calorie malnutrition (marasmus); Xeropthalmia (lack of vitamin A) and child blindness (DCPP 2007); worsened chronic respiratory illnesses, asthma, and increased risk for acute respiratory infections such as bronchitis and bacterial pneumonia due to the irritation of bronchial passage and lungs; gastrointestinal illnesses such as *E. coli* and Salmonella (CDC n.d.)
Earthquake	Crush injuries due to falling objects (DCPP 2007); trauma-related injuries and fatalities (Gunn 1995)
Fires	Poor air quality (build-up of smoke and pollutants in the air that leads to respiratory problems), burn injuries (DCPP 2007); aggravation of chronic heart and lung diseases, and fatalities (USEPA n.d.)
Floods	Drowning; diarrheal diseases; respiratory infections; dermatitis; and snake bites (DCPP 2007); asphyxia due to drowning, trauma-related injuries, hypothermia, animal bites, infected wounds, complications of injury, poisoning, poor mental health, communicable diseases, starvation, chronic disease, and other poverty-related diseases including malnutrition (Du et al. 2010)
Heat Waves	Hyperthermic illnesses; heat stroke (DCPP 2007); permanent disabilities, respiratory problems (CDC n.d.)
Hurricanes, Typhoons, Cyclones	Crush injuries such as superficial lacerations, trauma incidents (DCPP 2007); weather-related morbidity and mortality, water-borne and vector-borne diseases, water-related diseases, and mental health effects (Waddell et al. 2021; Veenema et al. 2017)
Tornadoes	Crush injuries due to collapsing structures; fractures, penetrating trauma, lacerations, abrasions (due to fine particles of soil, mud, sand that strike the body at very high speed) and other soft-tissue injuries (DCPP, 2007); mental health effects including posttraumatic stress disorder and depression symptoms (First & Houston 2020)
Volcanic eruptions	Acute respiratory distress syndrome, pulmonary edema, irritant conjunctivitis, joint pain, muscle weakness, and cutaneous bullae due to toxic volcanic gasses (i.e., carbon dioxide, carbon monoxide, and sulfuric acid) emitted into the atmosphere; burns due to steam coming from the volcano's vent; crush injuries, severe internal bleeding, multiple organ dysfunction syndrome a and asphyxiation due to mudslides and pyroclastic flows (DCPP 2007); infectious disease, burns, chronic lung problems (CDC n.d.)
Winter Storms	Frostbite (DCPP 2007); cardiovascular, lower respiratory diseases, respiratory infections, food- and water-borne diseases, injuries (Lin et al. 2021)

Table 4.5 SPEED data on communicable diseases post-typhoon Yolanda (2013), Pablo (2012), and Habagat (2012)

Diseases	Disasters events		
	Habagat (2012)	Pablo/Bophia (2012)	Yolanda (2013)
Hepatitis A/E	68	337	316
Leptospirosis	1447	562	531
Measles	1036	124	1534
Meningitis	141	69	1500
Dengue	–	–	3254
Tetanus	210	15	122

21 syndromes are provided which are divided into three groups of conditions: communicable diseases, non-communicable diseases, and injuries.

Salazar et al. (2016) made a study on post-disaster health impact of natural hazards in the Philippines in 2013 through an analysis of the 2013 SPEED data. In the study, a total of 4,645 SPEED reports are seen with 3,425 for typhoons, 609 for earthquakes, and 611 for floods. The 2013 SPEED data was divided into two groups to compare the syndrome rates per 10,000 individuals within 2 months and after 2 months post-disaster.

Within 2 months, communicable diseases had the highest number of syndrome rates for flood (N = 551), earthquake (N = 582), and typhoon (N = 1,614). For flood (N = 551), the syndrome rates are the following: communicable diseases are 18.8 (14.2–23.5); injuries are 1.7 (0.9–2.5); and non-communicable diseases are 0.7 (0.5–1.0) within 2 months. For earthquake (N = 582), the syndrome rates are the following: communicable diseases are 54.9 (36.5); injuries are 6; and non-communicable diseases are 9.2 (5.2–13.1) within 2 months. For typhoon (N = 1,614), the syndrome rates are the following: communicable diseases are 84.5 (69.45–99.4); injuries are 10.9 (8.3–13.5); and non-communicable diseases are 10.3 (8.4–12.2) within 2 months (Salazar et al. 2016).

After 2 months post-disaster, communicable diseases still had the greatest number of syndrome rates for flood (N = 60), earthquake (N = 27), and typhoon (N = 1,811). For flood (N = 60), the syndrome rates are the following: communicable diseases are 14.5 (12.4–16.7); injuries are 0.4 (0.1–0.6); and non-communicable diseases are 0.2 (0.1–0.2). For earthquake (N = 27), the syndrome rates are the following: communicable diseases are 14.7 (11.3–18.2); injuries are 1.2 (0.6–1.7); and non-communicable diseases are 1.5 (0.7–2.3). For typhoon (N = 1,811), the syndrome rates are the following: communicable diseases are 14.2 (12.5–15.8); injuries are 1.4 (1.1–1.7); and non-communicable diseases are 2.3 (1.9–2.6) (Salazar et al. 2016).

Typhoons account for most of the syndrome reports. Floods and communicable diseases also account for a large number of syndrome reports in the SPEED data. This is consistent with current literature regarding communicable diseases being a result of disruption of living conditions or displacement of populations during natural disasters (de Ville de Goyet 2004; Murthy and Christian 2010; Kouadio

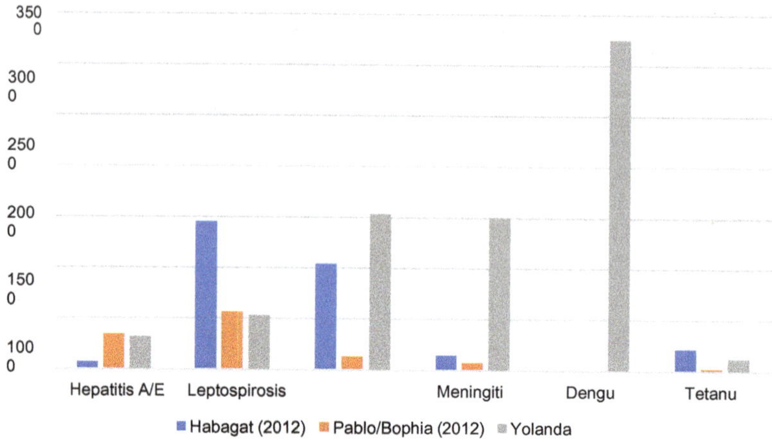

Fig. 4.1 SPEED data on communicable diseases post-typhoon Yolanda (2013), Pablo (2012), and Habagat (2012)

et al. 2012). Respiratory infection was seen to account for the highest consultation rates. In children under 5 years old, acute respiratory infection, open wounds, high blood pressure, skin disease, acute watery diarrhea, and fever account for the highest consultation rates, most of which are from under the communicable syndrome group (Salazar et al. 2016).

Communicable diseases appear to be the more prevalent conditions during occurrences of typhoon. SPEED data on post-disaster consultations for Typhoon Yolanda (2013), Bopha/Pablo (2012), and Southwest Monsoon Floods/Habagat (2012) resulted to increased number of Hepatitis A/E, Leptospirosis, Measles, Meningitis, Dengue, and Tetanus cases (Quiambao 2016). The table below summarizes the number of communicable disease cases on post-disaster periods. See Table 4.5 and Figure 4.1.

In this present decade, the Philippines saw the onslaught of Super Typhoon Yolanda in 2013 which resulted in massive losses in life, infrastructure, and resources. The Center for Research on the Epidemiology of Disasters Database (2019) reports a total of 7,354 deaths and 16,106,870 people affected at the end of the typhoon's visit. In health infrastructure alone, damage amounted to 19.2 million US Dollars (Health Emergency Management Bureau 2014). Total damages are estimated to amount to 14 billion US Dollars (Calamur 2013). Additionally, there was a total of 340,000 consultations reported through the SPEED system in the affected areas of the typhoon, and 58 percent (58%) of these consultations were for watery diarrhea, 13 percent (13%) for suspected measles, nine percent (9%) for hemorrhagic fever, six percent (6%) for acute bloody diarrhea, and five percent (5%) for leptospirosis; alerts for typhoid fever, dengue, rabies, and suspected meningococcemia were also noted in the disaster report (WHO 2014).

Typhoon Haiyan/Yolanda has been the subject of in-depth studies on post-disaster health impacts in the Philippines covering not only the direct physical effects brought about by the disaster but also the effects on the current management of people living with particular conditions as well as the mental health of the survivors and responders. During the typhoon, target client medications such as that of for Pulmonary Tuberculosis (PTB) were made unavailable resulting to a 6–8-week period where people with PTB left without continuous medication intake (Lew et al, 2015). These would mean possible development of antibacterial resistance in some clients and progression to a different category of treatment under the National TB Control Program.

Mental health is also affected during the course of natural disasters. In a study by Hugelius et al. (2017), assessing the health among survivors and responders to the Typhoon, 42 percent (42%) of the respondents noted that they have suffered mental health problems related to the typhoon at a particular point in time while 12 percent (12%) reported persistent mental health problems. Contributing factors to the occurrence of mental health problems were the damage and displacement from homes as well as losses of loved ones during the disaster.

The El Nino phenomenon has also caused droughts and in turn causing malnutrition, especially among children. This was reiterated in the study of Rojas et al in 2014 where they showed the link between drought with food insecurity and malnutrition cases. From 2000 to 2019, the Philippines saw four incidences of drought; there were no noted cases of deaths, however, a recorded estimate total of 197,687 people were affected with food shortage, among others (CRED, 2019). According to the National Nutrition Survey of the Food and Nutrition Research Institute (FNRI), there has been a noted increase in the cases of malnutrition in the country particularly in children from 30.3 percent (30.3%) to 33.4 percent (33.4%) in the 2013 report (Bermejo 2017). Hunger is seen to be more prevalent in areas where fishing and farming are the main sources of food and income (Felizco & Zapanta 2016). Malnutrition due to shortage in food also aggravates the increased risk to communicable diseases especially in areas with poor living and sanitary conditions. Food shortages can be attributed to either: (1) food stock destruction due to damage of crops or (2) disruption of food distribution systems. Malnutrition follows and exacerbates the occurrence/spread of diseases. The spread of diseases is also enhanced by lack of water supply and sanitation conditions. For instance, in Haiti, an outbreak of cholera was reported after the earthquake affecting 112,330 people, and eventually causing 2,478 deaths. The massive number of cases of cholera reported were attributed to unsanitary conditions (WHO 2010).

Likewise, the increasing risk of outbreaks of diseases and epidemics in the aftermath of natural disasters can be accounted by several factors such as mass population movement from the zone of danger or overcrowding in evacuation centers; inadequacies of health services and sanitary facilities; shortages of safe food, safe water, and food supplies. Disruption of surveillance and disease prevention services are also primary factors for spread or exacerbation of communicable diseases such as that of disruption in tuberculosis direct observed short-term treatments and the expanded program on immunization.

4.4 Conclusion

Communicable diseases remained the most commonly reported health conditions during post-disaster periods. The most common communicable diseases brought about by climate-related disasters are diarrhea, acute respiratory infection, measles, and malaria, which are aggravated by malnutrition. Internationally, the major epidemics occurring after a type of natural disasters are cholera, diarrhea, hepatitis A and E, measles, meningitis, tetanus, diarrhea, leptospirosis, acute respiratory syndrome, coccidiomycosis, and malaria. Population displacement, disruption in way of living and livelihood, and the inadequate health services contribute to the health burden during disasters.

There is a need for policymakers to actively investigate and address the vulnerability of communities against environmental disasters brought about by climate change, such as floods, cyclones, and droughts. Beyond pushing for disaster risk reduction interventions, policymakers should also shine light upon the actions needed to reduce and prevent further climate change. Studies should be done to look into sustainable practices such as renewable energy resources and recycling. Investments should be directed toward renewable energy and resources, while local government units should shift to the development of green cities.

References

Asian Disaster Preparedness Center (ADPC) (2001) The Philippine disaster management story: issues and challenges. Thailand

Association of Southeast Asian Nations (2016) AADMER Work Program. Retrieved from https://www.asean.org/wp-content/uploads/2016/02/AADMER-Work-Programme-2016–2020-v1.6.pdf on 5 Dec 2019

Bündnis Entwicklung Hilft (2011) World Risk Report 2011. Retrieved from https://reliefweb.int/sites/reliefweb.int/files/resources/WorldRiskReport-2018.pdf on 5 Dec 2019

Bündnis Entwicklung Hilft (2018) World Risk Report 2018. Retrieved from https://reliefweb.int/sites/reliefweb.int/files/resources/WorldRiskReport-2018.pdf on 5 Dec 2019

Calamur K (2013) World headline: the financial cost of typhoon haiyan. national public radio. Retrieved from https://www.npr.org/sections/parallels/2013/11/12/244701849/world-headlines-the-financial-cost-of-typhoon-haiyan on 9 Dec 2019

Cannon T (2007) Integrating disaster risk reduction into the millennium development goals: review of activities up to the present. international strategy for disaster reduction. Retrieved from https://www.unisdr.org/files/766_Desk_Review_Report_MDGs_HFA_Oct07_ActionAid%5B1%5D.pdf on 5 Dec 2019

Cappelli F, Costantini V, Consoli D (2021) The trap of climate change- induced "natural" disasters and inequality. Glob Environ Chang 70:102329

Castro MC (2017) Malaria transmission and prospects for malaria eradication: the role of the environment. Cold Spring Harb Perspect Med 7(10):a025601

Centers for Disease Control and Prevention (n.d.) Key Facts About Eruptions. Available from https://www.cdc.gov/disasters/volcanoes/facts.html

Centers for Disease Control and Prevention (2020) Food and waterborne diarrheal disease. Available from https://www.cdc.gov/climateandhealth/effects/food_waterborne.htm

Centers for Disease Control and Prevention. (n.d.). Malaria. Available from https://www.cdc.gov/malaria/about/faqs.html

Centers for Disease Control and Prevention (n.d.) Bacterial Meningitis. Available from https://www.cdc.gov/meningitis/bacterial.html

Centers for Disease Control and Prevention. (n.d.). Cholera – *Vibrio cholerae* infection. Available from https://www.cdc.gov/cholera/illness.html

Centers for Disease Control and Prevention (n.d.) Diphtheria. Available from https://www.cdc.gov/diphtheria/index.html

Centers for Disease Control and Prevention. (n.d.). Heat Stress. Available from https://www.cdc.gov/niosh/topics/heatstress/default.html

Centers for Disease Control and Prevention (n.d.) Epidemic Typhus. Available from https://www.cdc.gov/typhus/epidemic/index.html

Centers for Disease Control and Prevention (n.d.) Extreme heat can impact our health in many ways. Available from https://www.cdc.gov/climateandhealth/pubs/extreme-heat-final_508.pdf

Centers for Disease Control and Prevention (n.d.) Dengue and Dengue Hemorrhagic Fever. Available from https://www.cdc.gov/dengue/resources/dengue&DHF%20information%20for%20Health%20Care%20Practitioners_2009.pdf

Centers for Disease Control and Prevention (n.d.) Tetanus. Available from https://www.cdc.gov/tetanus/about/causes-transmission.html

Centers for Disease Control and Prevention (n.d.) Health Implications of Drought. Available from https://www.cdc.gov/nceh/drought/implications.htm

Centers for Disease Control and Prevention (n.d.) Tick-borne Relapsing Fever. Available from https://www.cdc.gov/relapsing-fever/index.html

Centers for Disease Control and Prevention (n.d.) Shigella – Shigellosis. Available from https://www.cdc.gov/shigella/symptoms.html

de Ville de Goyet C (2004) Epidemics cause by dead bodies: a disaster myth that does not want to die. Am J Public Health 15(9). Retrieved from http://iris.paho.org/xmlui/bitstream/handle/123456789/8238/22000.pdf?sequence=1&isAllowed=y on 6 Dec 2019

Delfin F, Gaillard JC (2008) Extreme versus quotidian: addressing temporal dichotomies in Philippine disaster management. Public Admin Dev 28:190–199. https://doi.org/10.1002/pad.493

Disease Control Priorities Project (DCPP) (2007) Natural disasters: coping with the health impact. Available from http://www.dcp2.org. Accessed 18 March 2011

Domingo S (2017) Institutional issues on disaster risk and management. Philippine Institute of Development Studies. Retrieved from https://pidswebs.pids.gov.ph/CDN/PUBLICATIONS/pidsdps1750. pdf on 5 Dec 2019

Du W, FitzGerald GJ, Clark M, Hou XY (2010) Health impacts of floods. Prehosp Disaster Med 25(3):265–272

Duque P (2005) Disaster management and critical issues on disaster risk reduction in the Philippines. International workshop on emergency response and rescue. Quezon City, Philippines

Dy P, Stephens T (2016) The Typhoon Haiyan response: strengthening coordination among Philippine Government, Civil Society, and International, Actors. Harvard Kennedy School Program on Crisis Leadership. https://www.hks.harvard.edu/sites/default/files/centers/research-initiatives/crisisleadership/files/Dy_and_Stephens.pdf on 9 December 2019

Economics of Climate Adaptation (ECA) (2009) Shaping climate-resilient development: a framework for decision-making. A Report of the Economics of Climate Adaptation Working Group.

Emergency Events Database (EM-DAT). 2019. The OFDA/CRED International Disaster Database. UCL – Brussels, Belgium. Available from http://www.em-dat.net

Esteban R and Fabian D (2004) The Philippine disaster management system. Malaysia. Available from http://www.pctc.gov.ph/updates/tpdms.htm. Accessed 18 April 2011

Felizco M, Zapanta J (2016) Drought and dry spell hits 58 provinces in the Philippines. Act Alliance. Retrieved from https://reliefweb.int/sites/reliefweb.int/files/resources/ACT-Philippine-Forum-Alert-El-Nino-Alert-20160504.pdf on 9 Dec 2019

First JM, Houston JB (2020) The mental health impacts of successive disasters: examining the roles of individual and community resilience following a tornado and CoViD-19. Clin Soc Work J 50(2):124–134

Gunn SW (1995) Health effects of earthquakes. Disaster Prev Manag An Int J

Health Emergency Management Bureau (2014) Heroes at the frontline: health responders' stories of haiyan. Department of Health

Hugelius K, Gifford M, Örtenwall P, Adolfsson A (2017) Health among disaster survivors and health professionals after the Haiyan Typhoon: a self-selected Internet-based web survey. Int J Emerg Med 10(1):13

Institute for Economics & Peace (2019) Global peace index 2019: measuring peace in a complex world, Sydney, June 2019. Retrieved from http://visionofhumanity.org/app/uploads/2019/06/GPI-2019-web003.pdf on 9 Dec 2019

Jha S, Martinez A, Quising P, Ardaniel Z, and Wang L (2018) Natural disasters, public spending, and creative destruction: a case study of the Philippines. ADBI Working Paper 817. Tokyo: Asian Development Bank Institute. Available: https://www.adb.org/publications/natural-disasters-public-spending-and-creative-destruction-philippines

Kahn ME (2005) The death toll from natural disasters: the role of income, geography, and institutions. Rev Econ Stat 87 (2):271–284. doi:https://doi.org/10.1162/0034653053970339

Kouadio I, Alijunid S, Kamigaki T, Hammad K, Oshitani H (2012) Infectious diseases following natural disasters: prevention and control measures. Retrieved from https://www.ncbi.nlm.nih.gov/pmc/articles/PMC4871893/#CIT0014 on 6 Dec 2019

Kreft S, Eckstein D, Junghans L, Kerestan C, Hagen U (2015) The global climate risk index 2015. Retrieved from https://germanwatch.org/sites/germanwatch.org/files/publication/10333.pdf on 6 December 2019

Lew W, Vianzon R, Garfin AMC, Hall JL (2015) Restarting the tuberculosis programme post-Haiyan. West Pac Surveill Response J 6(Suppl 1):91

Lin S, Zhang W, Sheridan S, Mongillo M, DiRienzo S, Stuart NA, et al. (2021) The immediate effects of winter storms and power outages on multiple health outcomes and the time windows of vulnerability. Environ Res

Lu J (2016) Impact of climate change on human health. Acta Medica Philippina 50(2)

March G (2002) Natural Disasters and the Impacts on Health. Available from http://www.eird.org/isdr-biblio/PDF/Natural%20disasters%20and%20the%20impacts.pdf. Accessed 18 March 2011

Mavrouli M, Mavroulis S, Lekkas E, Tsakris A (2021) Respiratory infections following earthquake-induced tsunamis: transmission risk factors and lessons learned for disaster risk management. Int J Environ Res Public Health 18(9):4952. https://doi.org/10.3390/ijerph18094952

Mori N, Takemi T, Tachikawa Y, Tatano H, Shimura T, Tanaka T et al (2021) Recent nationwide climate change impact assessments of natural hazards in Japan and East Asia. Weather Clim Extrem 32:100309

Murthy S, Christian M (2010) Infectious disease following disasters. Retrieved from https://www.ncbi.nlm.nih.gov/pubmed/21149220 6 Dec 2019

National Aeronautics and Space Administration (NASA) (n.d.) Overview: weather, global warming, and climate change. Retrieved from https://climate.nasa.gov/resources/global-warming-vs-climate-change/ on 6 Dec 2019

National Disaster Coordinating Council (NDCC) (2007) Philippine disaster management framework. Available from http://www.ndcc.gov.ph. Accessed 18 Apr 2011

National disaster risk reduction and management council (NDRRMC) (2011) Signing ceremony for the approval of the national disaster risk reduction and management framework. Retrieved from http://www.ndrrmc.gov.ph/attachments/article/1675/NDRRMC_Framework.pdf on 5 Dec 2019

NHS (2021) Causes typhoid fever. Available from https://www.nhs.uk/conditions/typhoid-fever/causes/

OCHA (2013a) Philippines (Mindanao) humanitarian action. Plan-mid-year review. Geneva, New York: OCHA

OCHA (2013e) Philippines: Typhoon Bopha. Situation Report. No. 19. 12 February

Office of Civil Defense (OCD) (2019) NDRRMC, partners to conduct summit on national disaster resilience. Philippine Information Agency. Retrieved from https://pia.gov.ph/news/articles/1025210 on 5 Dec 2019

Oxfam International (n.d.) 5 Natural disasters that beg for climate action. Retrieved from https://www.oxfam.org/en/5-natural-disasters-beg-climate-action on 6 Dec 2019

Philippine Atmospheric, Geophysical, and Astronomical Services Administration (PAGASA) (n.d.) Climate change in the Philippines. Retrieved from http://bagong.pagasa.dost.gov.ph/information/climate-change-in-the-philippines on 9 Dec 2019

Pan American Health Organization (n.d.) Climate change and health. Available from https://www.paho.org/en/topics/climate-change-and-health

Pedroso K (2010) From NDCC to NDRRMC. The Philippine daily inquirer. Available from http://newsinfo.inquirer.net/inquirerheadlines/nation/view/20101020-298696/From-NDCC-to-NDRRMC. Accessed 18 April 2011

Perez R (2009) Climate change in the Philippines: presented at symposium on the economics of climate change in Southeast Asia. De La Salle University

IOM Philippines (2013) Typhoon Bopha Emergency Response. Situation Report. 05 March

Porio E, Cruz F, Narisma G, Yulo-Loyzaga A (2019) Drought and urbanization: the case of the Philippines: methods, approaches and practices. DOI: https://doi.org/10.1007/978-981-10-8947-3_12. Retrieved from https://www.researchgate.net/publication/330023534_Drought_and_Urbanization_The_Case_of_the_Philippines_Methods_Approaches_and_Practices on 9 Dec 2019

PreventionWeb (2011) Disaster Statistics. Available from http://www.preventionweb.net/english/countries/. Accessed 9 March 2011

Prüss-Üstün A, Wolf J, Corvalán C, Bos R, Neira M (2016) Preventing disease through healthy environments: a global assessment of the burden of disease from environmental risks. World Health Organization. Available from https://www.who.int/publications/i/item/9789241565196

Munich Re (1998) Topics: annual review of natural catastrophes 1997, Munich: Munich Re

ReliefWeb – UN Office for the Coordination of Humanitarian Affairs (2009) Situation report Philippines – Typhoon Ondoy. Retrieved from https://reliefweb.int/report/philippines/situation-report-philippines-typhoon-ondoy on 9 Dec 2019

Rojas O, Li Y, Cumani R (2014) Understanding the drought impact of El Niño on the global agricultural areas: an assessment using FAO's Agricultural Stress Index (ASI). Food and Agricultural Organization of the United Nations. Retrieved from http://www.fao.org/3/a-i4251e.pdf on 9 Dec 2019

Salazar M, Pesigan A, Law R, Winkler V (2016) Post-disaster health impact of natural hazards in the Philippines in 2013. Retrieved from https://www.ncbi.nlm.nih.gov/pmc/articles/PMC4871893/#CIT0014 on 6 Dec 2019

Sato T, Nakasu T (2011) 2009 typhoon Ondoy flood disasters in Metro Manila. Natural Disaster Research Report of the National Research Institute for Earth Science and Disaster Prevention. Retrieved from https://dil-opac.bosai.go.jp/publication/nied_natural_disaster/pdf/45/45-04E.pdf on 9 Dec 2019

Surmieda MR, Lopez JM, Abad-Viola G, Miranda ME, Abellanosa IP, Sadang RA, et al. (1992) Surveillance in evacuation camps after the eruption of Mt. Pinatubo, Philippines. MMWR CDC Surveill Summ.41:963

The Climate Reality Project (2016) How is climate change affecting the Philippines, Retrieved from https://www.climaterealityproject.org/blog/how-climate-change-affecting-philippines on 6 Dec 2019

Thomas B (2016) The case of typhoon Yolanda (Haiyan). World Bank, Washington DC. Retrieved from http://agris.fao.org/agris-search/search.do?recordID=US2015601246 on 9 Dec 2019

United Nations International Strategy for Disaster Reduction (UNISDR) (undated) Good Practices. Available http://www.unisdr.org/eng/media-room/facts-sheets/fs-good-practices.htm. Accessed 18 Apr 2011

United Nations International Strategy for Disaster Reduction Secretariat (UNISDR).2009. ISDR (2009) Global Assessment Report on Disaster Risk Reduction: Risk and Poverty in a Changing Climate. United Nations, Geneva

United Nations Office for Disaster Risk Reduction (UNDRR) (2019) Hyogo Framework for Action. Retrieved from https://www.unisdr.org/we/coordinate/hfa on 5 Dec 2019

United Nations Office for Disaster Risk Reduction (UNDRR) (n.d.) Sendai Framework for Disaster Risk Reduction. Retrieved from https://www.preventionweb.net/sendai-framework/sendai-framework-for-drr on 5 Dec 2019

United Nations Office for Disaster Risk Reduction (UNDRR) (n.d.) SGDs with targets related to disaster risk. Retrieved from https://www.preventionweb.net/sendai-framework/sdg/target on 5 Dec 2019

United Nations (2013) A renewed global partnership for development. Retrieved from https://www.un.org/en/development/desa/policy/untaskteam_undf/glob_dev_rep_2013.pdf on 5 Dec 2019

United Nations (n.d.) Disaster Risk Reduction. Retrieved from https://sustainabledevelopment.un.org/topics/disasterriskreduction on 5 Dec 2019

United States Environmental Protection Agency (n.d.) How smoke from fires can affect your health. Available from https://www.epa.gov/pm-pollution/how-smoke-fires-can-affect-your-health

US Geological Survey. (n.d.). How can climate change affect natural disasters? Available from https://www.usgs.gov/faqs/how-can-climate-change-affect-natural-disasters#faq

US Geological Survey (n.d.) What are the long-term effects of climate change? Available from https://www.usgs.gov/faqs/what-are-long-term-effects-climate-change

Veenema TG, Thornton CP, Lavin RP, Bender AK, Seal S, Corley A (2017) Climate change-related water disasters' impact on population health. J Nurs Sch 2017(49):625–634. https://doi.org/10.1111/jnu.12328.[PubMed][CrossRef][GoogleScholar]

Waddell SL, Jayaweera DT, Mirsaeidi M, Beier JC, Kumar N (2021) Perspectives on the health effects of hurricanes: a review and challenges. Int J Environ Res Public Health 18(5):2756

Warner GS (2010) Increased incidence of domestic animal bites following a disaster due to natural hazards. Prehosp Disaster Med 25(2):188–190

Watson JT, Gayer M, and Connolly MA (2007) Epidemics after natural disasters. Emerg Infect Dis 13 (1)

Watts N, Adger WN, Agnolucci P, Blackstock J, Byass P, Cai W et al (2015) Health and climate change: policy responses to protect public health. Lancet 386(10006):1861–1914

Wendt U (2013) Typhoon Bopha and People Displacements in the Philippines. The State of Environmental Migration. Retrieved from http://labos.ulg.ac.be/hugo/wp-content/uploads/sites/38/2017/11/The-State-of-Environmental-Migration-2013-33-46.pdf on 9 Dec 2019

World Health Organization (n.d.) Afghanistan Post-disaster measles campaign completed in flood-affected provinces. Available from https://www.emro.who.int/afg/afghanistan-news/measles-campaign-flood-provinces.html

World Health Organization (2002) Control of communicable diseases and prevention of epidemics. Available from http://www.who.int/water_sanitation_health/hygiene/emergencies/em2002chap11.pdf. Accessed 18 March 2011

World Health Organization (2006) Communicable diseases following natural disasters: risk Assessment and priority interventions. WHO: Geneva. Available from http://www.who.int/diseasecontrol_emergencies/en/. Accessed 18 March 2011

World Health Organization (2010) Preventing the spread of disease after natural disasters. Available from http://pmmunc.synthasite.com/resources/PMMUNC3WHOPreventingDisease.pdf. Accessed 18 MARCH 2011

World Health Organization (2014) WHO Philippines Response to Typhoon Haiyan (Yolanda): the first six months. Retrieved from https://www.who.int/hac/crises/phl/sitreps/philippines_six_months_from_haiyan_may2014.pdf on 10 Dec 2019

World Health Organization (2021) Climate change and health. Available from https://www.who.int/news-room/fact-sheets/detail/climate-change-and-health

World Health Organization (2022) Hepatitis. Available from https://www.who.int/news-room/fact-sheets/detail/hepatitis-a

World Bank and the National Disaster Coordinating Council (2004) Natural disaster risk management in the Philippines: enhancing poverty alleviation through disaster reduction. Manila. Available from www.povertyenvironment.net/node/2031. Accessed 18 April 2011

World Bank (2005) Facts and figures on natural disasters. Available from www.worldbank.org/ieg/.../docs/natural_disasters_fact_sheet.pdf. Accessed 18 April 2011

Jinky Leilanie Del Prado Lu is a Research Professor and Scientist III at the National Institutes of Health, University of the Philippines Manila.

Chapter 5
Climate Change and Food Security in the Philippines: Impacts, Adaptation, and Climate Change Action

Maria Victoria Ortega-Espaldon and Celia dR. Medina

Abstract The climate emergency poses significant challenges to the agriculture sector which directly affects Philippine food security and economic growth. Climate change disrupts crop productivity, affecting domestic agricultural production and the development of agriculture as an industry. This situation coupled with the COVID-19 pandemic may result in food scarcity, hunger, and intensified poverty in the Philippines. Filipino farmers must rethink their ways of doing and integrate climate change in decision-making because "business as usual" is no longer an option. This chapter presents a brief review of the status of Philippine agriculture and its contribution to the national economy, an analysis of the profile of Filipino farmers relevant to climate and agriculture, and a discussion on how agricultural systems are connected and how they are influenced by the dynamics between the upland ecosystems (forests and watershed) and the downstream ecosystems (coastal and marine). This chapter deals with the multiple knowledge systems as sources of different climate change adaptation approaches and explores a holistic framework for managing climate risks to agricultural systems in the Philippines to ensure the well-being of Filipino farmers and society in general.

Keywords Philippines · Food security, climate smart agriculture · Climate change adaptations

Key Points

- The climate emergency threatens our farmers, our agriculture sector, and our economy. Climate change can result in substantial and further disruption in crop productivity affecting domestic agricultural production, the development of agriculture as an industry, and food security.

M. V. Ortega-Espaldon (✉) · Celia dR. Medina
University of the Philippines Los Baños, Los Baños, Philippines
e-mail: moespaldon@up.edu.ph

K. B. Berse et al. (eds.), *Climate Emergency in the Philippines*, Disaster Risk Reduction, https://doi.org/10.1007/978-981-99-7804-5_5

- Food security is measured not only in terms of availability of food products, but also access to affordable and safe food and utilization of sufficient and nutritious food.
- Adapting to impacts of a changing climate requires a holistic view of the role of ecosystems in reducing risks. Fortifying our watersheds, conserving our biodiversity resources, and protecting our coastal and marine ecosystems can secure water resources for irrigation, protect cultivated lands from sedimentation, and diversify food sources for common Filipinos.
- Forward-looking, risk-based agricultural planning is needed for climate adaptation to be effective.

Summary for Policymakers

Climate variability and extreme events threaten our agriculture and food security. Food security is measured not only in terms of availability of food products, but also access to affordable and safe food and utilization of sufficient and nutritious food.

- Agriculture is still the backbone of Philippine economy that can propel the transformation of our long-term economic growth as a nation.
- Climate change can result in substantial and further disruption in crop productivity and can affect domestic agricultural production, the development of agriculture as an industry, and food security in the long term.

Consistent with farmers' accounts on costs of disaster to agriculture and fisheries, it is estimated that damage cost to agriculture was about 37.2 billion pesos in 2018 alone (DA 2019). Various studies have also noted the adverse impacts of climate change on crop production using crop models and a general decline in crop production using climate scenarios in different time periods, 2030–2050.

- With increasing population of the country (annual growth rate at 1.14% in 2018) and base population at 100 million in 2015, we are looking at increasing domestic food demand. While we can complement our regular supply with imports, this policy of dependence is precarious since climate change impacts are global.
- For annuals like rice and corn, increased fertilization associated with temperature increases can result in increased yield. However, if the temperature threshold is reached particularly during specific growth stages, "the increased fertilization effect is negated" (Belder et al. 2005).

Economy-wide climate change may result in increase in prices of commodities like cereals (24%), fruits and vegetables (12.7%), meat (4.4%), sugar (7.5%), and roots and tubers (5.8%), based on modeling studies.

Below are some related results of studies on impacts of increase in temperature on crops in the tropics:

- Increase of 1 °C leads to about 8–14% decrease in rice yield during the dry season (Lansigan et al. 2007a).

- Increase at 2 °C mean air temperature could decrease rice yield by 5–12% (IPCC 2007).
- Increase up to 4 °C and CO_2 to about 350 ppm concentration can potentially reduce yield of maize by 30%; but at 700 ppm CO_2, yield can increase by 9% (Sahoo 1999).
- Temperature rise by 0.5–1.5 °C could reduce maize yield potentially by 2–5% (Aggarwal and Mall 2002).
- Increase in the minimum surface temperature in Los Baños and other areas in the Philippines led to a reduced rice yield (Lansigan 2007a).

Economy-wide climate change may result in increase in prices of commodities like cereals (24%), fruits and vegetables (12.7%), meat (4.4%), sugar (7.5%), and roots and tubers (5.8%), based on modeling studies (Rosegrant et al. 2016).

- The projected increase in consumer prices, especially cereals, will adversely impact prices of other food products like pork, beef, and chicken—major protein source of the country.
- Corn is the major ingredient of livestock feeds. While these increases in consumer prices seem positive for agriculture sector, this condition will not ensure that the economic condition of the farmers will improve.

Adapting to impacts of a changing climate requires a holistic view of the role of ecosystems in reducing risks to climate changes. A landscape approach to food security is our best bet. Fortifying our watersheds, conserving our biodiversity resources, and protecting our coastal and marine ecosystems can secure water resources for irrigation, protect cultivated lands from sedimentation, and diversify food sources for common Filipinos.

- For perennials, preliminary observations show that prolonged dry season resulted in observed increase in the yield of lanzones, mangoes, and rambutan. Studies to ascertain responses of major fruit crops to increases in temperature, changes in weather patterns, uncertain onset of rainy reasons, and prolonged dry and wet seasons are wanting and need to be examined closer to come up with better management strategies.
- For coconuts, impacts are not immediate but there is decreased production after extended period of drought, which can result in water stress during the stage of flowering and filling of the nuts (Comiso et al. 2014).
- In general, reports that the condition of agriculture sector, particularly smallholder farmers, would be worse under different climate scenarios.
- Pest outbreaks can be serious as in the case of cocolisap that had damaged coconut farms in the country and was associated to prolonged drought. The impact of infestation was considered monumental as 25 million Filipinos rely upon the industry, 3.5 million of whom are coconut farmers.

Because of the high vulnerability and risks faced by the agriculture sector to climatic impacts, the need to approach the future can be done through a risk-based adaptation planning in order to meet the country's goals. Risk-based

adaptation planning is rooted in the concept that impacts of climate change are widespread and responses to the risks should be systematically organized.

- Included in risk-based adaptation is risk transfer mechanism. Risk transfer involves the transference of risk or burden of financial loss to another party. For agriculture, this is crop insurance. Lansigan (2014) noted that science-based assessment of vulnerability and risks is imperative and is a sound basis for developing weather-based crop insurance suitable in the Philippine context.
- Integration of traditional knowledge and practices coupled with advances in science and technology to improve monitoring, forecasting, crop advisories, and practice can help build early warning systems for our farmers. There are already systems available to do this locally (e.g., Project SARAI of DOST-PCAARRD).
- The projected damage of climate change impacts on agriculture is huge. Estimates of damage costs from modeling studies have reached up to about 145 billion pesos per year until 2050 (Rosegrant et al. 2016).

> Integration of traditional knowledge and practices coupled with advances in science and technology to improve monitoring, forecasting, crop advisories and practice can help build early warning systems for our farmers.

While adaptation planning at the national level is imperative, place-based vulnerability and risks due to changing precipitation patterns, increasing temperature, and increasing and/or intensifying extreme climate events coupled with continuous environmental resources degradation need to be examined at the local level. Within the frame of landscape approach, vulnerability and risks analysis of local areas is imperative as basis for a risk-based strategic adaptation plan to ensure food security for the country.

5.1 Introduction

Climate change has undeniably been a major hazard and threat to the sustainability of agriculture production as well as to the well-being of Filipino farmers in the recent times. While there are many disparate factors affecting the conditions of farmers, climate change impacts such as extreme events like droughts, extreme rainfall causing massive flooding, changes in seasonality of precipitation, and increasing temperature exacerbate the socio-economic and political constraints that confront the sector and its stakeholders, especially the farmers. Coupled with what farmers have already experienced in the recent past, various studies have already noted out the adverse impacts of climate change on crop production (Lansigan et al. 2007b; Dait 2013; Comiso et al. 2014; Rosegrant et al. 2019). These studies consistently showed a general decline in crop production using different climate scenarios in different time periods, 2030–2050. With increasing population of the country (annual growth rate at 1.14% in 2018) and base population at 100 million in 2015, we are looking at increasing domestic food demand. While we complement our regular supply with imports, this policy of dependence is precarious. Climate change impacts are global and have also been wreaking havoc in crop production in developing countries like Thailand, Myanmar, and Vietnam, our main sources of rice imports. Hence, significant efforts must be done to protect our ability to produce food for domestic demand.

For the Philippine agriculture sector and the Filipino farmers, business as usual is not an option. Adapting to impacts of a changing climate requires re-evaluating the conventional approach to agricultural development including policy directions; traditional knowledge and practices of Filipino farmers; and the potentials of advances in science and technology to improve monitoring, forecasting, and crop advisories and practice.

The road to transforming agricultural systems in the Philippines, including transforming Filipino farmers' mindsets, is not a simple task. It requires a multidimensional understanding of the climate change context, the socio-economic context, policies, and institutions. With climate change, uncertainties in crop production are increasing risks of crop failures. Pushing science and technologies, and other extension services from research institutes and different agencies to a traditionally crippled agriculture sector can assist. More importantly, managing climate change impacts on agricultural systems requires a consideration of the interconnectivity of the ecosystems—from the watershed to the marine environment. This system's perspective is commonly known as ridge-to-reef approach, or landscape-seascape connectivity (Cruz and Bantayan 2011; Cruz-Trinidad et al. 2014). The complex interconnections of the various ecosystems in a landscape have been more complicated because of the archipelagic nature of the country. Major watersheds and larger island provinces straddle different administrative jurisdictions and this condition requires intricately linked management regimes—for which we are wanting up to the present.

This chapter aims to present four main concerns: (a) contribution of agriculture to the overall Philippine economy and its critical role in the survival and welfare of Philippine society; (b) the effects of climate change on agriculture, particularly on crop production; (c) potential adaptation strategies; and (d) consideration for general plan of action that mainstream climate change impacts.

5.1.1 Contribution of Agriculture in Philippine Economy

The first question we need to answer in the discourse on agriculture sustainability and climate change is this: How can the agriculture sector provide for the needs of the growing population of the country with climate change? Agriculture has been the traditional backbone of Philippine economy since the country earned its independence in 1945 and even before that. It has provided for the needs of the population in the past. With industrialization, urbanization, and increasing exodus of Filipinos from the farm to the cities, what is the future of agriculture in the country, given the changing climate? What could be our future?

To date, agriculture only accounts for measly 8% of the Philippines' domestic gross product (GDP), 57 percent (57%) of which is from the crop sub-sector (PSA 2019). Figure 5.1 shows the GDP from 2008 to 2018, which illustrates the gradual decline of the sector's contribution to the economy over a 10-year period (2008–2018) (PSA 2019). Service sector appears to be the major contributor to the country's GDP. In spite of this state, agriculture sector employs about 24 percent (24%) or about 10 million of the country's population (PSA 2018 cited by DA 2019). Among the agricultural sectors, crop production sub-sector exhibited the lowest growth in terms of GVA (gross value added) while livestock remains to be the strongest (Table 5.1). This may not be surprising because big private sector is the main player in the livestock industry while small and medium farmers, with a few large companies, are in the crop production. This lends crop production more vulnerable to climate extremes while big livestock companies have the capital to adjust their production system to cope with changes in climate. For example, advanced technologies and climate information are easily accessible to big poultry and livestock farms while small farmers remain to be business as usual. The decline in the growth rate of the agriculture sector may be attributed to climate change impacts.

DA (2019) reported that in 2017, damage due to typhoons has amounted to 8.2 billion pesos and damage has tremendously increased to 37.2 billion pesos in 2018. This translates to a 354 percent (354%) increase in damage between 2017 and 2018 alone (PSA 2019). Damages from Ondoy and Peping in 2009 had been estimated to

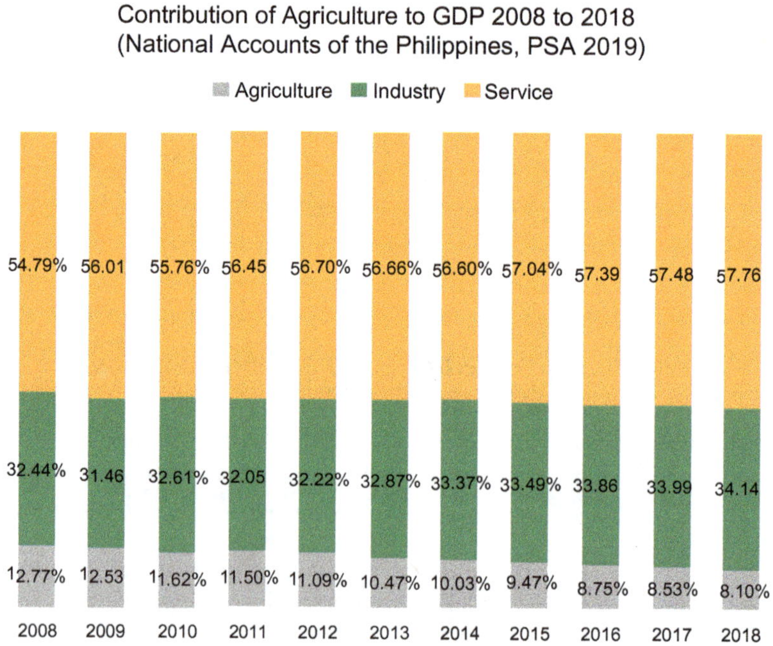

Fig. 5.1 Contribution of agriculture in the national economy (PSA 2019)

Table 5.1 Contribution of each sector of agriculture in the overall output (PSA 2019)

Agriculture sector (growth rate in %)	2008	2009	2010	2011	2012	2013	2014	2015	2016	2017	2018
Crops	22.24	1.20	6.61	17.72	−0.14	2.52	14.02	−8.15	3.00	9.77	3.82
Livestock	11.12	8.37	6.58	0.87	0.78	9.32	5.89	1.72	2.52	12.81	10.31
Poultry	11.91	5.54	9.78	4.97	5.43	5.68	11.20	2.12	3.69	6.26	7.62
Fishing	19.15	0.44	0.71	1.51	5.07	2.94	−1.14	−0.88	-4.43	6.48	8.96
Agricultural activities, services	16.71	5.98	4.75	12.73	0.14	3.70	12.50	0.26	8.18	8.44	6.06

amount to 206 billion pesos or about 1.8 percent (1.8%) of the country's GDP of that period while Super Typhoon Yolanda which devastated Leyte and other provinces on its path amounted to 571 billion pesos (Ravago et al. 2019). They added that the impacts of the climatic changes necessitate that the State should ensure that agricultural insurance are available to farmers. It is now evident that farmers' experience of weather events has actually been changing. This situation without full understanding renders this sector more vulnerable to changes in the frequency, duration, strength, and timing of rainfall. Hence, the need for more reliable weather information to supplement their memory and indigenous knowledge and practices and for them to be able to adapt is now gradually recognized by local stakeholders—farmers, cooperatives, and the local government units.

5.1.2 Profile of Filipino Farmers

A brief survey of national statistics noted that in 2015 approximately 11 million comprised the agricultural workers. They are mostly concentrated in rural areas (85%) in 2015. Of the total agricultural workers, 29.8 percent (29.8%) are wage earners and the rest are unpaid family labor and self-employed or about 70 percent (70%). It is unsettling to note the aging of the farming population with an average age of 57. Less and less young people are opting to go to agriculture.

Men and women are both active in farming, with varied responsibilities. Study conducted by Bordey et al. (2016) among Filipino rice farmers of major rice-producing area in the country noted that profiles of farmers can influence their capacity to manage their farms and how they deal with different stressors. Rice farming in this particular study showed that generally more men (87%) are engaged in rice farming than women. While poverty incidence grips about 25.2 percent (25.2%) of the total population (PSA 2019), the agriculture or the rural sector, which is estimated to comprise more than half of the total population of the country, the agriculture

Table 5.2 The impact of climate change on consumer prices of major agricultural commodities in % increase (Rosegrant et al. 2016)

Commodity	2030	2050
Cereals	14	24.3
Fruits and vegetables	6.1	12.7
Meat	2.1	4.4
Roots and tubers	2.5	5.8
Sugar	3.9	7.5

sector, concentrated in the rural areas, accounts two-thirds of poverty (Ravago et al. 2019).

In the study of poverty incidence in selected years from 1991 to 2015 (Ravago et al. 2019), it was evident that there was no significant reduction of poverty by regions. The National Capital Region has the lowest (3.9% share in 2015) followed by Region 4B (9.1% in 2015). Most of the regions continue to record double-digit poverty incidence. For instance, ARMM has a poverty incidence of 53.7 percent (53.7%) in 2015, now part of the Bangsa Autonomous Region of Muslim Mindanao or BARMM (Table 5.2). It is frustrating to note that while the country has reported economic growth, albeit small but steadily (4.7% in the average), poverty remains high, particularly in the rural areas. This is the main reason for the exodus of farm workers to the cities, which is spiraling the problems of crowding, degradation of social services, and vulnerable population in the urban areas.

With changing climate, the question becomes double edged: can agriculture be an engine to reduce poverty?

5.2 Impacts of Climate Change on Agriculture

Managing the climate change impacts on agricultural systems requires a consideration of the interconnectivity of the ecosystems—from the watershed to the marine environment. This system's perspective is commonly known as ridge-to-reef approach, or landscape-seascape connectivity (Cruz and Bantayan 2011). The complex interconnections of the various ecosystems in a landscape have been more complicated because of the archipelagic nature of the country. Major watersheds and larger island provinces straddle different administrative jurisdictions and require closer and intricately linked management regimes—for which we are wanting up to the present.

The impacts of a warming world on Philippine agricultural systems can be most profound. The deforestation of watersheds, land requirements for growing urban centers, and increasing needs for food supplies and industrial development of the country have summarily resulted to increased sedimentation of the rivers and pollution of waterways which eventually end up in the coastal and mangrove areas and marine ecosystems. Mangrove areas have also been converted to fishponds and coastal aquaculture. The impacts of these environmental degradation in the country

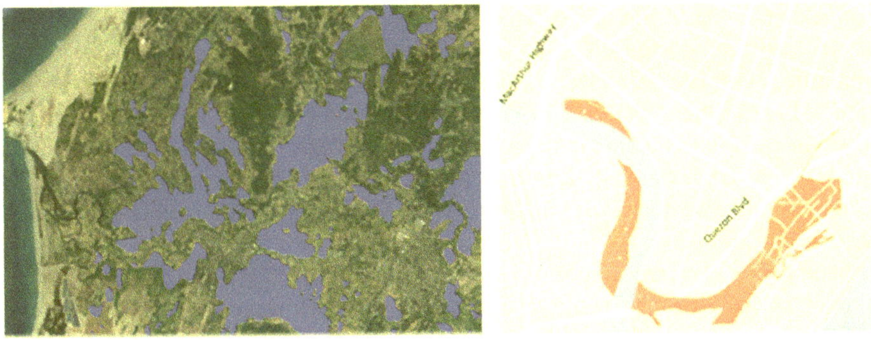

Fig. 5.2 Before and after typhoon Ineng (August 20 and August 23, respectively), Paoay, Ilocos Norte showing an estimated flooded rice area of 1223 ha (Project SARAI Enhanced Agriculture Monitoring Systems or SEAMS, 2019)

can exacerbate the impacts of climate change. The expected heavy precipitation on open and degraded lands in the uplands can translate to tons of top soils being eroded to the waterways and into the marine environment. This can lead to massive coral bleaching or the destruction of the known habitat and basis of rich fisheries of the country. The extreme events like droughts which are projected to occur more frequently in several areas can damage crop production due to insufficient irrigation water available, resulting from insufficient water supply from dams and rivers (Comiso et al. 2014), and poor and ill-managed irrigation facilities. For instance, typhoons and habagat or monsoon rains have caused massive floods both in terms of extent of area covered and duration of exposure, which resulted in agriculture damage amounting to about 37 billion pesos in 2017–2018 alone (DA Brief, ECCP Symposium August 2019), over 300 percent (300%) increase in damage from previous years.

Another example of the flood damages in rice-producing areas as a result of typhoon Ineng in 2019 is shown in Fig. 5.2. In 2018, NDRRMC reported crop losses of 8.96 billion pesos to rice and 4.49 billion pesos to corn due to typhoon Ompong (International name Mangkhut) alone.

5.2.1 Impacts on Agricultural Production

Crop productivity is dependent on climate. While degradation of natural resource base continued to adversely affect crop productivity, climate change impacts can be exaggerated and extensive. Temperature and other climatic attributes can significantly influence physiological productivity of major crops in the country. One of the well-studied crop is rice. Tibig (2001, cf: Comiso et al. 2014) noted that "changes in the onset of the rainy season, and the increase in nighttime temperatures… could significantly influence …overall productivity of farming system". Historically, gross value added or the difference between gross outputs and the intermediate outputs has the lowest dip during the most severe drought in the history of Philippine rice agriculture, 1998 El Nino. Decrease in the GVA is also noted in the El Nino events of 1972–1973, 1982–1983, and 1997–1998 (Comiso et al. 2014). The warm events' impact on corn is less compared with rice, as threshold for corn in terms of water availability is lower compared with rice (100 mm cumulated soil moisture versus 200 mm cumulative for rice), and so is with temperature (and associated ambient CO_2 concentration).

For annuals like rice and corn, increased carbon fertilization associated with temperature increase can result in increased yield. However, if the temperature threshold is reached particularly during specific growth stages, "the increased fertilization effect is negated" (Peng et al. 2005). His team noted that for every 1 °C increase in growing season night time temperature, a corresponding 10 percent (10%) decrease is expected. Lansigan et al. (2007a) corroborated to include the effects of 8 to 14 percent (8–14%) yield reduction depending on location. Below are some related results of studies on impacts of increase in temperature on crops in the tropics:

- Increase of 1 °C leads to about 8 to 14 percent (8–14%) decrease in rice yield during the dry season (Lansigan et al. 2007a).
- Increase at 2 °C mean air temperature could decrease rice yield by 5 to 12 percent (5–12%) (IPCC 2007).
- Increase up to 4 °C and CO_2 to about 350 ppm concentration can potentially reduce yield of maize by 30 percent (30%), but at 700 ppm CO_2, yield can increase by 9 percent (9%) (Sahoo 1999).
- Temperature rise by 0.5 to 1.5 °C could reduce maize yield potential by 2 to 5 percent (2–5%) (Aggarwal and Mall 2002).
- Increase in the minimum surface temperature in Los Banos and other areas in the Philippines led to a reduced rice yield (Lansigan 2009).

While these scientific studies, particularly from Sahoo (1999) and Aggarwal and Mall (2002), may be criticized to have been conducted in different environmental conditions and have not yet yielded a gross estimate of agriculture damage and losses due to climatic changes specific in the Philippines, it is evident that the rise in temperature by 1.5 °C could translate to a reduction in rice yield of a maximum of 14 percent (14%) per ha, excluding potential damages to pest which could be serious. Lansigan (2009) noted that in Los Banos and other areas in the Philippines, increase in the minimum surface temperature led to a reduced rice yield. For corn, the effect of temperature increase varies. For a 0.5–1.5 °C increase, potential yield can be reduced by 2 to 5 percent (2–5%) (Aggarwal and Mall 2002), but findings by Sahoo (1999) showed that for an increase of up to 4 degree centigrade at 700 ppm CO_2, yield can increase by 9 percent (9%). Dait (2015) reported that a climate scenario with increased 1 °C can result to a decrease in gross production value in agriculture by 19.21. This climate projection is not a farfetched reality. PAGASA (2018) has already noted an increase in temperature by 0.6 °C based on a long-term database on weather in the country and is expected to continue to increase by 2030, 2040, and 2050 projections using compounded climate scenarios. Gross rice production in the

Philippines (January–March) is about 4.6 MT in 2018 (PSA 2018). Implication of the results of the crop simulation model by Dait would be an estimated reduction of about 1 MT of rice production for first quarter. This calculation is done via DSSAT which considers the basic needs of crop physiological processes—soils, water, temperature, and genetic coefficient. Climate change is also expected to influence the emergence and intensity of occurrences of pests and diseases and the frequency and intensity of drought events. In general, the Philippine agriculture losses to climatic changes in production can be more than what can be expected. Agriculture uses about 80 percent (80%) of surface water for irrigation and occurrences of droughts could be devastating.

At the global scale, in the 2015 Assessment Report of IPCC, scientists recommend to leaders of the world to go for no less than 1.5 °C increase in global temperature. This target requires a call to climate action that will push for drastic reduction of emissions of countries, especially the industrialized countries, otherwise referred to as the G7. This has implications on the patterns of industrialization, use of fossil fuel, and transformation to clean energy. However, observations have shown that transformation of global economies to reduce GHG has not been in place to realize the goal of not going beyond 1.5 °C target. This is expected to create massive impacts to agriculture in developing countries like the Philippines, who do not have the means to undertake strategic climate adaptations actions. For changed patterns of precipitation, adverse impacts are expected. Modeling study (Dait 2015) has shown that while gross production value in agriculture is expected to increase if number of rainy days increase, gross production value can decrease by 0.24 for every 1 mm increase in precipitation.

More comprehensive study was made by Rosegrant et al. (2016) using four global climate models (GCM) and employing biophysical models to estimate the impacts of climate change and crop management regimes on crop yields such as DSSAT (Decision Support System for Agrotechnology Transfer developed in 2003) and WaNulCAS (Water, Nutrients, Soils). Results of the biophysical models were then inputted to macroeconomic models like IMPACT developed by IFRI to project food supply, demand, prices, impact of water supply, and climate change. Results of their study showed that a decrease in yields by an average of 1.7 percent (1.7%) across four GCMs can be expected in 2050 compared with baseline levels. Rosegrant et al. (2016) also noted that cereal production is expected to fall by 2050 compared with baseline levels. Corn, for example, is projected to decline by 13 percent (13%) while rice, 3.2 percent (3.2%). Because corn is mainly used for animal feeds, livestock production is likewise expected to be adversely affected. Declining cereal production coupled with increasing demand by increasing population (current annual growth rate at 1.7%, base population at 104M in 2015), agricultural commodity prices are expected to skyrocket, making the already poor people even more marginalized. Results of Rosegrant studies are alarming since increase in consumer prices by 2050 can increase by 24 percent (24%) (cereals) and fruits and vegetables by 13 percent (13%). Table from Rosegrant study shows the expected increase in consumer prices in 2030 and 2050 based on averages of the 4 GCMs runs.

The projected increase in consumer prices, especially cereals, will adversely impact prices of other food products like pork, beef, and chicken—major protein source of the country. Corn is the major ingredient of livestock feeds. Prices of fruits and vegetables are going to increase significantly after cereals. While these increase in consumer prices seem positive for agriculture sector, this condition will not ensure that the economic condition of the farmers will improve.

For perennials, preliminary observations show that prolonged dry season resulted to observed increase in the yield of lanzones, mangoes, and rambutan. Studies to ascertain responses of major fruit crops to increase in temperature, changes in weather patterns, uncertain onset of rainy reasons, and prolonged dry and wet seasons are wanting and need to be examined closer to come up with better management strategies. For coconuts, impacts are not immediate but there is decreased production after extended period of drought which can result in water stress during the stage of flowering and filling of the nuts (Comiso et al. 2014).

Based on all future climate scenarios, the condition of agriculture sector, particularly the smallholder farmers, would be worse (Sebastian et al. 2019). Agriculture as a livelihood and economic has become highly vulnerable and risky, with the physiological impacts on crops, pests and diseases, extreme events, and changes in the weather patterns. With increasing demand of increasing population, the future of the country can be grim.

5.2.2 Pest and Diseases Outbreaks

Outbreak of pests and diseases has also been projected to be more frequent because of climate changes. The devastation of coconut farms in different parts of the country due to coconut scale insect (CSI) or cocolisap (2010–2014) that nearly wiped out standing coconut palms has been partly attributed to prolonged dry season of 2010. In CALABARZON alone, CSI has affected 1.2 million coconut palms. The impact of infestation was considered monumental as 25 million Filipinos rely upon the industry, 3.5 million of whom are coconut farmers. In addition, 27 percent (27%) of the country's agricultural land are planted to coconut while 68 out of 79 provinces in the country grow coconuts (PCA 2010 cf. Villareal 2014).

Commonly known as cocolisap, this coconut scale insect or CSI has been identified as an introduced species (Aspidiotus rigidus) and not the native A. destructor as originally identified. Villareal noted that the spread can be attributed to two factors: changing wind patterns due to climate change and/or unintentional introduction of the pest through human economic activities (Villareal 2014). Its infestation was observed to spread to larger areas in the country due to temperature, relative humidity, wind direction, and planting density.

While science continues to understand the nature of cocolisap and look for solutions, gradual reduction of cocolisap infestation was observed after typhoon Glenda (international name: Rammasun), a Category 5 super typhoon with a highest wind speed of 259. It was deduced that the gradual disappearance of the cocolisap population is due to the combined effect of varied human interventions that include both organic and inorganic measures of pest management and the super typhoon Glenda.

Notable in the discussion of climate change and agriculture is the occurrence of pest and disease infestations. Infestation of farms and plantations in the country can reach emergency proportion as in the case of cocolisap that nearly destroyed the coconut industry. The event was attributed to introduction of new species (A. rigidus) to the Philippines and the prolonged dry season (Villareal 2014) which provided the new species favorable condition for reproduction. The other destructive disease linked to climate change is Fusarium wilt on banana. Modeling study which compared the current and future scenarios showed that the 21 percent (21%) susceptible areas to Fusarium under baseline climate condition in the Philippines is expected to increase to 27 percent (27%), which cover 91.2 percent (91.2%) and 28.5 percent (28.5%) of the country's highly and moderately suitable areas for banana, respectively (Salvacion et al. 2019). This is equivalent to approximately 67 percent (67%) of the country's total harvested area for banana.

First spotted in June 2019 in Isabela, corn farmers are now observing and concerned about a new pest emerging known as fall army worm or FAW (Spodoptera frugiperda). This has infected about 66 municipalities and 17 cities in the Philippines, as of October 2019 (DA Report). While this has been believed to be introduced inadvertently through economic trades and regular trade routes, the threat of the pest to proliferate in the country's food-growing areas is not farfetched given the country's favorable temperature. Early description of FAW in the US corn-growing areas (Sparks 1979) noted that the masses that contain few to hundreds of its eggs can hatch in 2 to 4 days with mean temperature between 70 and 80 °F and complete life cycle can take an estimated 4 weeks during the warm months of June to August. He also noted that life cycle can take 80–90 days in colder temperature. With the Philippines' more consistent warm weather and prolonged dry season due to changes in climate patterns, FAW outbreak of disastrous proportion is expected. As early as October 2019, Department of Agriculture has alerted corn farmers on the need for early detection and reporting of occurrence of FAW in the corn-growing areas in the country.

5.3 What Can Be Done: Toward Risk-Based Climate Change Adaptation Strategies in Agriculture

The impact of climate change on agriculture and the Filipino farmers in general is devastating and no action at this juncture is not a viable option. Rosegrant et al. (2016) highlighted the need for the country to develop adaptation and growth strategies to ensure agricultural production and "to underpin broader economic growth and structural transformation". Because of the high vulnerability and risks faced by the agriculture sector to climatic impacts, the need to approach the future can be done through a risk-based adaptation planning and meet the country's goals for food security, inclusive growth, and poverty reduction. Risk-based adaptation planning is rooted in the concept that impacts of climate change are widespread and responses to the risks should be systematically organized.

Agriculture in developing countries like the Philippines is highly vulnerable to climate variabilities and extreme events. Irrigation water, Emergence and occurrence of pests, uncertain weather conditions, and extreme weather events are climate risks confronted by the sector, and these risks are increasing. Vulnerability and risks in crop production vary from crop to crop and depending on location. Lansigan (2014) noted that science-based assessments of vulnerability and risks are imperative and are sound basis for developing risk transfer mechanisms suitable in the Philippine context. This is one strategy that is unfolding which is aimed to contribute to climate resilience of the country's agriculture sector. Risk transfer involves the transference of risk or burden of financial loss to another party. The significance of expansion of coverage of crop insurance cannot be underestimated in this respect. Consensus among policy and resource managers at the national level is already attained but efforts to put this in place remain to be seen. As a risk management strategy, crop insurance instruments have the potential to reduce climate change impacts for farmers. However, crop insurance products in the country have not reached far enough to make a difference. Studies showed that the low coverage of crop insurance in the Philippines is due to the premium cost and "apparent subjectivity and bias" in the damage assessment (Lansigan 2014).

With the advent of increasing risks due to climate variability and extreme events, the development of weather indexed-based insurance products is gradually taking shape. It has been proposed to address the weaknesses of the conventional crop insurance products—its subjectivity and bias in risk assessment. Lansigan in 2014 forwarded an approach to determine the threshold level or index critical in rice production at different phases of growth. In 2018, Lansigan and his team in Project SARAI (Smarter Approaches to Reinvigorate Agriculture as an Industry), a research project funded by the Department of Science and Technology (DOST), started to generate through crop simulation modeling, on-site-specific climate risk profiles of various locations that can be used by the Philippine Crop Insurance Company (PCIC) to improve the current crop insurance products. These new products are hope to address the lack of appropriate threshold levels for quick and objective damage assessments. The paucity of automatic weather stations to gauge or base the threshold level in the country is still something we need to struggle with. Along this concern, a silver lining is being expected as the Philippine weather agencies known as PAGASA (Philippine Atmospheric, Geophysical and Astronomical Services Administration) and DOST have already began to pursue modernizing weather monitoring. One of the constraints in the implementation of WIBI is in making the protocol more efficient. Smooth cooperation between agencies is pre-requisite for cutting the bureaucratic process of government-sponsored crop insurance program.

5.3.1 Integrated Early Warning System

Increased frequency of extreme events and climate variability in the country has been observed to result in catastrophic loss of lives, and for the agriculture sector, devastation of farms and plantations. These climate change impacts vary in scale and location—specificity of events and impacts is expected. Adaptive capacity of the sector to cope also varies by location and by commodity. Agrometeorological early warnings then must be commodity and place specific and must be based on local conditions relevant to agricultural activities (Yin 2014). Weather information from reputable sources must be utilized to inform farming communities of the best planting dates for different crops, or when is the best time to fertilize, or when to conduct pest management and harvest. Weather forecast and outlooks can be used to plan the activities of farmers and reduce potential damage of not knowing amount of rainfall to expect in during the crop-growing season.

The integrated early warning system designed for agriculture can be adapted from the DRR framework of the WMO. The framework is anchored on understanding the vulnerability and risks faced by agriculture sector—rainfall, temperature and humidity and the implications of these weather attributes to the different phases of growth of crops. Continuous monitoring and processing of both real-time and historical weather data can generate forecasts and outlooks that are needed by farmers in advance. In the Philippines, a classic example of how farmers use climate information in farming activities is the Climate Field School of the Municipality of Dumangas in Iloilo City. It is managed and operated by the LGU Agriculture Office, in partnership with PAGASA. The weather data is sent automatically to PAGASA for processing and relay back to the LGU weatherman and disseminated to the farmers in the dialect, with analysis relevant to major farming activities. Currently, Project SARAi is working with the municipality to advance the combined use of remote sensing and automatic weather stations to generate rainfall and drought outlook in 5–6 months. This is sufficient information to inform the farmers of the best dates and crops and varieties to plant to reduce risks to flood and drought or optimize the potential availability of moisture in soils during the duration of crop growth. This protocol is now being set up in some selected areas in the country with assistance from different agencies—DOST, DA, universities, DILG, and LGUs (Fig. 5.3).

5.3.2 Integrated Crop Monitoring and Forecasting

While space technology has been used traditionally in national defense of large and developed nations, its application in agriculture of developing countries is emerging. For example, India has set up a system using remote sensing and satellites in monitoring 15 major crops through the Mahalonobis National Crop Forecast Center (NCFC) of the Ministry of Agriculture (www.ncfc.gov.in). It uses its own satellites to monitor their major crops and reports to the Ministry of Agriculture. Crop

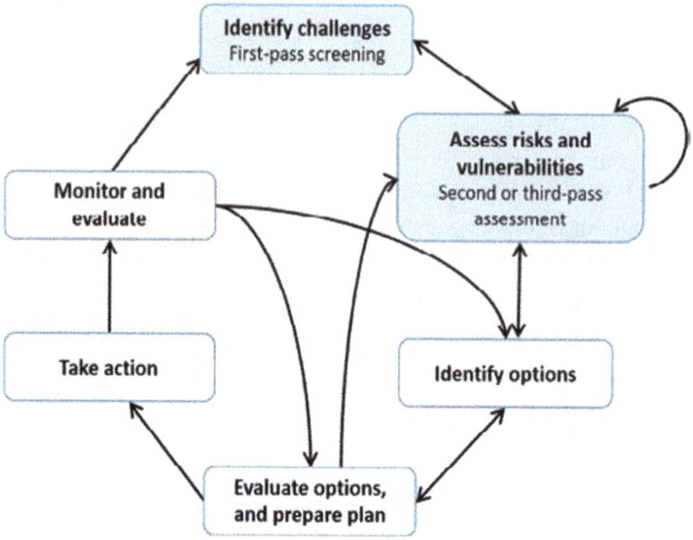

Fig. 5.3 Protocol for early warning system designed for agriculture that can be adopted from the DR Framework of the WMO

advisories emanating from the center are given to the extension office of the Ministry for dissemination nationwide.

The use of remote sensing and satellites in the Philippines has been gradually gaining grounds, including its utility in agriculture.

For instance, the Philippine Rice Information System or PRISM, the first satellite-based rice monitoring system in the Philippines, is developed by CGIAR researchers at the International Rice Research Institute or IRRI in collaboration with the Philippine Department of Agriculture. In operations since 2014, PRISM collects data from satellites and employs various advanced technologies of crop modeling, cloud computing, and statistics to produce data and maps about rice.

Another system that uses remote sensing and satellites is coming from the Department of Science and Technology or DOST-related efforts known as SARAI or Smarter Approaches to Reinvigorate Agriculture as an Industry. Project SARAI is a research and development program that aims to establish a real-time crop monitoring and forecasting system for crops other than rice. It covers major crops like corn, banana, coconut, coffee, cacao, tomato, soybean, and sugarcane. The system is based on developed crop models and different crop phenologies, ensemble of weather and climate data, real-time weather data via automatic weather stations built in strategic areas, remote sensing, geographic information system (GIS), and field monitoring reports. The knowledge products near real-time crop monitoring information and seasonal crop forecasts. Some examples of this include site-specific crop growth stage, crop health, weather and crop relationships, possible pest and disease infestations, and drought forecasting for agriculture. Mobile applications for pest and disease identification and harvesting advisories were also developed for use by different stakeholders in agriculture such as farmers, local government executives, traders and entrepreneurs. Examples are—SPIDTech for pest and disease identification, CAPHE for coffee harvesting advisories, BanaTech for banana yield and harvesting estimates, Bantay SARAI for online-based seasonal crop, livestock and poultry monitoring (Espaldon et al. 2018; Cabangbang et al. 2019).

In October 21, 2022, President Ferdinand Marcos, Jr. issued a directive dated PBBM-2022-086-90 for the nationwide implementation of SARAI as the digital platform of agriculture, to wit "to implement, in coordination with the DOST the SARAI Project nationwide and ensure that areas with no or limited internet connectivity will be updated with the information provided by the project."

5.3.3 Traditional Ecological Knowledge in Climate Change Adaptation

The discussion of climate change adaptation is not complete without putting in the centerfold of the discourse the contribution of traditional ecological knowledge. While scientific movement is gradually making impacts on how agriculture concerns due to climate risks are managed at the national level, the local and indigenous ecological knowledge systems are still found in the practice among indigenous people. In 2003, Monsalud et al. recorded that indigenous practices for crops and livestock to reduce the impacts of El Niño are still used in provinces such as Bulacan, Pampanga, Catanduanes, and Camarines Sur. These include the practice of smoking or pausok, use of herbal pesticide, lighting of lamps at night to scare insects, and putting of poison frogs in the field as plant pest controls. In Catanduanes and Camarines Sur, farmers practice the smudging, soapy water spray or water mixed with small pepper juice extract. Documentation of the Philippine indigenous knowledge on climate change adaptations has been replete with wisdom as shown in various studies done at homefront (Coxhead and Rola 1998; Bornales 2004; Monsalud and Montesur 2003; Peras 2005; Ducusin et al. 2019; Nelson et al. 2019; Zamora et al. 2020). These indigenous knowledge systems which are already documented have shown the rich well spring of knowledge that local communities can learn from and can be supported by the agriculture department who is in charge of ensuring that farms and farmers can reduce the damage that climate risks can cause. How these qualitative studies and information can enhance the adaptive capacity of the farming communities, along with advances in technologies of weather and climate information, crop advisories emanating from systems like PRISM and SARAI require that communities be equipped not only with their own knowledge tested through time, but also the use of advancing in science and technology.

5.4 Conclusion

The projected damage of climate change impacts on agriculture is huge. Estimates of damage costs from modeling studies have reached up to about 145 billion pesos per year until 2050 (Rosegrant et al. 2016). Consumer prices of major food products are expected to increase based on the average of all climate scenarios. Reduction in crop yields, especially corn and other cereals, is expected to result in increase in the prices of animal feeds. Emergence and re-emergence of pests and diseases can be expected. Without doing anything would be similar to riding a speeding train without a brake—it could mean an economy-wide collapse. Global analysts as well as local decision-makers have reached a consensus that we need to implement a concerted effort to avert the impending global and local systems collapse due to climate changes. The urgency for climate action has never been more real than the present.

A changing climate system in the Philippines results in compounding impacts to agriculture and well-being of small-scale farmers in the country.

Based on studies, climate change impacts can possibly result in 10 to 15 percent (10–15%) reduction of potential yield of rice, our staple crop. Climate change impacts also include potential increased damage to crops due to emerging pest and diseases with increasing temperature, prolonged drought, and intense humidity. Crops are also highly vulnerable to flooding which are expected impacts of more intense tropical cyclones. Without systems of precaution and technical support, our farms are left to the intensifying onslaughts of extreme weather events and increasing weather variabilities.

To address the threat of intensifying climate impacts, considered one of the greatest threats to human kind, especially in the developing countries, a national risk-based adaptation planning is called for. While adaptation planning at the national level is imperative, place-based vulnerability and risks due to changing precipitation patterns, increasing temperature, and increasing and/or intensifying extreme climate events coupled with continuous environmental resources degradation need to be examined closely to lay out site-specific adaptation strategies. Based on location-specific vulnerability and risk analysis, appropriate adaptation strategies can now be in place.

Risk transfer mechanism is an integral part of risk-based adaptation planning for agriculture. Although crop insurance has been in the Philippines for many years, low coverage and low acceptance among farmers are still its major implementation concern. The crop insurance sector has also been criticized severely as too bureaucratic, expensive, tedious, and subjective in its damage assessment protocol. With climate change, the need for an improved crop insurance system has become more pronounced. Because agriculture is highly dependent on weather and climate, timing of release of support after climate-related disasters happen in a region. In this concern, crop insurance product weather-based indices are seen to be highly relevant toward making farmers more resilient. Aside from using weather-based indices, the coverage must include not only rice and corn, but other major crops like banana, coconuts, and sugarcane. Subsidy for crop insurance maybe considered at first phase until farmers can fully recover from damages.

Another feature of a national risk-based adaptation planning is the nation-wide expansion of application of information technology, space technology, crop modeling, and automatic weather stations in agriculture. The digitalization of agricultural and market information can help in making these accessible to ordinary farmers. Infrastructure support, massive capacity building, and changing mindsets are requisites to implement climate change adaptation strategies. With climate change, collaboration of different institutions like Department of Agriculture, Department of Science and Technology, DILG, DICT, NEDA, State Universities and Colleges (SUCs), CCC, and OCD with resource allocation from the Department of Budget will be an important feature of "bayanihan" if we are to succeed in our target to make agriculture a major prime mover of the country's economy in the future.

Rice is our major staple, and it has become the icon of food security. It is no wonder that huge attention is given to rice. According to Rosegrant et al. (2016), increasing rice productivity and increasing irrigated rice area may increase the Gross Domestic Product or GDP. Increasing rice productivity entails use of new crop management strategies such as crops and varieties that are flood resistant or drought resilient, depending on the risks of a place. It also entails efficient crop protection strategies. Development of new sources of irrigation water, e.g., more groundwater than surface water, and rehabilitation of watershed as traditional water sources should be part of the adaptation plans—integrated into the comprehensive land use plans of each municipality.

To transform agriculture into an engine of poverty reduction, this requires a multi-scale approach of adaptation and a combination of many epistemological forms of knowledge. One emanates from wisdom of local knowledge and experience and the other one comes from the science and technology branch of knowledge production. At this juncture of the Philippine agriculture's history, it would be wiser to pool the resources coming from many sources.

Strengthening the adaptive capacity and preparedness of our farming communities is imperative to enable us to deal with the impacts of climate change. Scientific knowledge admittedly can be very useful in this regard. It has produced tools and processes that can reduce the risks in farming. This list is getting longer—we have the crop advisories coming out of automatic weather stations located in different parts of the country, crop modeling for crop forecasting, GIS and remote sensing to monitor and forecast potential harvests, and losses in cases of extreme events.

The challenges of climate change and food security are global in scale. It requires the agriculture sector to be adaptive and able to mitigate the effects of changing climate to agricultural production. Risk-based adaptation planning involves making agriculture climate resilient, hence regular vulnerability and risks must be in place so strategies can be made more adaptive. The risk-based adaptation planning generally aims to (1) increase agricultural productivity and income through sustainable agriculture; (2) build an adaptive and resilient agriculture sector; and (3) exploring ways to reduce greenhouse gas emissions. The goal is to identify the most suitable combination of strategies in terms of local climate, physical characteristics, and socio-economic condition because it will be the basis of a successful intervention strategy. Large-scale implementation demands an immense amount of investment and requires the support from various stakeholders from the national down to the farm level where impacts are felt by the households.

The Philippines is not behind any other nations who incorporates climate change adaptation as well as contributing to reduction of GHG emission in their development agenda for 2030 through 2050. Efforts have already been initiated at the central Department of Agriculture with the establishment of climate-resilient communities called AMIA villages, coordinated by the DA System-wide Climate Change Office (SWCC).

On the other hand, the DA Field Operations Office started mainstreaming the use of remote sensing and GIS protocol being advanced by SARAI Enhanced Monitoring System first in Region 4B (Mindoro, Marinduque, Romblon, and Palawan). In 2019, eight more regions in the country were added as pilot sites for the enhanced system. The system includes setting up a crop monitoring system for major crops such as corn, coffee, cacao, and banana using remote sensing. The end goal is to set up a nationwide crop monitoring system that can make use of the climate information being provided by PAGASA and different agrometeorological stations located all over the country for crop advisories. The system is also expected to deploy nationwide the SARAI-enhanced monitoring system (SEAMS) protocol, including drought assessment and forecasting system, developed under the auspice of the Department of Science and Technology (DOST) through the Philippine Council for Agriculture, Aquatic and Forestry Resources Research and Development (PCAARRD). Other technologies in the pipeline of R&D are expected to be mainstreamed as they become available like crop models and pest management tools for major crops (corn, banana, coffee, cacao, tomato, sugarcane, and soybean). The information generated here are useful in generating crop advisories as to: when to plant, what alternative crops to plan, irrigation management, pest management, and damage assessment on larger scales (barangay, municipal, provincial). Capacity building among the agriculture sector is ongoing to ensure that the stakeholders are able to understand and able to use the information to complement their own traditional practices. Adaptation is a process, and monitoring must be undertaken to determine the best measures of living with a new climate.

References

Aggarwal PK, Mall RK (2002) Climate change and rice yield in diverse agro-environments in India: II. Effect of uncertainties in scenario and crop models on impact assessment. Clim Change 52:331–343

Belder P, Bouman BAM, Spiertz JHJ, Peng S, Castaneda AR, Visperas RM (2005) Crop performance, nitrogen and water use in flooded and aerobic rice. Plant Soil 273:167–182

Bordey FH, Beltran JC, Litonjua AC, Launio CC, Manalili RG, Mataia AB, Malasa RB, Relado RZ, Tanzo IR, Paran SJC, Yusongco CG (2016) Profile of an Asian rice farmer, competitiveness of

Philippine rice in Asia. Philippine Rice Research Institute: International Rice Research Institute, pp 172
Bornales RG Jr (2004) Adaptation and vulnerability of subanen community to environmental conditions in Mt. Malindang National Park, Philippines, UPLB PhD Thesis
Cabangbang P, Espaldon MVO, Mendoza HD, Lacson JAA, Ducusin RJC, Eslava DF, Dorado MA, Ballaran VG, Ebuenga ME, Khan CL, Salazar BM, Protacio CM, Aguilar EA, Bato VA (2019) Paving the pathway for climate smart agriculture among small scale farmers in the Philippines. In: Shirato Y, Hasebe A (eds) Climate smart agriculture for the small scale farmers in the Asian and Pacific region. Japan and Taiwan, pp 227–240
Comiso JC, Balance CA, Sarigumba TI, Espaldon MVO, Lansigan FP, Baguinon NT, Birosel RC, Coladilla JO, Cruz RVO, Florece FM, Guerrero III RD, Lasco RD, Perez RT, Pulhin JM, Tibig LV (2014) Changing Philippine climate: impacts on agriculture and natural resources. The University of the Philippines Press, Dilliman, Quezon City, pp 72–76
Coxhead I, Rola AC (1998) Economic development, agricultural growth and environmental management: what are the linkages in Lantapan. In SANREM CRSP-Philippines conference, Malaybalay, Bukidnon, Philippines pp 18–21
Cruz RVO, Bantayan NC (2011) Climate change adaptation plan for INREM river basins in the Philippines. TA-6420 (REG): promoting climate change adaptation in Asia and the Pacific. Asian Development Bank, Pasig City
Cruz-Trinidad A, Aliño PM, Geronimo RC, Cabral RB (2014) Linking food security with coral reefs and fisheries in the coral triangle. Coast Manag 42(2):160–182
Dait JMG (2013) Effect of climate change on Philippine agriculture. Int J Sci Res
Dait JM (2015) Effect of climate change on Philippine agriculture. Int J Sci Res 4(9):1922–1924
Department of Agriculture (DA) (2019) Rebuilding, reforming, digitializing the Philippine agriculture and fisheries sector. A Briefer on Philippine Agriculture. Presented in European Chamber of Commerce Philippines (ECCP) Forum, 2019. (www.da.gov.ph)
Ducusin RJC, Espaldon MVO, Rebancos CM, De Guzman LEP (2019) Vulnerability assessment of climate change impacts on a Globally Important Agricultural Heritage System (GIAHS) in the Philippines: the case of Batad Rice Terraces, Banaue, Ifugao, Philippines. Clim Change 153(3):395–421
Espaldon MVO, Mendoza HD, Quiray AE, Dorado MA (2018) Using digital technology in enhancing climate resilience of Philippine agriculture: the SARAI-enhanced monitoring system. In: Sajise PE, Lasco RD, Cadiz MCH, Bantayan RB (eds) Policy-Enabling environment for climate change adaptation: some experiences in Southeast Asia. Southeast Asian Regional Center for Graduate Study and Research in Agriculture (SEARCA), Los Baños, Laguna, Philippines
Espaldon et al (2019) Project SARAI enhanced agriculture monitoring systems or SEAMS. Project report submitted to PCAARRD DOST
IPCC 2015 (2007) Climate change working group report, pp 137–154
Lansigan FP (2014) Implementation issues in weather index-based insurance for agricultural production: a Philippine case study. SEARCA
Lansigan FP, De Los Santos WL, Coladilla JO (2000) Agronomic impacts of climate variability on rice production in the Philippines. Agric Ecosyst Environ (Elsevier Science B.V.) 82:129–137
Lansigan FP, Cruz RVO, Lasco RD (2007a) Linkages of forest and water resources use and management for sustainable rice production in the Philippines. In: Banta SJ (ed) Rice, water and forests. Asia Rice Foundation, Los Banos
Monsalud FC, Montesur J (2003) Stabilization of Upland agriculture under El Nino-induced climatic risk: impact assessment and mitigation measures in the Philippines. A CGPRT Centre working paper no. 71, Philippines
Nelson GL, Zamora OB, de Guzman LEP, Tatlonghari RV, Espaldon MVO, Brillon JA (2019) J Environ Sci Manag 22–1:87–98
Oscar M. Lopez climate change center. Working group II

Peras J (2005) Vulnerability and adaptation strategies of Pantabangan-Caranglan Watershed communities to climate variability and extremes, Philippines. MS Thesis. University of the Philippines Los Banos

Philippine Statistical Authority (PSA) (2019) Performance of Philippine agriculture report. www.psa.gov.ph

Ravago MJV, Balisacan AM, Sombilla MA (2019) Current structure and future challenges of the agricultural sector. In: Rosegrant MK, Sombilla MA (eds) The future of Philippine agriculture under a changing climate policies, investments and scenarios

Rosegrant MW, Perez N, Pradesh A, Thomas TS (2016) The economywide impacts of climate change on Philippine agriculture. Climate change policy note 1. International Food Policy Research Institute

Rosegrant MW, Sombilla MA (2019) The future of Philippine agriculture under a changing climate: policies, investments and scenarios (No. 144101). ISEAS

Sahoo SK (1999) Simulating growth and yield of maize in different agroclimatic regions. MS thesis, Indian Agricultural Institute (Unpublished)

Sebastian L, Bernardo EB (2019) Making the Smallholder Farmers in Southeast Asia Climate Smart - The CCAFS R4D Thrust. In: Shirato Y, Hasebe A, eds. Climate Smart Agriculture for the Small-Scale Farmers in the Asian and Pacific Region. Taipei, Taiwan: Food and Fertilizer Technology Center (FFTC) for the Asian and Pacific Region and National Agriculture and Food Research Organization (NARO) of Japan, pp 201–226

Sparks AN (1979) A review of the biology of fall army worm. Fla Entomol 62(2):82–87

Tibig LV (2001) Responding to the threats to climate change on Philippine agriculture. In: Villarin R (ed) Disturbing climate. Manila Observatory, Ateneo de Manila University, Manila

Villareal RV (2014) The coconut scale insect infestation in the Philippines. StarScience. Philippine Star. 28 August 2014. Downloaded: 7 Feb 2020

Yin JI (2014) Agrometerological early warning system: a service infrastructure for climate smart agriculture. Korean J Agric For Meteorol (Korean Society of Agricultural and Forest Meteorology) 6(4):403–417

Zamora OB, de Guzman LEP, Tatlonghari RV (eds) (2020) Stories of adaptation to climate change. Department of Agriculture-Bureau of Agricultural Research and College of Agriculture and Food Science, UPLB, 130 pp. ISBN 978-971-0347-64-3. https://drive.google.com/file/d/1kvfIG6ZiCByWlwn2jfrtJjEy17xxNJ0G/view?usp=sharing

Maria Victoria Ortega-Espaldon Ph.D. is a human geographer, UP Scientist, and Professor at the School of Environmental Science and Management (SESAM), University of the Philippines Los Baños. She served as Vice Chancellor for Research and Extension and as Dean of SESAM Currently, she serves as Chair of the UPLB Climate and Disaster Risks Study Center and as Program Leader of Project SARAi or Smarter Approaches to Reinvigorate Agriculture as an Industry in the Philippines, funded by the Philippine Council for Agriculture, Aquatic and Natural Resources (PCAARRD) under the Department of Science and Technology (DOST).

Celia dR. Medina Ph.D. is a Professor in entomology at the University of the Philippines Los Baños. She is an insect ecologist and a pest management specialist pursuing innovations that provide solutions to pest problems that beset country. She has turned her work on the non-destructive detection of mango pulp weevil into the establishment of a soft X-ray based mango screening center in Palawan. Her studies on population ecology were rolled into a coconut scale insect monitoring software adopted by the Philippine Coconut Authority, and her ecological studies of mango pests were converted into a web application for insecticide resistance management.

Chapter 6
Water Security in the Context of the Global Climate Emergency

Alejandro N. Ciencia Jr., Lorelei C. Mendoza, Gladys A. Cruz, and Rainier Ric de la Cruz

Abstract In March 2019, many Metro Manila residents were shocked to find that their water faucets had suddenly run dry. The water supply crisis that hit Metro Manila, the worst it had experienced in a decade, was indicative of "water stress" which is symptomatic of "water insecurity," a term which describes the condition of people when they have limited access to adequate quantities of potable water. Climate change exacerbates water insecurity manifesting in the form of unpredictable occurrences of heavy rainfall and severe flooding or drought. This chapter discusses how water insecurity in the Philippines now constitutes an emergency situation linked to climate change. Addressing water insecurity ought to be an utmost concern for national leaders and institutions because water is essential to human survival and well-being. This chapter argues that to make Filipinos water secure, policymakers must address the major problem that plagues the country's water resources sector: poor management. The sector is regulated by several legal frameworks, with each law separately prioritizing some aspect of the sector and assigning responsibilities to an assortment of departments and agencies at various levels of government. The resulting situation is that of conflicting legal provisions, discordant policies, piecemeal accomplishments, and overlapping mandates. To address the problem of poor management, a single, holistic, and integrated framework to regulate all aspects of the water resources sector is needed.

Keywords Water security · Water stress · Climate change · Climate emergency

---

A. N. Ciencia Jr. (✉) · L. C. Mendoza · G. A. Cruz · R. R. de la Cruz
University of the Philippines Baguio, Baguio, Philippines
e-mail: anciencia@up.edu.ph

L. C. Mendoza
e-mail: lcmendoza@up.edu.ph

G. A. Cruz
e-mail: gacruz@up.edu.ph

R. R. de la Cruz
e-mail: rbdelacruz1@up.edu.ph

K. B. Berse et al. (eds.), *Climate Emergency in the Philippines*, Disaster Risk Reduction, https://doi.org/10.1007/978-981-99-7804-5_6

Key Points

- The problem of water insecurity is also attributable to the poor management or governance of the water resources sector.
- Government must recognize that there is a water insecurity problem linked to the climate emergency and treat this with utmost concern and urgency.
- Policymakers at the national level must allocate funds for R&D and the construction and maintenance of water sewerage systems and wastewater treatment plants.
- Policymakers at the local level must faithfully monitor the quality of water sources located/available in the city or municipality and the granting of permits allowing water extraction activities.

Summary for Policymakers

The problem of water insecurity in the Philippines is generally attributed to the concurrence of dwindling water supply and rapid population growth. Climate change exacerbates water insecurity by threatening the quantity and quality of water that people can draw from available water sources. Water insecurity is experienced by people when their access to adequate amounts of safe and clean water becomes limited or uncertain. The poor in Philippine society are the most vulnerable to the problem of water insecurity on account of their limited resources which prevent them from responding well to constraints and unanticipated circumstances.

The problem of water insecurity is also attributable to the poor management or governance of the water resources sector.

- The sector is regulated by several legal frameworks, with each law separately prioritizing some aspect or aspects of the sector, and assigning responsibilities or mandates to an assortment of departments and agencies at various levels of government. The resulting situation is that of conflicting legal provisions, discordant policies, and overlapping mandates—which, in turn, lead to inefficient operations, uncoordinated undertakings, high coordination costs, poor enforcement or non-enforcement of regulation, piecemeal accomplishments, governmental inaction, etc.
- Poor management of the water resources sector can also be explained by other problems of the Philippine bureaucracy—e.g., meager budgets, poor staffing, limited manpower, lack of field presence for some national agencies, limited opportunities for staff development and upgrading, etc.
- Poor management of the sector is explained by the absence of a single overarching body which can harmonize, synchronize, coordinate, and rationalize water-related policies and processes in the country. Water security is a multi-dimensional concept.

Climate change exacerbates water insecurity by threatening the quantity and quality of water that people can draw from available water sources. The poor in Philippine society are the most vulnerable to the problem of water insecurity on account of their limited resources which prevent them from responding well to constraints and unanticipated circumstances.

To address the problem of poor management, a single holistic and integrated framework to regulate all aspects of the sector is needed.

To address water insecurity, policymakers at the national level must first of all recognize the problem as a matter of utmost concern. The following policy recommendations are humbly offered:

a. For **Congress** to prudently enact but with dispatch a law rationalizing the country's water laws. It is preferable for such a law to contain provisions that recognize

   - The threat of climate change on our country's water supply.
   - The need to rationalize existing economic regulations and the financial aspects of providing water services to Filipinos, and to make potable water more accessible and affordable.
   - The importance of reviewing the various legal frameworks that govern the water resources sector in the country.
   - The urgency of addressing the sector's governance problems.

b. For the **Office of the President** to see to it that in preparing the country's annual national budgets there is ample funding for the following:

   - Research and development (R&D) in the areas of water desalination, water filtration, water purification, water recycling, rainwater harvesting, wastewater treatment, etc.
   - The construction and maintenance of water sewerage systems and wastewater treatment plants all over the country as part of the government's infrastructure program.

To address water insecurity, policymakers at the local level, in coordination with national agencies and their field offices, must faithfully implement water-related regulations—e.g., those pertaining to:

- The monitoring of the quality of water sources located or available in the city or municipality.
- The granting of permits allowing water extraction activities within the city or municipality.

The construction and maintenance of water sewerage systems and wastewater treatment plants all over the country must be part of the government's infrastructure program.

6.1 Introduction

On July 28, 2010, the United Nations through its General Assembly adopted Resolution 64/292 to officially recognize the "right to safe and clean drinking water and sanitation as a human right" (UN 2010). By doing so the international body explicitly made it the duty of states to provide people with potable water. But if one examines the resolution and other pertinent UN documents, one finds that, in addition to providing water that is safe and clean, states are also obligated to ensure that the water provided to people is sufficient, accessible, and affordable. For the UN, the human right to water and sanitation is to be understood as the right to be "water secure." The UN defines "water security" as "the capacity of a population to safeguard sustainable access to adequate quantities of acceptable quality water for sustaining livelihoods, human well-being and socioeconomic development, for ensuring protection against water-borne pollution and water-related disasters and for preserving ecosystems in a climate of peace and political stability" (UN Water 2013).

But what are the practical implications of pursuing policies that promote water security and are grounded on the claim that access to safe and clean drinking water is a human right? What specifically does it mean for government to pursue water security-oriented policies, especially when we recognize that factors like population growth and climate change seriously challenge government's ability to provide water to people? How can policymakers address the dilemma of ensuring that water is accessible and affordable to all without undervaluing it and while protecting it as a precious resource?

This chapter draws attention to the importance of water as a resource and examines the concept of water security and how it relates to the climate emergency. It will highlight the need to pay greater attention to the problem of water insecurity in the Philippines. It will also review the findings of studies conducted on water security in Baguio City as a jumping off point for drawing out lessons and insights that can inform pertinent policies and interventions in the Philippines. A list of policy recommendations will be provided.

6.1.1 Some Facts and Water-Related Concepts

Water is a vital resource. It supports life and is essential to human survival. According to some estimates, an average person "can go without food for about three weeks but would typically only last three to four days without water" (Spector 2018). In an archipelago like the Philippines, water is abundant. It is, in fact, an abundant chemical compound on earth. More than 70 percent (70%) of the earth's surface is covered by water (USGSa).

Water on our planet actually comes in three forms or states: liquid, solid, and gas. Ocean water is perhaps the best example of water in liquid form. Water in the oceans and inland seas constitute around 97 percent (97%) of all the earth's water. Ocean water is liquid water in its most abundant form. Glaciers comprise 2.1 percent (2.1%) of the earth's water—that is, in the form of ice or solid water. Water in gaseous form consists of invisible water vapor that is found in the earth's atmosphere. It constitutes less than 1 percent (1%) of the total amount of water on the planet (USGSa).

Now, when we speak of water as essential to human survival, we are not simply talking about water in its liquid form but about water as a potable liquid—that is, water that is safe to drink. Ocean water is not safe for human consumption because of its high salinity or high concentration of dissolved salts. The other general type of water is fresh water. Unlike ocean water, fresh water has low levels of salt content. Lakes, rivers, and streams are common sources of fresh water. They are examples of surface water. Fresh water can also come from aquifers, which are, underground layers of rock that store water. Water from aquifers is an example of groundwater. Fresh water—from lakes, rivers, streams, and aquifers—is commonly used for drinking and irrigation.

Precipitation, in the form of rain or snow, replenishes the fresh water that is found in lakes, rivers, streams, and aquifers. The replenishment of water in aquifers is known as groundwater recharge. The process of groundwater recharge is very important in communities that rely mostly on groundwater for their drinking water and irrigation because constant replenishment results in a steady supply of water for such communities. In places where water is scarce, one can expect that surface water is mostly not an available or dependable source of fresh water. In such places, the people that live there rely on groundwater as their main fresh water source. It is important for them that it does not take too long for their underground sources of water to be recharged by precipitation, mostly in the form of rain. Otherwise, drought occurs. It is also critical that groundwater remains uncontaminated.

At this point, we need to clarify that fresh water is not the same as potable water. As mentioned earlier, potable water is water that is safe for drinking and for preparing food. The basis for distinguishing ocean water from fresh water is salinity or salt content. Fresh water is water with low salt content. For the Philippines, fresh water is water with less than 500 parts per million (ppm) of dissolved common salt (Philippine Clean Water Act of 2004).

Low salinity is only one of the many characteristics of potable or safe water. Potable water is safe to drink on account not only of its low salt content but also because of its being free from physical, chemical, biological, and radiological contaminants, in addition to having an acceptable taste, smell, and appearance (WHO 2017). Potable water is perhaps best viewed as fresh water that has undergone some treatment to make it safe for human consumption and "not represent any significant risk to health over a lifetime of consumption" (WHO 2006, 1).

Only 2.5 percent (2.5%) of the earth's water is fresh; the rest of it is saline. We need to note that only 1 percent (1%) of the earth's water is usable and easily accessible for human survival, with much of its fresh water trapped in snowfields and glaciers. Interestingly, it has been reported that "in essence, only 0.007 percent of the planet's water is available" for its inhabitants (National Geographic 2019). Clearly, there is no shortage of water per se when we see the abundance of ocean water on the planet. But as suggested earlier, humans cannot survive on abundant ocean water. The current population of the earth, which is estimated to be 7.7 billion, relies on a tiny fraction of the earth's accessible fresh water. It bears noting that the earth is a closed system in the sense that very little water enters or leaves its atmosphere (NGWA 2019), even when we take into account the various states of matter that water assumes when it undergoes change. The number of humans on the planet, however, has significantly increased over time.

6.2 Climate Emergency and Water (In)security

On November 5, 2019, the Washington Post reported that more than 11,000 scientists from 153 different countries and from diverse disciplinal backgrounds signed a declaration proclaiming that the planet "clearly and unequivocally faces a climate emergency." Invoking the moral obligation of scientists to "clearly warn humanity of any catastrophic threat and to 'tell it like it is'" (Ripple et al. 2019), the declaration spelled out the premises for announcing a climate emergency which are as follows: (1) the average temperature of the earth has risen significantly since the 1800s; (2) the total yearly global greenhouse gas (GHG) emissions continue to rise at a fast rate; (3) there is a link between GHG emissions and the occurrence of extreme weather events, particularly the alarming magnitude and frequency of such events; and (4) current efforts of the global community to address the problem of climate change have not been enough to arrest climate change (UNEP 2019).

The declaration of a climate emergency is essentially a call for greater urgency, more commitment, and better coordinated global actions against climate change. It is worth stressing that the declaration also noted the potentially irreversible effects of excessive consumption and unsustainable production on natural ecosystems, societies, and economies and the catastrophic chain reactions that can result from continuing with "business as usual" (Ripple et al. 2019). Undeniably, the declaration is just the latest in a long line of dire warnings about the urgency of the climate crisis, but the reference to excessive consumption and unsustainable production as fueling the process of climate change should remind us that human behaviors and mindsets are largely to blame for the emergency. Framed as such, the situation actually gives us some reason to be optimistic. Hopefully, humans can discard their irresponsible behaviors and mindsets before the effects of climate change become irreversible.

That climate change in the form of extreme weather events raises people's exposure to hazards, risks, and disasters is now accepted by many people as a given. But we need to point out that while the effects of the climate crisis are felt globally, certain regions of the planet are more at risk than others. The Asia–Pacific region, in particular, is home to some of the countries that are most vulnerable to climate change (ADB 2016). The latest Global Peace Index, which links climate change with peacefulness and stability, indicates that 6 of the 9 countries with the highest risk of multiple climate hazards are found in the region. Alarmingly, the Philippines topped the list with the highest overall hazard score of 8.5 (out of 10), followed by Japan, Bangladesh, Myanmar, China, and Indonesia, India, Vietnam, and Pakistan

(IEP 2019). According to ABS-CBN News in a report that appeared on 15 June 2019, around 47 percent (47%) of the Philippine's population live in areas of high exposure to climate hazards. If left unchecked, climate change will mean more lives and properties lost to disasters, diseases, and other risks in the coming years. Climate change poses real and serious problems for the Philippines.

In 2011, the World Economic Forum (WEF) outlined how water security, food security, energy resources, and the climate are intimately linked. In the book *Water Security: the Water-Food-Energy-Climate Nexus*, the WEF asserted that "(w)ater security is the gossamer that links together the web of food, energy, climate, economic growth, and human security challenges that the world economy faces over the next two decades" (WEF 2011, 1). Addressing water insecurity is vital to addressing the other challenges that confront the global economic system.

The WEF (2011) declared that the way water is managed on the planet is fundamentally unsound. Failure to manage water properly and to address water insecurity will endanger the various sectors of the global economy. Economic growth in different parts of the globe will be accompanied by the intensified demand for the earth's available freshwater resources mostly from the industrial, energy, and urban sectors—thus hastening their depletion. Addressing water insecurity then is crucial but it must be done in a manner that does not result in the emission of greenhouse gases and contribute to climate change. The situation entails systemic change—one that requires urgent collective and coordinated action.

According to the WEF (2011), the consistent underpricing of water, which leads to its overconsumption and wasteful use, is largely to blame for water insecurity in different parts of the planet. Reliance on groundwater to satisfy current demand has contributed to its depletion, threatening the ability of future societies to meet their water needs. Water, as an irreplaceable vital resource, clearly needs to be properly managed. In the words of the WEF (2011, 2), "We simply cannot manage water in the future as we have in the past, or else the economic web will collapse."

How did we reach this current level of water resource management? In regions worldwide, water is often excessively utilized leading to wastage. Groundwater reserves continue to be depleted, threatening their ability to meet future needs. As the global economy grows, the demand for water of future populations will inevitably surge. Given that water is an irreplaceable resource, it is imperative for us to rethink our water management behaviors, so as not to risk collapse of the global economic system (WEF 2011, 1–2).

As to the link between climate change and water insecurity, the WEF writes: "Much needs to be done now to improve the reliability of water supply in the face of a highly variable climate, while recognizing the social and environmental factors associated with development and management" (WEF 2011, 197). Just recently, the WEF reiterated the prediction that water supply crises are most likely to occur in the future in different parts of the planet, noting that a water (supply) crisis is one of the top five global risks with a high likelihood of occurrence and with the biggest impact to society and business. A water supply crisis has consistently been part of the list of the most urgent global concerns since 2012, along with climate change and extreme weather events, with a water crisis becoming the top global risk in 2015

(WEF 2019). Water insecurity on a global scale threatens humankind. The world of the future is predicted to be a water insecure world.

6.2.1 Water Stress

The earth's population is projected to reach 9.7 billion by 2050 (UN 2015), only 30 years from now. We can expect that in the coming years, the demand for fresh water will significantly increase with the projected growth in the world's population and will outpace the supply of fresh water that is available to humans, especially in the absence of discoveries of new sources of fresh water and/or radical technological innovations in the production and/or extraction of potable water. Among other things, our predicament should inculcate in our minds the importance of investing in research and development to discover technological solutions to the problem of water insecurity.

History tells us that the great civilizations of the past were propelled by the need to harness and control water as a resource. Modern cities are now marked by the existence of extensive piped water systems that are meant to ensure the uninterrupted delivery of abundant quantities of potable water to urban residents on a daily basis. Alongside the growth of national populations, urbanization is also predicted to characterize modernizing countries in the coming years with the transformation of rural areas into cities.

But if recent developments in a number of cities around the world are any indication of the general condition of life in the future, life in the cities of tomorrow does not appear enticing. Quite recently, the City of Flint in the state of Michigan, United States, found itself in the middle of a water crisis. On October 4, 2015, the Guardian reported that drinking water delivered through the city's pipelines was found to contain dangerous levels of lead. On January 1, 2018, the New York Times reported that Cape Town in South Africa had announced that within 3 months the city's water supply would run out largely as a consequence of three consecutive years of very low levels of rainfall. "Day Zero"—the designated day of the actual stoppage of the city's water supply system—may be regarded as foreshadowing the water crises that await big urban centers around the world.

Interestingly, apart from Cape Town, eleven (11) other big cities in different parts of the world are predicted to experience "water stress" within the next 20 years. Water stress occurs when the demand for water in cities "exceed(s) the available amount during a certain period or when poor quality restricts its use" (European Environment Agency 2019).

The 11 cities are Sao Paulo, Bangalore, Beijing, Cairo, Jakarta, Moscow, Istanbul, Mexico City, London, Tokyo, and Miami. It bears noting that water stress can be viewed not only as a phenomenon that strikes cities but one that also affects the entire planet. Citing UN-sanctioned studies, BBC News on February 11, 2018 reported "global demand for fresh water will exceed supply by 40 percent (40%) in 2030, thanks to a combination of climate change, human action and population growth."

Water stress is also projected to hit countries. The World Resources Institute (WRI) (2015) has identified thirty-three (33) countries that are mostly likely to face "extremely high water stress" in 2040. Perhaps not surprisingly, most of the countries on the list are found in the Middle East. But there are non-Middle Eastern countries on the list like San Marino, Singapore, Chile, and Spain. In a report that appeared on August 29, 2015 the Straits Times clarified that water stress is generally associated with "limited surface water and high demand." Considering the size of Singapore's territory, it is quite understandable then why it was included on the list.

The WRI rankings, however, should not be taken to mean that only the countries on the list will face water crises in the future. The large number of countries—i.e., 33—should indicate that water stress will not be an isolated occurrence for a few countries in the future. It is also worth noting that countries in the future are expected to face not only water stress but also inter-state tensions or intra-state civil wars as a consequence of water scarcity and the anticipated increased demand for water. Now, we need to remind ourselves that these projections are not to be taken as saying that some countries are "predestined" to experience water stress. Instead, the projections ought to be regarded as warnings that if appropriate steps are not taken to address anticipated challenges, water stress will indeed be the destiny of ill-prepared countries or cities. As disclosed by the Straits Times in the report mentioned earlier, Singapore is already investing in "innovative methods to manage (its) water resources." In countries that are less endowed with financial resources, there is always the option of modifying water consumption and utilization behaviors. In addition to seeking technological solutions to water insecurity, we ought to promote responsible water consumption and utilization behaviors.

Now, we really do not need to look far to find other examples of countries or cities that are experiencing "water stress." In March 2019, many Metro Manila residents were shocked to find that their water faucets had suddenly ran dry. A water supply crisis had hit the cities of Metro Manila (Yap and Calonzo 2019). It may confound the casual observer why a country surrounded by water experiences disruptions in the delivery of water supply.

But then again, as mentioned earlier, water scarcity is mostly about the scarcity of potable water.

It also needs to be noted that water shortages are actually a fact of life, for quite some time now, in some parts of the Philippines. Baguio City is perhaps the best example. The city is not surrounded by ocean water but it "has the highest average rainfall in the entire Philippines, and is cited… as one of three places in the country that receive the greatest amount of annual rainfall" (Mendoza et al. 2020, 1). But Baguio City residents experience water shortages on a regular basis. Water insecurity in the case of Baguio and the coping behaviors of its residents will be discussed in the section below. But at this point we would like to reiterate that we, the inhabitants of this planet, are facing a future where shortages in potable water and water conflicts are likely to be persistent features of our lives, especially if we ignore the warnings and fail to make the appropriate preparations.

6.2.2 *Water Security*

Scrutiny of the definition of water security will suggest that the concept has at least three (3) dimensions, namely, (a) quantity, (b) quality, and (c) accessibility. In relation to quantity, the UN instructs populations and their governments to secure for themselves adequate amount of water for their consumption and use. In relation to quality, the international body directs people to secure for themselves water "of acceptable quality," on top of the requirement of adequate quantity. "Acceptable quality" actually refers to potability.

The dimension of accessibility, meanwhile, enjoins populations and their governments to ensure that water resources are "within reach." Note, however, that accessibility to water can be further conceptualized to refer to (a) physical accessibility, and (b) economic accessibility or affordability. Physical accessibility requires that water resources are physically present in the places where people are found and, accordingly, are physically obtainable.

Economic accessibility or affordability, meanwhile, raises the possibility that water resources may indeed be physically present but people are actually barred from obtaining water on account of lack of money. Note here that economic accessibility mostly imagines water access in an economic arrangement where cash or some medium of exchange is used to procure water and that, accordingly, water is a commodity. At this point, it is important to highlight the fact that, in the real world, poverty prevents many people from having access to the water they need for their survival and well-being. That lack of access to water, either in terms of physical or economic access/affordability, can lead to "unpeace" and "instability" certainly deserves some comment. It is clear that water insecurity can lead to resentment and feelings of injustice among people who are denied access to it, and to political conflict between groups within a state or between neighboring states. Water insecurity can ignite violent conflicts between and within countries.

Now, we probably also need to remind ourselves that the right to sanitation and the right to clean and safe drinking water go hand in hand. Access to water also means access to sanitation and hygiene. In fact, there are water security advocates or practitioners who prefer to use the label WASH for water access, sanitation, and hygiene (see UN Water (b)). UN Resolution 64/292, which recognizes the human right to water, also acknowledges the right to sanitation as a human right.

6.3 Water Insecurity, Climate Change, and the Philippines

Climate change has particularly dire consequences for water security in the country. Climate variability, "including changes in temperature, evaporation, and precipitation, will impact regional water resources and is expected to increase the frequency of floods and droughts and reduce river flows, particularly during low flow periods"

(ADB 2016, 10). Water scarcity remains to be one of the country's most pressing challenges, with changing precipitation patterns, increasing temperatures, and heightened severity of droughts posing additional threats to existing water resources on top of rapid population growth, increasing demand for food production, urbanization, pollution, and excessive and inefficient use of water (ADB 2013). Projected changes in rainfall rates in the Philippines in the next 50 years have the potential of further heightening the risks of extreme flooding and severe droughts, which, in turn, threaten the availability and/or quality of existing water resources.

The Climate Change Commission (2011) anticipates that highly urbanized cities and metropolitan centers will be more vulnerable to worsening water shortages from increasing demand and deteriorating water supplies, both in terms of quantity and quality, while other areas face problems brought about by "increasing runoffs and flooding that spread pollutants, contaminate water sources and overwhelm water infrastructures" (CCC 2011, 47). Rising sea levels may also result in saltwater intrusion in fresh water supplies.

To summarize this section, we anticipate that climate change will continue to bring about extreme weather events in the Philippines, with severe flooding and droughts occurring unpredictably at different times of the year, threatening the lives, property, and livelihood of many Filipinos. We further anticipate that urban areas will experience periodic occurrences of water shortages while sources of fresh water in the country will become more vulnerable to contamination, if not depletion.

6.3.1 National Level Data on Water Security in the Philippines

Discussions on the issue of water security necessitate the use of reliable baseline data. However, as numerous studies and reports have noted in the past, data on water security indicators in the Philippines remain to be fragmented and lack standardization, reflecting for the most part the many agencies involved in the management of water resources in the country (ADB 2013; Alikpala and Ilagan 2018). With this in mind, we will now discuss some statistical figures which will nonetheless give us some idea about the current state of water insecurity in the Philippines.

Water Availability per Capita: Quantity. We can calculate water availability per capita which is just total available supply divided by the population. Note that calculating water availability per capita gives the impression that the available water supply is physically accessible and affordable to all Filipinos, which is rarely the case in the real world. Using data from the 2015 census, water availability per person stands at 1,446 m^3 per year (PWSSMP 2018). If we estimate further and use the projected population of more than 107 million Filipinos in 2019, then per capita availability falls to 1,361 m^3 annually. These figures indicate that the country is actually experiencing water stress since the values are way below the threshold of 1,700 cubic meters per person (NEDA, 8). If annual supplies further drop below 1,000 m^3 per person, then the population faces water scarcity, while below 500 m^3 entails absolute water scarcity (UN Water cited in PWSSMP 2018). The Philippines actually breached the water stress threshold as early as 2007, more than a decade ago, when per capita availability dipped to 1,650 m^3 (Alikpala and Ilagan 2018). If we assume a constant supply of renewable fresh water resources, continued population growth, and rapid development are expected to worsen the country's state of "water stress."

Access to safe drinking water: Quality and Accessibility. Data on access to safe drinking water reveal a more positive situation. The Philippine Statistics Authority's (PSA) 2017 National Demographic and Health Survey (NDHS) reports that 95 percent (95%) of Filipino households use an improved source of drinking water, a significant jump from 90 percent (90%) in 2008. Note that "improved sources of water" are defined as those sources of water that "protect against outside contamination so that water is more likely to be safe to drink" (PSA 2018). The NDHS data show that 98 percent (98%) of urban households use improved sources of drinking water as opposed to 93 percent (93%) of rural households (PSA 2018). Bottled water or water from refilling stations continue to be the most common source for both urban and rural households but is more commonly used for urban (58.2%) compared to rural (31.2%) residences. Water piped into the dwelling/yard/plot or piped to a neighbor is the second largest source for both household types. Other improved sources account for minimal contributions except for tube wells or boreholes which remain to be a common source for rural dwellers, with almost 20 percent (20%) of households still sourcing their drinking water from these, as opposed to around 4 percent (4%) for urban types (PSA 2018). In terms of the physical or geographical accessibility of water, around 80 percent (80%) of households reported the availability of drinking water within their premises, including 91 percent (91%) of urban and 71 percent (71%) of rural households. Meanwhile, around 3 percent (3%) of local households still have to travel 30 min or longer just to fetch water (PSA 2018).

Disparities in Access to Improved/Safe Sources of Water: Accessibility. In terms of regional disparities, the latest data from the NDHS indicate that the National Capital Region has the highest percentage of the population with improved sources of drinking water at 99.5 percent (99.5%), with only a minimal segment left with unimproved sources (PSA 2018). Meanwhile, 29 percent (29%) of the population in Autonomous Region in Muslim Mindanao (ARMM) still lack access to improved sources, the highest proportion in the country. This is followed by the Cordillera Administrative Region (CAR) with 15.4 percent (15.4%) of the populace still relying on unimproved sources of drinking water. We can also divide the population into wealth quintiles, or into five different parts according to the level of wealth. The first quintile corresponds to the poorest 20 percent (20%) of the population while the fifth quintile is comprised of the wealthiest 20 percent (20%) of the population. The data from the PSA indicate a clear disparity between the poorest and the wealthiest segments of the population. Among the poorest Filipinos, around 16 percent (16%) still rely on unimproved sources and 84 percent (84%) use improved sources, while almost all (99.5%) of those who belong to the wealthiest quintile rely on improved sources (PSA 2018).

The data presented above show that the country currently faces water stress in terms of the water availability per capita. They show that while there are considerable improvements in people's access to potable water in the country, there are significant regional and socioeconomic/class disparities in Philippine society in terms of access to safe or improved sources of water. National data indicate the need for our national leaders and institutions to take the lead in responding to a host of problems associated with water insecurity in the country, including that of equitable access to safe water.

In October 2019, Representative Joey Salceda of the House of Representatives filed House Bill No. 4944 which aims to create a Department of Water Resources and a Water Regulatory Commission. HB 4944 deserves attention because we believe it provides a reliable description of the current state of the Philippines' water resources sector. HB 4944 informs us that the sector is characterized, among other things, by (a) the dwindling supply of renewable water resources vis-a-vis increasing demand of a growing population; (b) the limited access of people to an uninterrupted 24 hours a day, 7 days a week) adequate supply of safe and clean—i.e., potable—water distributed through a piped water system; (c) the deteriorating quality of available water resources; (d) a grossly inadequate sanitation and sewerage sub-sector, and the limited access of people to improved sanitation facilities, particularly wastewater treatment services and sewerage systems; (e) inequitable water resource access and utilization; and (f) water tariffs that are not set rationally; water is not properly priced. Needless to say, we believe Salceda is actually describing water insecurity in the Philippines.

As to the factors that contribute significantly to the water resources sector's poor condition, Salceda squarely puts the blame on "fundamental governance and institutional issues" (HB 4944: 2). There is not enough space here to discuss in detail Salceda's depiction and assessment of the water resources sector but the broad description is that the sector is regulated by several legal frameworks, with each law separately prioritizing some aspect or aspects of the sector, and assigning responsibilities or mandates to an assortment of departments and agencies at various levels of government. The resulting situation is that of conflicting legal provisions, discordant policies, and overlapping mandates—which, in turn, lead to inefficient operations, uncoordinated undertakings, high coordination costs, poor enforcement or non-enforcement of regulation, piecemeal accomplishments, governmental inaction, etc. The issue of overlapping mandates negates the concept of governmental accountability. If the performance of a task is assigned to several agencies, which agency then will take full responsibility when an important task is left unaccomplished?

For Salceda and others—e.g., Rola et al. (2015)—who studied the matter, the current setup is marked by the continued fragmented, uncoordinated, and overlapping management of water resources in the country. Coupled with the other problems that tend to beset the Philippine bureaucracy—e.g., meager budgets, poor staffing, limited manpower, lack of field presence for some national agencies, limited opportunities for staff development and upgrading, etc., the current state of affairs precludes the realization of a comprehensive, integrated, doable, sustainable, long-term, and science-based solution to the problems of the sector.

Governance problems are largely responsible for the water insecurity of Filipinos. For Salceda, the country's water resources sector clearly lacks "a strong apex body" to harmonize and synchronize water-related policies and processes in the country (HB 4944). Water security is a multi-dimensional concept. Addressing water insecurity is not simply about providing people with water. As such, the challenge of making adequate quantities of clean and safe water physically accessible and affordable to all Filipinos requires a holistic, integrated, and coordinated approach or framework. Clearly, addressing the problems of the water resources sector in the country requires a thorough review of the relevant governing legal frameworks and the adoption of a coherent and harmonizing approach to the formulation of a water resources policy. We strongly believe HB 4944 is a big step in the right direction for the country. It bears noting at this point that HB 4944 actually cites climate change as a major threat to our water resources and one of the reasons why addressing the management problems of the water resources sector is crucial.

6.4 Review of Findings of Water Security Studies on Baguio

Now, we would like to draw attention to the state of water security in Baguio City. In late 2014, the faculty of the Department of Economics and Political Science of UP Baguio secured a grant from the United Kingdom-based International Institute for Environment and Development (IIED) to conduct a study on water security in Baguio City. More than a year after, a second grant was obtained from ICLEI-Southeast Asia Secretariat to pursue another water security study on Baguio City but with special focus on poor households. (For the discussion that follows, Ciencia et al. 2015 and Ciencia et al. 2017 will refer respectively to the first and second publications resulting from the IIED study, while Mendoza et al. 2016 will refer to the ICLEI study). We believe the findings of these studies are worth mentioning here for two (2) reasons. First, the study of water security in the case of Baguio reveals that a city that receives significant amounts of rainfall on a yearly basis can in fact experience water shortages. Second, the case of Baguio illustrates how residents of the city have adapted to the reality of water shortages. Readers are however warned not to take the case of Baguio as representative of all cities in the Philippines. Accordingly, policymakers are encouraged to pay attention to the place-specific features of their respective jurisdictions. We believe policymaking should suit the peculiarities and particular conditions of places and people. While we acknowledge that climate change, as a global phenomenon, affects all parts of the country, we are also aware that the effects may differ from place to place. In any case, we now discuss the key findings of our studies on Baguio City.

Increasing water demand amidst lagging water supply. Baguio City is experiencing rapid population expansion like many growing cities in the Philippines. Initially designed for 25,000 inhabitants, Baguio is now home to a population of 356,000, making the delivery of social services, including water, a challenging enterprise.

Rapid urbanization, expanding economic activity, and increased consumption all contribute toward raising water demand levels at a rate much faster than the current water system will allow. Water shortage in the city is reflective of the Baguio Water District's (BWD) inability to meet the water needs of the population. Substantial shortfalls in water supply are actually experienced as anticipated in the local water utility's estimates of demand and supply from 2013 to 2021 (Ciencia et al. 2015, 23). While the BWD aims to expand its service to 100 percent of households in its coverage area by 2020 as part of its long-term plans and mandate (Royeca 2015), its own projections of supply show that it will be insufficient for the growing population's needs. BWD attributes water shortage to declining pumping water levels due to fast depleting recharge areas of water sources resulting from urban development, the unregulated proliferation of private deep wells, and the loss of forest cover caused by human encroachment (Royeca 2015). Unless such processes are controlled, BWD supply will remain inadequate. Consequently, water shortage is expected to worsen, not improve, in the coming years.

Low water coverage and inefficient water distribution. Since its establishment in 1975, the service coverage area of the Baguio Water District has expanded and is now, according to BWD information, at 94 percent (94%) of households in the city. Only 122 of the 129 barangays in the city are connected to the BWD. Comparatively, the data on coverage that we derived in our first IIED study (Ciencia et al. 2015) from a city-wide survey of households are much lower at 68.6 percent (68.6%) of households having a private BWD connection. Lack of access to water is more pronounced among poor households given that only 27.7 percent (27.7%) of them have a BWD line (Ciencia et al. 2017).

Non-Revenue Water. Expanding BWD's coverage is not the only challenge. BWD also needs to address the problems of leaky pipes, illegal connections, and water pilferage. Also referred to as non-revenue water, BWD's "unaccounted for water" has nonetheless steadily declined from 46.46 percent (46.46%) of total output in January 2010 to 29.29 percent (29.29%) in December 2014. Reducing such leakages makes economic and ecological sense. Non-revenue water not only raises production costs but tariff rates to concessionaires as well.

Scheduled Water Distribution/Rationing Scheme. Another aspect of water supply provision is hours of service. Because of Baguio's mountainous topography, BWD distributes water following a rotational schedule in order to serve households in steep and higher elevation locations. The average household typically receives BWD water two to three times a week as reported by 21 percent (21%) and 35 percent (35%) of households, respectively. Uninterrupted water supply is a luxury enjoyed by only 10 percent (10%) of households in the city. Providing longer hours of service and expanding coverage can only be done if BWD can significantly increase its output. The water rationing scheme that is followed by the BWD is suggestive of BWD's limited capacity to produce more water for city residents. We anticipate that climate change will exacerbate water shortage in the city.

Vulnerability, inadequate infrastructure, and limited social services. Baguio City is vulnerable to disasters as attested to by the July 1990 earthquake, the regular occurrence of typhoons and monsoon rains yearly, and the occasional landslides that maroon the city from the outside world. While local residents generally demonstrate the "ability to adapt to disruptions and to thrive in times of instability" (Ciencia et al. 2017, 149), poverty and the lack of infrastructure and social services hinder successful adaptation. Poor households do not have access to BWD water and rely instead on free water sources like springs and creeks. They are adversely affected when water systems become murky or silted after rainy episodes or when these dry up during the dry season. They are more vulnerable to water-borne diseases due also to inadequate sanitation facilities and the lack of means to obtain cleaner but more expensive water for cooking and drinking. It is also important to note that only 24 percent (24%) of the city's population is connected to the sewerage system. The city needs to actively find ways to manage its wastewater more effectively. Incidentally, Baguio City's inadequate sewerage system brings to mind the Philippine Supreme Court's August 2019 decision in *Maynilad versus Secretary of Environment and Natural Resources.* The court case is significant because it draws attention to the serious need to build more sewerage systems and wastewater treatment facilities all over the country.

Institutional weaknesses hamper water security. Baguio City's water insecurity problem emanates mainly from BWD's failure to fulfill its mandate to "provide affordable water for all." While in our ICLEI study (Ciencia et al. 2017, 6) we recognize that this is affected by engineering, technical and topographical factors, we also note that "institutional weaknesses and technical inadequacy (rather) than (the) physical availability of water" are responsible for the water insecurity of residents. That Baguio receives the highest amount of rainfall in the country tells us that the problem is not the absolute lack of water. Policy instruments that govern people's access to and control over water resources—e.g., the National Water Code (P.D. 1067) and the Baguio Water Code of 2007—are in place. What is clearly missing however is effective enforcement. For instance, while the power to award water rights rests exclusively on the National Water Resources Board (NWRB), that same board lacks the capacity to effectively monitor actual water use at the local level. Just exactly how many of the estimated 1,000 private deep wells (Royeca 2015) operating in the city actually have water permits cannot be ascertained. As a result, we fear that the city's water supply runs the risk of over-extraction. Meanwhile, the Baguio Water Code contains a provision requiring new construction to install water harvesting facilities. This provision however remains unimplemented.

Engineering, technical, and topographical factors constrain BWD's ability to expand its water system but it is perhaps the huge financial cost of expansion, in addition to replacing leaky pipes, that actually hinders it from doing so (Ciencia et al. 2017, 159). Legal impediments on giving water connection to informal settlers also explain why not all households have BWD lines. Another major institutional weakness lies in the lack of monitoring mechanisms to ensure that water from water refilling stations, delivery services, and springs/creeks pass quality standards. As these water sources comprise a major segment of the water sector of Baguio City, the spread of water-borne diseases can be averted by requiring mandatory water testing conducted at regular intervals. Our studies on water security in Baguio City have revealed that the monitoring of water quality and water extraction activities in the city is limited.

6.5 Efforts to Address the Challenges of Water Insecurity

We will now discuss some of the coping behaviors and initiatives that Baguio residents have adopted through the years to address the long-standing problem of water insecurity in the city. We will also mention some of the significant initiatives that local government officials, the water district, and community associations have pursued to enhance water security in the city ever since the problem became apparent.

Households-based Behaviors. In response to the city's water shortage problem, Baguio households have invested in the purchase of water tanks, drums, or containers which allow them to store and have enough water to sustain their needs in between water schedules or water deliveries by private water concessionaires. Thirty-one percent (31%) of all Baguio City households, whether BWD-connected or not, rely on water delivered for a fee by private water concessionaires. A large percentage (79%) of Baguio households practice water recycling. An example of water recycling is using water from rinsing clothes to flush the toilet, water plants, clean the house, etc. Such behavior can be observed even during the rainy season. Rainwater harvesting is also practiced by almost half of Baguio households (49.4%). This is mostly done with the use of drums, pails, and basins rather than cisterns. The quality of harvested rainwater however could be unreliable and the practice actually poses a serious risk in light of the threat of the spread of dengue. In any case, rainwater harvesting is supported by local legislation although it is not effectively enforced.

There are Baguio households that rely on natural sources of water like springs and creeks found in the city. Forty-seven percent (47%) of poor households in the city report using spring water for their daily needs. The utilization of water from natural sources is generally free although manually fetching water from such sources normally entails some investment of time. If a resident can afford it, he can pay for installing PVC (polyvinyl chloride) pipes from the natural water source to his place of residence. Spring water, however, becomes murky during the rainy season.

An interesting adaptive behavior of Baguio households is the sharing of a BWD connection with those who do not, oftentimes for a fee. As practiced, water is normally sold at an average price of 29 pesos per drum. The cheapest price is reported to be 10 pesos while the highest is 50 pesos. Now, the cost of water in this arrangement may be considered exorbitant when compared to the BWD rate of 38 pesos per cu m (or PHP7.60 per drum) but it is actually competitive when compared to the price of 28 pesos per drum charged by private water delivery services. When we realize that most of the households that enter into these BWD connection sharing arrangements are poor, we cannot help but ask questions about the affordability and the "correct" or "proper" price of water in the country.

Community-based solutions. Local communities or barangays in Baguio City have also devised ways to address the problem of water insecurity. Perhaps the most noteworthy initiative taken by communities in Baguio City is the establishment and operation of community water systems. A community water system is one that utilizes a water source such as a spring or deep well, stores the water in tanks, and delivers it through water distribution pipes to individual houses. In Baguio, 8.9% of households not connected to the BWD report their reliance on community water systems. The creation of such systems however usually involves huge infrastructure investment made possible through grants or loans. Moreover, the quality of water from such systems could be unreliable.

Commercial Initiatives. Water delivery services and water refilling stations have filled the market gap that is created when BWD water is not available or is inadequate. Purified water from water refilling stations actually comprises the drinking water of a very significant percentage (85%) of Baguio households. The quality of water from these commercial establishments however is not guaranteed as there is no government body that consistently and effectively monitors this.

Initiatives of the City Government. The local government in Baguio has adopted a number of legislative initiatives to address water insecurity in the city. Around 15 years ago, the City Government with support from external partners undertook the Sustainable Water-Integrated Management and Governance (SWIM) project from September 2004 to August 2005, resulting in the passage of the city's Water Code of 2007 which provided the policy framework for the holistic management of water resources and pertinent facilities in the city. The City also passed its Environment Code in 2016. Among the important provisions of the Code is the one requiring owners of new buildings to obtain a certification that a water capture facility has been installed prior to actual occupancy of the building. While the passage of key measures may seem like major accomplishments for the city, monitoring of their enforcement must also be given equal importance (Mendoza et al. 2020). Harmonizing its legislative initiatives with national laws may also pose serious challenges for the city.

Initiatives of the Local Water District. The BWD has pursued a number of initiatives to improve water security in the city. Among its most recent accomplishments is the rehabilitation of a two-hectare rain basin. In the summer of 2018, the BWD completed the rehabilitation of the Sto. Tomas Rain Catchment Basin which stores rainwater for distribution during the summer months. A major undertaking for the BWD is the proposed adoption of a Bulk Water System, an idea it has actively pursued as early as 1997. As proposed, the project will require the BWD to enter into a partnership agreement with a private water supplier. Thus far however no partnership between BWD and a private water supplier has materialized.

6.6 Analysis: Lessons and Insights from Previous Studies

This section summarizes some lessons and insights that may be drawn from the literature on water security and from studies such as the one about Baguio City. We believe policymakers can pick up some ideas from this section which hopefully can contribute to the formulation and implementation of sound interventions in addressing water insecurity in the Philippines.

The climate emergency foreshadows the arrival of a more water insecure world. Climate change in the form of extreme weather events such as severe flooding and droughts is predicted to threaten our water supply in terms of its quantity and quality. Failure to arrest the climate emergency now while there is still time will mean a more water insecure world in the future.

Water security is an economic issue. As an economic problem, water insecurity can be analyzed using a demand and supply framework which views the water sector as consisting of resources and agents that interact in a system of relations. As a system, the sector seeks to achieve water security, an outcome that characterizes its performance as efficient. Economic efficiency is "the effectiveness with which a system utilizes its available resources (including knowledge) at a particular time (static efficiency) or through time (dynamic efficiency)" (Gregory and Stuart 2014, 58). Economic efficiency in the water sector means minimizing the amount of water for a particular use.

Water has many economic uses; it is an input in agricultural processes, crop and livestock production, energy production, and even tourism. But to all of us, water is first and foremost, a consumption good, a basic commodity essential to daily survival. In our research, the focus was on water as a consumption good. Hence, the concern for water security is viewed as an issue about how persons and communities have "access to adequate quantities of acceptable quality water" (UN Water 2013). The water security problem requires an approach that uses a water demand management (WDM) strategy which aims at "reducing losses in the (water delivery) system, improving operational efficiencies, promoting rational use of water resources, ensuring equitable distribution of resources, and exploring alternative sources such as recycling of wastewater for non-potable uses" (Abansi et al. 2018, 234–235).

The WDM strategy complements the traditional approach that tends to focus on the supply side of the water sector and is largely concerned with the availability and adequacy of water sources. A "demand-viewpoint" directs the attention of water policy and its instruments toward understanding better the water users—their needs, their goals, and their constraints. Paying attention to water users means taking into account the nature of livelihoods, the pattern of social networks, and the character of governance that define the behavior of people's use of water.

Water security means giving the poor access to water. Related to the economic concepts of supply and demand is the idea of equitability. Within communities, specific attention must be made to ensure that the poor have adequate access to potable water. Often water providers like local water districts focus on water adequacy defined in terms of achieving a total level of water volume for a target population per unit of time. They do not establish technical and organizational mechanisms that are directed to ensuring that water supply reaches poor households. One might remember the "trickle-down" metaphor that once described the "growth" perspective where an increasing real gross domestic product per capita was sufficient as the goal of economic production; the concern for an equitable distribution of output can come second because it was believed that the "fruits" of economic growth will naturally reach all strata of society at some future time without fail. Today, the prevailing opinion is that equitable distribution of output or any critical product like water services must be pursued as a distinct goal. In the case of water providers, achieving a level of total water supply is insufficient as a goal of their operation. They must also ensure that poor households living in their coverage areas have access to water services.

Water is a human right because it is precious. The issue of equitability brings us back to the concept of access to safe and clean water as a human right. We have just argued for equitability and paying attention to the needs of the poor as water consumers. Needless to say, there is a position that is critical of this pro-poor, pro-user "water-is-a-human right" point of view. Proponents of this critical position worry that promulgating the idea that water is a human right—and the accompanying idea that water ought to be free—will only lead people to use water irresponsibly and not give it the value it actually possesses as a precious and un-substitutable resource. We can label this the "water-is-a-precious-resource" position. Its proponents uphold the idea of not only putting a price on water but to give it the high price it deserves. Consistent with this idea is the argument that putting a high price tag on water would enable water districts to collect considerable profits for maintaining piped water systems, sewerage systems, and wastewater treatment facilities. Upholding water's value and putting a high price on it would improve the delivery of water services.

The "water-is-a-precious-resource" position clearly has compelling arguments but we also see problems with it. As things stand, many poor people already lack financial access to water. Setting a higher price for water would only aggravate the poor's lack of access to water. In short, many poor people would be priced out of the experience of accessing water. Moreover, putting a high price tag on water only encourages the management of water for financial gain—i.e., the "commercialization" of water. This will only make the poor and society in general more vulnerable to the profit-seeking motives of business interests, and exacerbate existing economic equalities in society. Perhaps it needs to be mentioned that there is the persistent suspicion among some observers that for some time now big corporations and billionaire businessmen have been quietly buying rights to various water resources in the country and other parts of the world in anticipation of the huge financial windfall that they will receive when water becomes more scarce and more precious as the demand for water resources rises considerably in the coming years (Global Research 2012).

There are also compelling reasons to acknowledge access to water as a human right. Foremost among this is the undeniable fact that water is essential to human survival. To deprive people of water is to deny them the right to live. Treating access to water as a human right seems consistent with the idea of treating people equitably and acknowledging that they have the right to live as people. We are convinced that providing potable water as a basic need and human right is consistent with the idea of equitability, and the reason why water ought to be provided to all people is because water is indeed a precious and invaluable life-sustaining resource. The core idea of "water-is-a-precious-resource" is not denied by proponents of "water-is-a-human right." It is precisely because it is precious and essential to life that it ought to be declared a human right. We can endorse the view that people have a right to clean and safe water and at the same time argue that people have an obligation to value water and to protect the integrity of water resources; that irresponsible users of water ought to be penalized; and that responsible users of water and custodians of water resources ought to be rewarded. We have drawn attention to a debate which we believe will characterize discussions about water utilization in the coming years. Our position is to reconcile the two seemingly rival positions. The "water-is-a-human-right" versus "water-is-a-precious-resource" is a false dichotomy because water can be viewed as both. The challenge for people then, and for policymakers, is to find how to concretize the reconciliation of the two positions. Incidentally, it might be beneficial to revisit the water demand management (WDM) strategy which was discussed earlier. It just might provide the reconciliation of the "water-is-a-human-right" and "water-is-a-precious-resource" positions.

Addressing the climate emergency and water insecurity requires the proper utilization and management of resources. We raise an important point alluded to by the WEF as discussed earlier: water (in)security—and the climate crisis—are essentially "management" (or "governance") issues; addressing these twin problems will require managing people's resource utilization-related behaviors and actions. Water insecurity and the climate emergency are more than technical or engineering problems. Making the country water secure and climate-smart is not simply a matter of designing and building new environmentally-friendly structures, machines, water storage facilities, etc. Addressing these problems will also involve molding people to be responsible users of the planet's resources, including water. Moreover, it will essentially require the crafting of good—i.e., comprehensive, integrated, doable, sustainable, long-term, and science-based—policies and their effective enforcement. The improper and irresponsible utilization and management of the earth's resources by humans actually account for both the global climate emergency and water insecurity that people now face. Acknowledging this while there is still time is critical because it tells us that modifying the way we utilize and manage the planet's resources holds the key to addressing the climate emergency and water insecurity.

Addressing Water Insecurity Requires Coherent and Coordinated Management. Water security involves the provision of water to people as a prerequisite for their survival and well-being. Societies often become water insecure when their water supply dwindles while the population grows. On its face, water insecurity is a consequence of the combination of uncontrolled population growth and depleting water resources. But water security is a multi-dimensional concept and is not simply about ensuring that society has an adequate supply of water; the water must also be safe and accessible to all people in society. To address the problem of water insecurity in the Philippines, we need to adopt a coherent and coordinated approach to managing the various aspects of the water resources sector—e.g., quantity, quality, physical accessibility, and affordability. As things currently stand, the water resources sector in the country is regulated by several legal frameworks. The situation is not conducive to the coherent and coordinated management of the water resources sector. A thorough review of water-related laws is needed as a prelude to drafting a unified, harmonized, coherent, and comprehensive policy for the water resources sector. The formulation of a unified and enforceable legal framework for the sector is a major step in addressing water security in the country.

6.7 Conclusion

The chapter looked into the link between water security and the climate emergency. We view the climate emergency as a call for urgent global action to address the threat of climate change before its effects become irreversible. We endorse the view that climate change and water insecurity are closely linked.

We recognize water as a life-sustaining resource. In its potable form, water is invaluable to humans. Now, if it is indeed the prime constitutional duty of government to serve and protect the people, to protect life and the general welfare, to provide adequate social services, and to promote people's right to health, then it is but logical for the government to prioritize the provision of potable water to all its citizens and to guarantee that people are water secure. In light of the indispensable life-sustaining quality of water and the credible projections of worsening global water scarcity in the foreseeable future, we humbly implore our top elected national government officials to prioritize the matter of ensuring the water security of the Filipino people, in line with the principles of democratic governance and public accountability. It is humbly recommended that addressing water insecurity within the context of the climate emergency be made a matter of utmost importance by our country's top national officials and institutions. We implore our national leaders to lead all Filipinos in a war against water insecurity and climate change which exacerbates it.

We urge Congress to prudently enact but with dispatch a law rationalizing the country's water laws. With its recognition of (a) the threat of climate change on our country's water supply, (b) the need to rationalize existing economic regulations and the financial aspects of providing water services to Filipinos, (c) the importance of reviewing the various legal frameworks that govern the water resources sector in the country, and (d) the urgency of addressing the sector's governance problems, House Bill No. 4944 appears to be a promising starting point for such an endeavor. The bill deserves the serious consideration of our lawmakers. We respectfully push for provisions that will make potable water more accessible and affordable, especially for the poor.

We urge the Office of the President to see to it that in preparing the country's annual national budgets there is ample funding for (a) research and development (R&D) in the areas of water desalination, water filtration, water purification, water recycling, rainwater harvesting, wastewater treatment, and the like. Investment in technological innovations resulting in the overall efficient and responsible utilization of our water resources will contribute significantly to making the Philippines a water secure nation, and (b) the construction and maintenance of water sewerage systems and wastewater treatment plants all over the country as part of the government's infrastructure program. Such facilities will help protect our water supply.

Finally, we urge LGUs, in coordination with national agencies and their field offices, to faithfully implement water-related regulations. We urge LGUs and agencies to regularly provide the public with (a) a list of water sources—e.g., water district, water refilling stations, water delivery services, community water systems, deep wells, natural springs, etc.—located or available in the city or municipality, alongside periodic information on the quality of water drawn from such sources, and (b) a list of all deep wells located in the city or municipality, alongside periodic information on water permits issued to applicants. An aim of these activities is to instill a sense of rule compliance among the people and government personnel, in addition to promoting the regular monitoring of water quality and water extraction activities.

References

Abansi CL, Hall RA, Siason IML (2018) Water demand management and improving access to water. In: Rola AC et al (eds) Water policy in the Philippines: issues, initiatives, and prospects. Global issues in water policy, vol 8. Springer International Publishing, pp 233–259

Alikpala R, Ilagan C (2018) A policy brief on the Philippine water sector. Arangkada Philippines policy brief no. 7: 2018. http://www.investphilippines.info/arangkada/wp-content/uploads/2015/09/A-Water-Policy-Brief-on-the-Philippines-July2018-SEPT8.pdf. Accessed 1 May 2020

Asian Development Bank (ADB) (2013) Philippines: water supply and sanitation sector assessment, strategy, and road map. ADB, Mandaluyong City, Philippines

Asian Development Bank (ADB) (2016) Asian water development outlook 2016: strengthening water security in Asia and the Pacific. ADB, Mandaluyong City, Philippines

Ciencia AN, Mendoza LC, Cruz GA, Calde NL, Cabalfin MR, Penalba MA (2015) Towards establishing water security and urban resilience in the City of Baguio. Asian cities climate resilience working paper series 27: 2015. https://pubs.iied.org/pdfs/10754IIED.pdf?. Accessed 1 May 2020

Ciencia AN, Mendoza LC, Cruz GA, Penalba MA, Calde NL, Cabalfin MR (2017) Water security and urban resilience: the case of Baguio City, Philippines. In: Archer D, Colenbrander S, Dodman D (eds) Responding to climate change in Asian cities: governance for a more resilient urban future. Routledge, pp 146–164

Climate Change Commission (CCC) (2011) National climate change action plan: results-based monitoring and evaluation system

European Environment Agency (2019) Water stress. https://www.eea.europa.eu/archived/archived-content-water-topic/wise-help-centre/glossary-definitions/water-stress. Accessed 24 Nov 2019

Explained: The world's water crisis. Season 1, episode 19. Produced by Frank Matt, featuring Kyle MacLachlan as Narrator. First aired 12 September 2018. Netflix, 2018

Global Research (2012) The new "water barons": wall street mega-banks are buying up the world's water. https://www.globalresearch.ca/the-new-water-barons-wall-street-mega-banks-are-buying-up-the-worlds-water/5383274. Accessed 26 Nov 2019

Gregory PR, Stuart RC (2014) Economic systems and economic outcomes. In: The global economy and its economic systems. South-Western CENGAGE Learning, Australia, pp 54–78

House Bill No. 4944: An act establishing the national framework for water resource management and creating the Department of Water Resources and the Water Regulatory Commission, and for other purposes. Introduced by representative Joey Sarte Salceda on October 1, 2019. http://www.congress.gov.ph/legisdocs/basic_18/HB04944.pdf. Accessed 27 April 2020

Institute for Economics and Peace (IEP) (2019) Global peace index 2019: measuring peace in a complex world. IEP, Sydney

Mendoza LC, Ciencia AN, Cruz GA, Penalba MA, Calde NL, Cabalfin MR (2016) Engaging communities in addressing water security, sanitation, and urban resilience challenges in Baguio City. A report submitted to the ICLEI-Southeast Asia Secretariat. http://icleiseas.org/wp-content/uploads/2017/03/ICLEI-REPORT-UP-Baguio_ICLEI-SEAS_clean-version-2.pdf. Accessed 25 Nov 2019

Mendoza LC, Cruz GA, Ciencia AN, Penalba MA (2020) Local policy and water access in Baguio City, Philippines. Int J Soc Ecol Sustain Dev (IJSESD) 11(1):1–13

National Economic and Development Authority (NEDA) Philippine water supply and sanitation master plan (PWSSMP) 2019–2030. https://neda.gov.ph/wp-content/uploads/2021/09/120921_PWSSMP_Main-Report.pdf. Accessed 5 January 2024.

National Geographic (2019) Freshwater crisis. https://www.nationalgeographic.com/environment/freshwater/freshwater-crisis/#close. Accessed 24 Nov 2019

National Groundwater Association (NWGA) (2019) Information on earth's water. https://www.ngwa.org/what-is-groundwater/About-groundwater/information-on-earths-water. Accessed 24 Nov 2019

Philippine Statistics Authority (2018). National demographic and health survey 2017

Philippine Water Supply and Sanitation Master Plan (PWSSMP) Databook (2018). http://www.pwssmp.com/databook. Accessed 1 May 2020

Presidential Decree No. 11067, s. 11976. A decree instituting a water code, thereby revising and consolidating the laws governing the ownership, appropriation, utilization, exploitation, development, conservation and protection of water resources (The Water Code of the Philippines). https://www.officialgazette.gov.ph/1976/12/31/presidential-decree-no-1067-s-1976/. Accessed 27 April 2020

Republic Act No. 9275: an act providing for a comprehensive water quality management and for other purposes (Philippine Clean Water Act of 2004). Signed into law on 22 March 2004. https://www.officialgazette.gov.ph/2004/03/22/republic-act-no-9275/. Accessed 27 April 2020

Ripple WJ, Wolf C, Newsome TM, Barnard P, Moomaw WR (2019) World scientists' warning of a climate emergency. https://academic.oup.com/bioscience/advance-article/doi/10.1093/biosci/biz088/5610806. Accessed 24 Nov 2019

Rola AC, Pulhin JM, Tabios III GQ, Lizada JC, Dayo, MHF (2015) Challenges of water governance in the Philippines. Philipp J Sci 144(2):197–208

Royeca S (2015) Baguio city water supply situationer

Spector D (2018) Here's how many days a person can survive without water. Business Insider. https://www.businessinsider.com/how-many-days-can-you-survive-without-water-2014-5. Accessed 24 Nov 2019

Supreme Court of the Philippines, G.R. No. 202897 (Maynilad Water Services, Inc. Vs. The Secretary of the Department of Environment and Natural Resources (DENR)) Decision, 6 August 2019. http://sc.judiciary.gov.ph/6445/. Accessed 27 April 2020

United Nations Environment Programme (UNEP) (2019) Facts about the climate emergency. https://www.unenvironment.org/explore-topics/climate-change/facts-about-climate-emergency. Accessed 25 Nov 2019
UN Water (a). Human rights to water and sanitation. https://www.unwater.org/water-facts/human-rights/. Accessed 24 Nov 2019
UN Water (b). Water, sanitation and hygiene. https://www.unwater.org/water-facts/water-sanitation-and-hygiene/. Accessed 24 Nov 2019
UN Water (2013) What is water security? Infographic. https://www.unwater.org/publications/water-security-infographic/. Accessed 24 Nov 2019
United Nations (2010) Resolution 64/292. The human right to water and sanitation. https://undocs.org/A/RES/64/292. Accessed 24 Nov 2019
United Nations (2015) World population projected to reach 9.7 billion by 2050. https://www.un.org/en/development/desa/news/population/2015-report.html. Accessed 24 Nov 2019
United States Geological Society (USGSa) (n.d.) How much of the Earth's water is stored in glaciers? https://www.usgs.gov/faqs/how-much-earths-water-stored-glaciers?qt-news_science_products=0#qt-news_science_products. Accessed 24 Nov 2019
United States Geological Society (USGSb) (n.d.) Saline water and salinity. https://www.usgs.gov/special-topic/water-science-school/science/saline-water-and-salinity?qt-science_center_objects=0#qt-science_center_objects. Accessed 24 Nov 2019
World Economic Forum (2011) Water security: the water-food-energy-climate nexus. Island Press, Washington, DC
World Economic Forum (2019) Global risks report 2019. https://www.weforum.org/reports/the-global-risks-report-2019. Accessed 25 Nov 2019
World Health Organization (WHO) (2006) Guidelines for drinking—water quality first addendum to third edition, Volume 1 Recommendations. https://www.who.int/water_sanitation_health/dwq/gdwq0506.pdf. Accessed 24 Nov 2019
World Health Organization (WHO) (2017) Guidelines for drinking water quality, 4th edition incorporating the first addendum. https://apps.who.int/iris/bitstream/handle/10665/254637/9789241549950-eng.pdf;jsessionid=43B52CCDF8E4162ADC039E5DC3242A47?sequence=1. Accessed 24 Nov 2019
World Resources Institute (2015) Ranking the world's most water-stressed countries in 2040. https://www.wri.org/blog/2015/08/ranking-world-s-most-water-stressed-countries-2040. Accessed 24 Nov 2019
Yap C, Calonzo A (2019) The worst water crisis in nearly a decade hits Manila. Here's why. Bloomberg. https://www.bloomberg.com/news/articles/2019-03-18/worst-water-crisis-in-decade-hits-philippine-capital-here-s-why. Accessed 27 April 2020

Alejandro N. Ciencia Jr. Ph.D. is Professor of Political Science at the College of Social Sciences, University of the Philippines Baguio. His research interests include water security, urban resilience, indigenous peoples' rights, justice, and judicial politics.

Lorelei C. Mendoza is a Professor of Economics at the College of Social Sciences, University of the Philippines Baguio. She has completed researches in natural resource management and gender and household economics.

Gladys A. Cruz is an Associate Professor of Economics at the University of the Philippines Baguio. She teaches statistics, mathematical economics, and quantitative economics. Her research interests include urbanization, population, and resource management.

Rainier Ric de la Cruz is a former Assistant Professor of Economics at University of the Philippines Baguio and is currently an economist at the Business and Economics Division of the Philippine Competition Commission. He received his BA in Social Sciences (magna cum laude and batch valedictorian) from UP Baguio and his MA in Economics from UP Diliman.

Part II
Local Experiences from Selected River Basin Ecosystems and Coastal Communities

Chapter 7
The Need for a Clear Rehabilitation Roadmap for Manila Bay Coastal Ecosystems: A Microcosm of Climate Emergency in the Philippine Context

Rene N. Rollon, Giannina Marie G. Albano, Green Ann A. Cruz, Maia Patricia D. Garcia, Kristine Camille V. Buenafe, Anisette Monique G. Albano, and Betty May R. Villamayor

Abstract Manila Bay is plagued by various environmental challenges. This chapter revisits the changes in mangrove occurrence and distribution relative to the spatio-temporal changes and history of the 200 km coastline of the entire Manila Bay. Using topographic maps, brackish water ponds, and toponyms, we found that about 75,000 to 90,000 hectares of mangroves apparently existed in the 1900s, comprising ca. 10 percent of the river basin. At present, only barely 1 percent (<1,000 ha) is left, reflecting massive deforestation largely due to conversion into brackish water fishponds, salt pans, and rice paddies as well as development into residential and industrial settlements. Knowing the bay's history, it is imperative to protect the ecological integrity of Manila Bay and to restore former mangrove areas as the government moves forward in its rehabilitation programs, projects, and activities. The Manila Bay story seems to be a microcosm of national and global significance, exemplifying how we deal with development projects which transform our coastal ecosystems into other uses. We stress the need for more environmental scientists looking through broader spatio-temporal lenses in our work toward conserving the Philippine coastal ecosystems. This chapter discusses specific policy recommendations in setting a clear rehabilitation roadmap for Manila Bay Coastal Ecosystems considering present and projected climate scenarios.

Keywords Manila Bay mangroves · Rehabilitation roadmap · Highlands-to-oceans perspective

---

R. N. Rollon (✉) · G. M. G. Albano · G. A. A. Cruz · M. P. D. Garcia · K. C. V. Buenafe · A. M. G. Albano · B. M. R. Villamayor
University of the Philippines Diliman, Quezon City 1101, Philippines
e-mail: rnrollon@up.edu.ph

K. B. Berse et al. (eds.), *Climate Emergency in the Philippines*, Disaster Risk Reduction, https://doi.org/10.1007/978-981-99-7804-5_7

Key Points

- Mangroves likely abound in the deltas and coastal floodplains of rivers, more particularly those that drain significantly large basins. The Pampanga River discharges annually some 6 billion m^3 of river water contributing to about 49 percent of freshwater input into the 180,000-ha Manila Bay and forming a vast area of coastal plain.
- About 75,000–90,000 hectares of mangroves apparently existed in the 1900s but at present, only barely 1 percent (<1,000 ha) is left reflecting massive deforestation.
- Knowing Manila Bay's history, it is imperative to restore former mangrove areas, as they are crucial to mitigate climate crisis projections.
- The DENR, BFAR, and other national and local agencies and decision-makers must seek scientific advice and agree on a clear rehabilitation roadmap for Manila Bay.

Summary for Policymakers
There is an urgent need for policymakers to craft a clear restoration-development roadmap for Manila Bay that rationalizes our desire for further socio-economic development while minimizing the impact on the biophysical integrity of the large environmental system.

We further argue that a restoration target even just a meager 10% reversion of the coastal floodplain of the Pampanga River basin back to mangroves would be crucial to mitigating climate crisis projections, to sustaining the viability of the coastal productivity of Manila Bay, and to revitalizing the nutrient and fisheries connectivities with the open ocean. The findings of our investigation to the circumstances of the coastal ecosystems and associated resources of Manila Bay in relation to the following:

- Historical transformations, land-use changes, and urbanization perspectives;
- Highlands-to-oceans continuum perspective in sustaining the vitality of the biophysical system;
- The importance of conserving and rehabilitating the system, and;
- Imminent climate threats vis-à-vis functions of these coastal ecosystems on coastal bioshield protection, productivity of our coastal fisheries, and biodiversity conservation and restoration.

DENR and BFAR must agree on a science-advised mangrove restoration roadmap that is site-specific, map-based, and modeled using robust inputs, including various climate change scenarios.

We then urgently call on a number of key national agencies to act accordingly, namely:

- For the **Department of Environment and Natural Resources (DENR)** and **Bureau of Fisheries and Aquatic Resources (BFAR)** to collaborate closely on the inventory and GIS-mapping of aquaculture ponds with fishpond lease agreements (FLAs), and their respective status (i.e., active/productive, or abandoned, underutilized, and undeveloped [AUUs]), as well as, those aquaculture ponds operating without proper FLAs (illegal);

- For **DENR** and **BFAR** to agree on a science-advised mangrove restoration roadmap (i.e., site-specific, map-based, modeled using robust inputs, including various climate change scenarios), considering the food security mandate of BFAR, while optimizing the forest extents and the services (e.g., biodiversity, nutrient and sediment fluxes, trophic functions, etc.), as well as climate resiliency function, that these coastal ecosystems provide;
- For **DENR, BFAR, and other national agencies** in the economic cluster of the cabinet to agree on a short-, medium-, and long-term development roadmap that ensures conservation, enhancement, and protection of the non-negotiable areas of key coastal and upstream ecosystems;
- For all national agencies concerned, with support from **higher education institutions (HEIs)**, to revisit existing laws and regulations (including the current Philippine Environmental Impact Assessment [EIA] system), and when necessary, propose a new law for such roadmaps, and;
- For Congress to prioritize enactment of laws with provisions for funding specifically supporting these roadmaps, but more generally, policies governing the conservation, rehabilitation, and protection of the coastal ecosystems of the country in the context of highlands-to-oceans connectivity, and their relevance to the climate emergency.

Local HEIs must be tapped to provide a platform for information, education, and communication (IEC) campaigns, as well as assist in the assessment and optimization of ecosystem restoration targets.

Aligned with the efforts of the national agencies, the following should be pursued at the local level:

- The formation of local government unit (LGU) alliances, closely coordinating with
- The concerned national agencies in pursuing ecosystem conservation, rehabilitation, and protection roadmaps, mindful that ecosystem boundaries are not the same as political/administrative boundaries, and thus, ecosystem management needs inter-LGU collaboration;
- The initiation of local efforts on an integrated upland and coastal resource management planning (IUCRMP), and the rationalization of inter-LGU coordinated targets;
- Strengthening partnerships with non-governmental organizations (NGOs) in continuing local initiatives pursuant to roadmap targets;
- Stronger role of local HEIs in providing platform for information, education, and communication (IEC) campaigns, as well as assisting in the assessment and optimization of ecosystem restoration targets;
- More proactive involvement and education of the youth sector, ensuring continuity of efforts, and long-term sustainability of roadmap implementation, and;
- Legislation of local ordinances, with provisions for funding schemes, in pursuit of the baywide-coordinated restoration-development goals.

7.1 Introduction

Manila Bay: Highlands-to-oceans Biophysical Setting

Manila Bay is a 180,000-ha embayment, with about a 200 km coastline, connecting to the West Philippine Sea through the 18 km wide southwest bay entrance (Fig. 7.1). West of the bay lies the province of Bataan, a peninsula with two adjacent dormant volcanoes (i.e., Mounts Mariveles and Natib), forming generally steep slopes along its coastline, and thus, also having narrow intertidal zones. The southeast coastline of the bay has also generally narrow intertidal areas, sloping (ca. 0.5% at the lowland) toward Tagaytay Ridge (~600 m above mean sea level). Along the eastern coastline lies the Metro Manila megacity of about 13 million inhabitants, and also where Pasig River discharges, draining the sub-basins of San Juan and Marikina rivers, as well as the Laguna de Bay and its multiple tributaries, altogether comprising about 1/3 of the total sub-watersheds discharging into Manila Bay. The largest of these watersheds is the basin of the Pampanga River (PRB), a complex network of tributaries, with a total catchment area of about 1 million hectares, encompassing four provinces (i.e., Bulacan, Pampanga, Tarlac, and Nueva Ecija), discharging some 6 billion m^3 annually which comprises half of the total freshwater flowing into Manila Bay. As the fifth longest river in the country (i.e., > 250 km stretch along the main channel), the Pampanga River forms a vast coastal floodplain north of Manila Bay, creating a large estuary with shorelines apparently prograding seaward over the millennia (Fig. 7.1). At present, the PRB coastal floodplain covers about 100,000 hectares, with tidal influence still detectable in areas more than 20 km upstream of the present shoreline. Such a large nutrient-rich estuarine area (where mangrove forests are also likely natural components of this coastal formation) must be playing a crucial role in the overall fisheries productivity of the bay (see also Dorenbosch 2006 for factors affecting the degree of inter-habitat exchange), and to the broader open ocean (West Philippine Sea), linked by monsoon-influenced surface water current system (Fig. 7.1).

Coastal Land-use and Fisheries within the Bay

The total area of the basins discharging to Manila Bay is about 2 million hectares, with some 32 million total inhabitants (Table 7.1). Although PRB comprises 1/2 of this total watershed area, the human population inhabiting the southeast quadrant (i.e., Metro Manila, Rizal, Cavite, and Laguna) is about 22 million (71%), with the densely-packed Metro Manila (22,219 people per km^2 population density) besting other areas of the southeast quadrant (population density: 1,000–3,000 per km^2). Elsewhere, human population density is less than 1,000 per km^2, except in Angeles City (Pampanga) where 6,496 inhabit per km^2 (Table 7.1). With the expanding urbanization trend much more intense in this quadrant, development pressures on the coastal areas north (Bulacan) and south (Cavite) of the metropolis is also highest, although pockets of industrial development booms also exist elsewhere (e.g., Bataan coastline). In the vast coastal floodplain of Pampanga and Bulacan areas, the proliferation

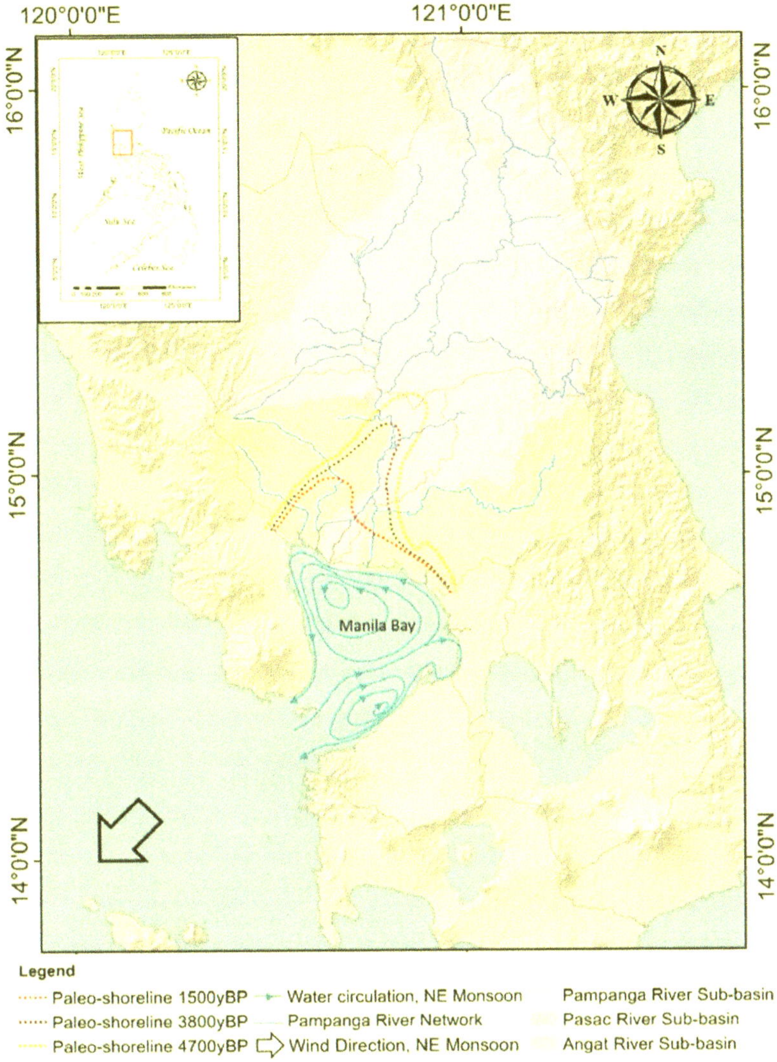

Fig. 7.1 Manila Bay and its shaded relief environs, highlighting: (1) generalized surface water circulation during NE monsoon (de las Alas and Sodusta, 1985); (2) river network and sub-basins of the Pampanga River basin, and; (3) paleo-shorelines (Soria 2009), extending > 50 km upstream of the present coastline

of aquaculture ponds also accelerated during the second half of the twentieth century, apparently converting extensive areas of mangrove forests (some 54,000 hectares in 1890 to only 794 hectares in 1995; BFAR Report 1995 as cited in PEMSEA and MBEMP TWG-RRA 2004).

Table 7.1 Density and distribution of human population within the basins of river systems draining into Manila Bay

Manila Bay Section	Province	Provincial Total Population (million)*	Population Density (n persons per km^2)	Manila Bay Population (million)**	Percent of Total Manila Bay Population***
North	Pampanga	2.61	3,797****	2.20	7.1
	Tarlac	1.37	434.12	0.68	2.2
	Nueva Ecija	2.15	376.25	1.94	6.2
	Bulacan	3.29	1,236.60	3.29	10.6
Southeast	Metro Manila	12.88	22,219.73	12.88	41.4
	Rizal	2.88	2,188.16	2.88	9.3
	Cavite	3.68	2,849.37	3.68	11.8
	Laguna	3.03	1,084.28	3.03	9.8
West	Bataan	0.76	555.53	0.51	1.6
Total (Average)		**32.85**	**(3,542.43)**	**31.50**	**100.0**

Legend *, 2015 population projection (PSA 2018); **, estimated number of inhabitants within the river basins draining into Manila Bay; ***, proportion relative to the 31.50 million people within Manila Bay watershed; *** *, with Angeles City being the densest (i.e., 6,496 persons per km^2) while the rest of the Pampanga province would otherwise have $< 1,000$ person per km^2 on average

Despite the multitude of environmental challenges besetting Manila Bay (e.g., harmful algal bloom [Siringan et al. 2008; Wang et al. 2008]; organic contaminants [Carvalho et al. 2009]; PCBs in sediments and oysters (Villanueve et al. 2010); hypoxia (Jacinto et al. 2011), sedimentation (Sta Maria et al. 2009), polycyclic aromatic hydrocarbon (Santiago 1997), overfishing (Armada 1994); habitat conversion and degradation (PEMSEA and MBEMP TWG-RRA 2004 etc.), the bay remains important to fisheries production from coastal waters (Fig. 7.2; 1980–2018), yielding ca. 8,000 metric tons annually (FSP 2016–2018). In fact, until now, the total number of fish corrals (i.e., among the various fishing gear contributing to the municipal fisheries production of the bay) would amount to ca. 1,600 (this study, 2018 imagery; Fig. 7.6), mostly distributed in the vicinity of the Pampanga River delta (n = 1,338; 84%), and in the Bacoor Bay area (n = 254; 16%). Even more significantly, the production from brackish water aquaculture yields some 220,000 metric tons annually over the last five years (Fig. 7.2).

Manila Bay Rehabilitation Initiatives and Further Challenges

Responding to the Supreme Court continuing mandamus (GR Nos. 171947–48, 2008 ordering concerned government agencies to clean up, rehabilitate, and preserve Manila Bay, and restore and maintain its waters to SB level to make them fit for swimming, skin-diving, and other forms of contact recreation), the Department of Environment and Natural Resources (DENR, the primary government agency implementing the order) formulated the "Operational Plan for Manila Bay Coastal Strategy 2017–2022" (OPMBCS 2018), providing comprehensive details on how such an order

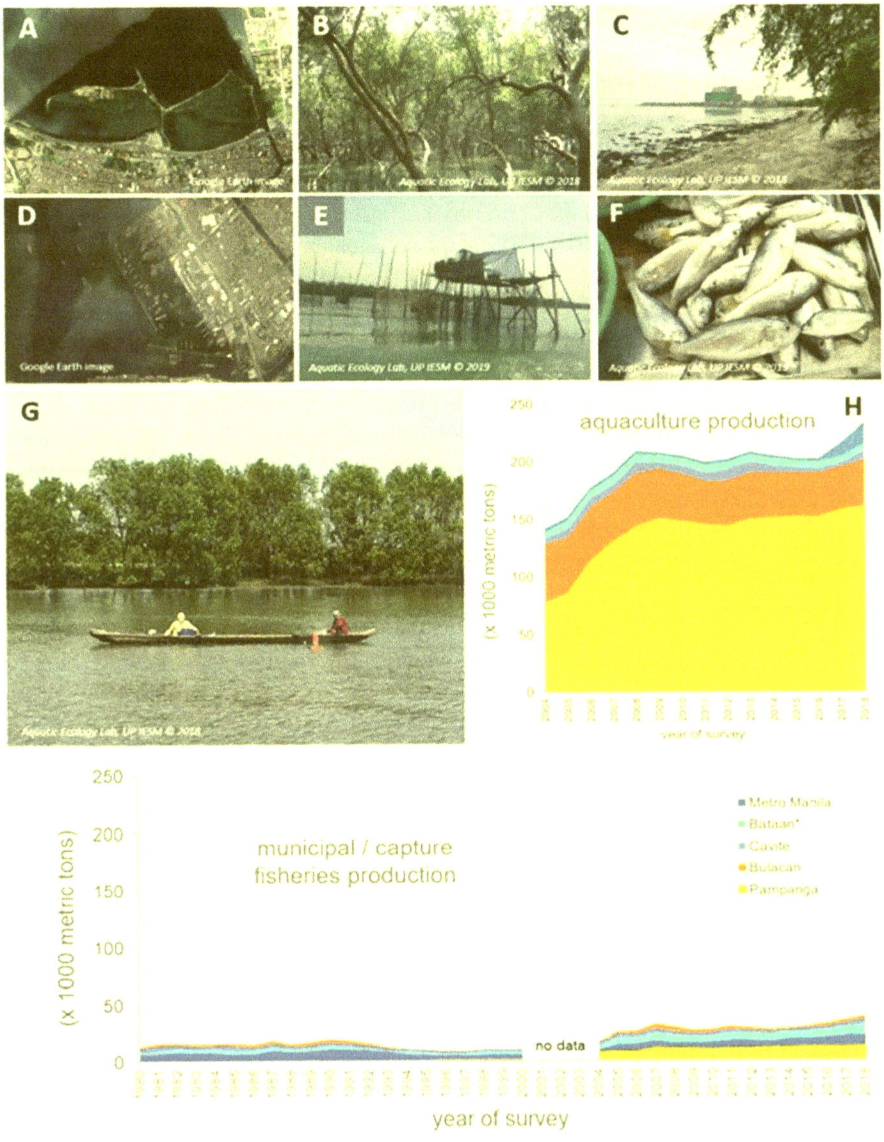

Fig. 7.2 Coastal land-use, fisheries and challenges within Manila Bay: **a** mangroves in LLPCHEA (Google Earth 2018); **b** *Avicennia marina*-dominated mangrove forest near the mouth of the Pampanga River; **c** vertical infrastructure development post-reclamation (Pasay area);**d** Navotas Fish Port (Google Earth 2018); **e** "baklad" or fish corral (mouth of Pampanga River, but see also Fig. 7.6 for comprehensive distribution); **f** fish catch of the day at the Masantol Wet Market in Pampanga; **g** gill net fishing ca. 10 km upstream of the Pampanga River mouth; **h** annual brackish water aquaculture production in Manila Bay (FSP 2004–2006, 2007–2009, 2010–2012, 2013–2015, 2016–2018); **i** annual municipal/capture fisheries production in Manila Bay (1980–2018; FSP 2000 as cited by PEMSEA and MBEMP TWG-RRA 2004)

would be complied with, identifying among others, information gaps, stakeholders, institutional arrangements, roles and responsibilities, and assigning specific Outputs to the 13 mandamus agencies (DA, DBM, DENR, DepEd, DILG, DOH, DPWH, LWUA, MMDA, MWSS, PCG, PNP-MG, and PPA). More recently, the clean-up activities in Manila Bay had been intensified, following similar rehabilitation efforts in Boracay.

As further deliberations on the court order progressed (i.e., 2008–2015), a national reclamation plan was approved in 2012 at the cabinet cluster level (PRA 2011–2012), targeting some 38,272 hectares (a total of 102 projects) within coastal areas all over the country, but 26,234 (69%) of which are along the Manila Bay coastlines. This plan actually led to a series of commentaries (e.g., Rodolfo 2013a, 2013b, 2014) and protest actions, making the discussions on the mandamus and implications to increased vulnerability to climate change impacts even more intense. In 2017, the National Economic Development Authority (NEDA) commissioned the formulation of the Manila Bay Sustainable Development Master Plan (NEDA 2017), yielding a Report (MBSDMP 2019) articulating an 8-measure development agenda, consistent with the supreme court mandamus. Despite such a Report, uncertainties as to the ecological rehabilitation goals for Manila Bay may still be huge.

Chapter Objectives

Although the past initiatives of restoring Manila Bay may already have highlands-to-oceans connectivity perspective (e.g., OPMBCS 2018; MBSDMP 2019), studies focusing on the spatio-temporal changes of the distribution and abundance of coastal habitats are scarce. Such studies are crucial benchmarks in setting restoration targets and establishing overall strategic directions.

This chapter aims to (1) revisit the present and historical spatio-temporal distribution of mangroves in Manila Bay, (2) rationalize the extent and spatial distribution of conservation and enhancement goals, (3) consider the threats of further development plans which may be incongruent to rational targets, and (4) discuss intervention options for the coastal habitats of Manila Bay in the broader context of climate emergency. Hopefully, this work may serve as crucial systems-approach inputs to refine and/or reorient programs, activities, and projects (PAPs; OPMBCS 2018; MBSDMP 2019) toward ecological restoration of Manila Bay. More broadly, the Manila Bay story is apparently a microcosm of national and global significance, exemplifying how we deal with development projects which transform our coastal ecosystems into other uses.

7.2 Revisiting Extent and Distribution of Coastal Habitats

Mapping the Present Extent of Mangroves

To determine the present extent and distribution of mangroves, a number of available basemaps (e.g., Long and Giri 2011; MBSDMP 2019) were used initially. The initial basemaps were superimposed on recent high-resolution imageries (Google Earth;

2015 coverage). Initial existing mangrove polygons were refined to reflect more accurately the present image-based extents. For image sections which were obviously mangrove vegetation (experts' inference) but were not classified as mangroves, additional polygons were created.

Validation of these mangrove polygons was done by (1) ground surveys, and; (2) Google Street View [360; recent, various dates] scanning. The ground surveys were carried out in connection with the PCARI Project ("Developing Information Infrastructure for Managing Antibiotics and Endocrine Disrupting Substances in Pampanga River Basin and its Coastal Environs: Maps, Transport Models and Bioindicators of Ecological and Public Health Risks"), assessing, among others, the habitats within the coastal floodplains of the Pampanga River delta. Specifically, some 20 km from the main river mouth (Macabebe-Hagonoy; Fig. 7.4) going upstream (Apalit area) were assessed for mangrove occurrence and distribution, with plot surveys (English et al. 1994) carried out at sampling points where mangrove extents were found significant. Using the 360 degree street view facility of Google Earth (2015–2019), the validation of the mangrove polygons was also immensely augmented. Pre-delineated polygons along 360-degree street view tracks were carefully browsed through, and when the image resolution would so permit, mangrove stands were identified at least at the general level, e.g., *Rhizophora*. In most cases, however, mangrove stands could be identified at the species level, e.g., *Nypa fruticans*, *Sonneratia caseolaris*, and *Avicennia marina*. In total, this exercise yielded 197 validation points (Fig. 7.5).

From these image-based mapping with ground validation exercise, we estimated that the mangroves in Manila Bay cover a total of 1,350 hectares (2015 imageries), the bulk (86.6%) of which are distributed only within the northern Manila Bay (i.e., Bulacan-Pampanga-Bataan stretch, *sensu* Rodolfo 2014; Fig. 7.3; Table 7.2), with some forest fragments still remaining in the Metropolitan Manila (4.5%; LPPCHEA and Navotas-Malabon) and Cavite (8.8%; mainly Bacoor Bay). From our survey plots near the mouth of Pampanga River (i.e., Macabebe-Masantol area), we recorded the occurrence of at least 9 mangrove species: *Sonneratia alba*, *S. caseolaris*, *Avicennia marina*, *Rhizophora apiculata*, *R. mucronata*, *R. stylosa*, *Lumnitzera racemosa*, *Excoecaria agallocha*, and *Nypa fruticans*. Remarkably, some 15 km upstream of the main river mouth (i.e., Masantol-Apalit area), forest fragments of the apple mangrove S. caseolaris were found in high frequency and widely. Elsewhere in the country, such a high abundance of *S. caseolaris* may not be observed as commonly, relative to its congeneric *S. alba*. For the LPPCHEA mangroves, we reckoned the total extent to be about 30 hectares (2018 imagery; Google Earth), predominantly *Avicennia marina* with occasional presence of six other mangrove species, namely, *Lumnitzera racemosa* (largest patch near Freedom Island boardwalk), *Rhizophora mucronata*, *Bruguiera cylindrica*, *Excoecaria agallocha*, *Sonneratia alba*, and *Acrostichum aureum* (December 2018 ground survey). The presence of five other mangrove taxa (i.e., *Rhizophora stylosa* ["bakauan bato"], *Xylocarpus granatum* ["tabigi"], *Aegiceras* sp. ["saging-saging"], and *Scyphiphora hydrophyllacea* ["nilad"]) had been reported earlier as well (DENR-NCR n.d.).

Reconstructing Historical Extent of Mangrove Areas

To reconstruct the historical extent of mangroves, all available old maps (e.g., NAMRIA topographic charts reflecting aerial photographic surveys in the 1950s, and all other literature materials with information on the distribution of mangroves in Manila Bay) were geo-referenced, and digitized for mangrove areas. To calculate for maximum mangrove extent potential, the existing brackish water (salinity-influenced) aquaculture ponds were similarly digitized and considered intuitively as areas vegetated with mangroves in the past. As salinity would be crucial in the mangrove distribution limits, salinity assessment along the coastal transect of the Pampanga River was carried out to infer on the likely upstream mangrove distribution boundary. As supporting ground evidence of the past geographic distribution of mangroves in Manila Bay, we evaluated the name(s) given to each locality (sitio, barangay, and municipality) along the coastal areas of Manila Bay as such may be derived from certain mangrove taxa and/or mangrove-associated flora and fauna (toponyms), probably indicative of their abundance in the past.

Indeed, our conservative approximations of historical potential distribution of mangroves in Manila Bay yielded vast extent, i.e., up to 93,000 hectares, mostly (> 86%; > 66,000 hectares) occurring at the Bataan-Pampanga-Bulacan section (Table 7.2; Fig. 7.3), with some 54,000 hectares still present in 1890 (Fig. 7.3c). This

Table 7.2 Past and present distribution of mangroves, reflecting massive loss over the years

Manila Bay Province	Past (pre-1890 Maximum Potential)		Present (2015 Distribution) (%)		% Overall Loss
	Area (hectares)	Provincial contribution (%)	Area (hectares)	Provincial contribution (%)	
Bataan	3,583	4.9	202	16.1	94.4
Bulacan	25,356	34.9	657	52.3	97.4
Cavite	1,490	2.1	111	8.8	92.5
Metropolitan Manila	4,963	6.8	57	4.5	98.9
Pampanga (less Candaba)	37,205	51.2	229	18.2	99.4
Pampanga (Candaba* only)	(21,209*)		(0)		
Total for Manila Bay	**72,596 (93,795*)**	**100**	**1,256**	**100**	**98.3**

Legend *, Candaba swamp (see Figs. 7.3 and 7.5) may be included in the calculation of the maximum potential mangrove extent by virtue of the paleo-shoreline history (4500 yBP; Soria 2009). However, for mangrove restoration purposes, this paper excludes this part

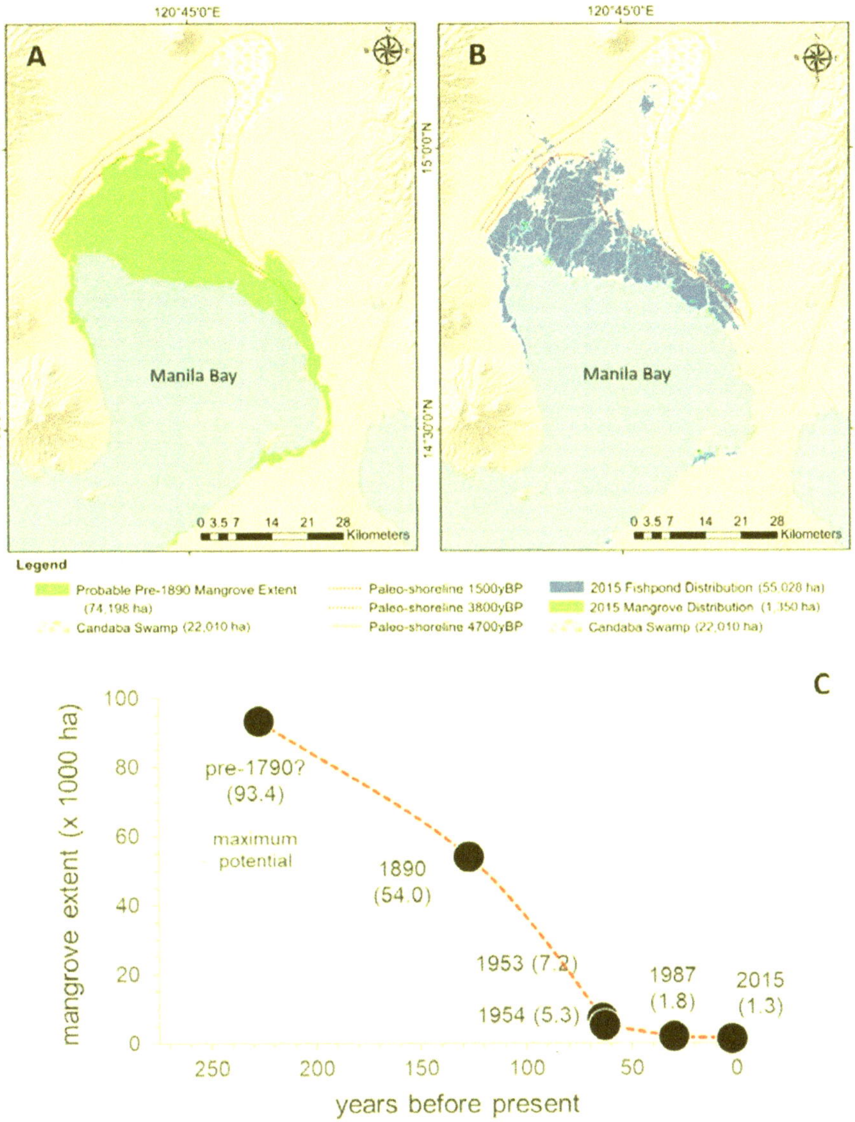

Fig. 7.3 Spatio-temporal changes in the distribution and abundance of mangroves in Manila Bay: **a** the historical distribution potential of mangroves prior to the year 1890, with paleo-shorelines (Soria 2009) also superimposed; **b** the present (2015) mangrove distribution vis-a-vis aquaculture fishponds, and; **c** decline of the Manila Bay mangroves over the years

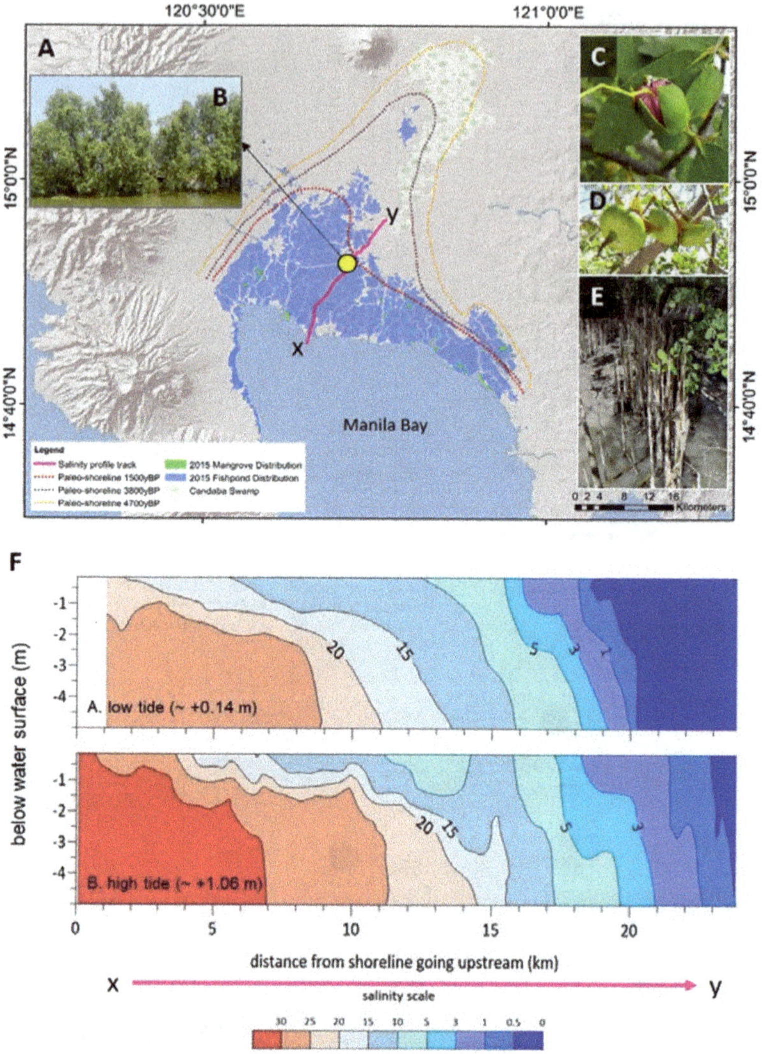

Fig. 7.4 Ground surveys confirming reconstructed mangrove distribution: **a** a map showing the 25 km survey transect (x to y); **b** remnant forests of the apple mangrove *Sonneratia caseolaris*, occurring in high frequency albeit fragmented along the x–y transect, with taxonomic identification confirmed by flowers **c**, fruits **d**, and pneumatophores **e**; **f** low and high tide salinity profiles (Surfer 16 Plots, integrating surface-to-bottom conductivity-temperature-depth measurements at 200–500 m intervals across x–y stretch), providing evidence of saltwater influence up to 25 km landward

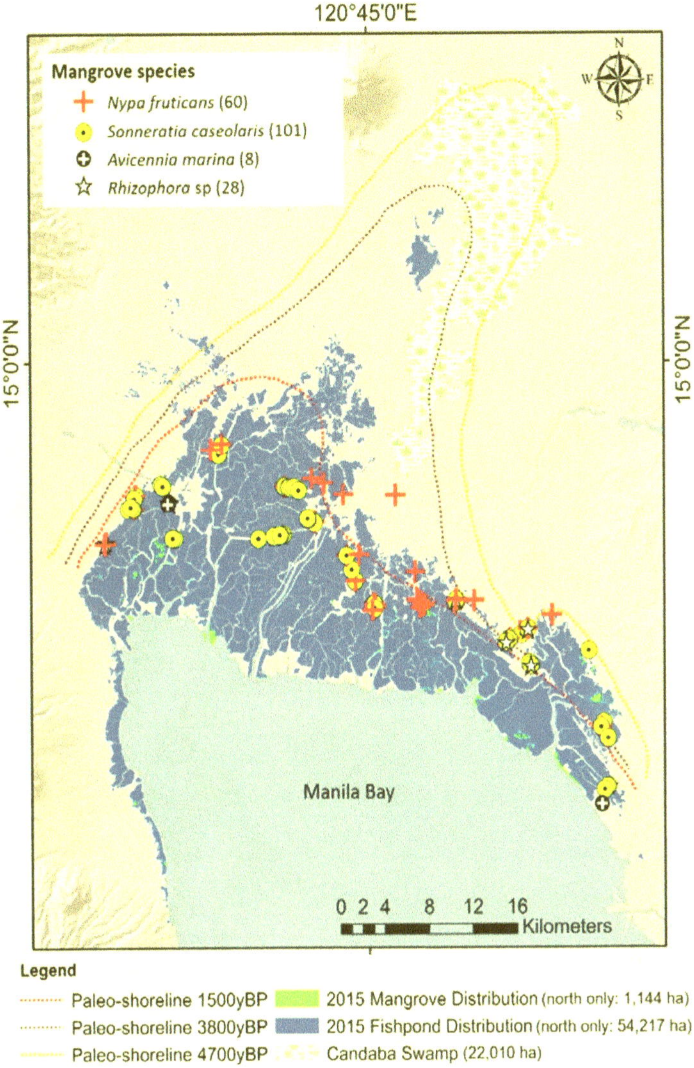

Fig. 7.5 Photograph-based occurrences (Google Earth Street View 2018; wherever available) of four common mangrove taxa (*Nypa fruticans*, *Sonneratia caseolaris*, *Avicennia marina*, and *Rhizophora* sp.), providing strong evidence of the expansive historical distribution of mangroves in the coastal floodplains of Pampanga River Basin

value is ca. 1/3 of the present national total distribution of the remaining mangroves (~300,000 hectares; FMB 2018). Mangrove loss therefore had been massive in this area, with barely one percent (1%) left today (Fig. 7.3b). Such historical reconstruction of the vast intertidal coastal plain dominated by mangrove forests in the past is actually strongly corroborated by the salinity profiles (Fig. 7.4; but see also

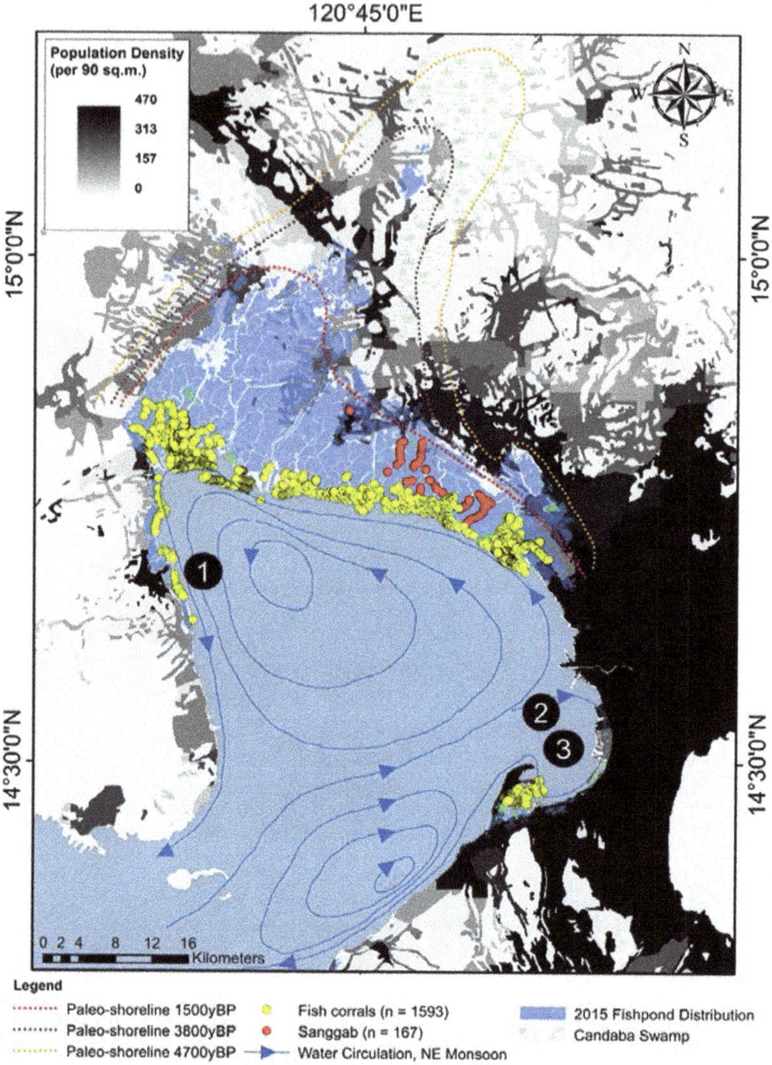

Fig. 7.6 Distribution of baklad (yellow circles, fish corrals) and sanggab (red circles, filter mesh nets), two types of a suite of fishing gear used in the coastal areas of Manila Bay. Sightings of marine mammals (**1**, ca. 500 dolphins near Bataan coastline in 2009; **2**, dwarf sperm whale near Mall of Asia in 2006; **3**, ca. 12 melon-headed whales in 2018) are also shown

Tan et al. 2020 for further corroborative isotope-based salinity evidence covering wider geographic extent), reconstructed paleo-shorelines (Soria 2009; Fig. 7.3), ground surveys of mangrove occurrence (Fig. 7.4), Google Earth 360 photographic documentation (Fig. 7.5), and toponyms. More particularly, the high frequency of fragment stands of the apple mangrove *Sonneratia caseolaris* (Fig. 7.4a) in

Masantol area as well as elsewhere (Fig. 7.5), some 15 km upstream of the mouth of Pampanga River, together with the salinity profiles (Fig. 7.4f) and the "open sea versus upstream" tide level synchrony measurements, provided the strongest evidence of such vast mangrove forest coverage in the past. In fact, today, this species (locally called "palapat" which incidentally sounds close to the common name "pagatpat" of the congeneric *Sonneratia alba*) was not known to our survey guides and local collaborators as a mangrove species. This is an apparent manifestation that not only had mangroves been decimated from their expansive abundance in the past, but also that the remnant forest fragments (e.g., *S. caseolaris*) had been actually forgotten as mangroves by the locals. Interestingly, an existing barangay "Palapat" (Hagonoy, Bulacan), located some 11 km upstream of the Binangbang River—Labangan Channel from the Manila Bay coastline, appears to be named after this mangrove *S. caseolaris*, which was probably predominant in this area in the past.

Land-use Changes and Decline of Mangrove Ecosystems

The decline of mangroves in Manila Bay over the years may not be strikingly different from that of the national trend, in this case also mostly attributable to the conversion of vast areas of mangroves into brackish water fishponds, saltwater evaporators, and/or coastal rice paddies (80%; Fig. 7.3; but see also parallel cases elsewhere in Samson and Rollon 2008, 2011). In Manila Bay however, other factors such as urbanization, human settlements, and other coastal developments also accounted substantially (19%; Fig. 7.2) for the decimation of mangroves over the recent centuries.

7.3 Rationalizing Roadmap to Conservation and Enhancement of Coastal Ecosystems

Considering the past and present distribution of mangroves in Manila Bay vis-à-vis challenges of further horizontal development of the mega metropolis, a climate-smart and system-rationalized roadmap to conservation and enhancement of coastal ecosystems should be urgent. Such a roadmap may put premium on the: (1) significance of these mangroves to the broader coastal biodiversity, fisheries, blue carbon dynamics, and nutrient fluxes, (2) bioshield function against climate-related threats (e.g., storm surge, and sea-level rise), and (3) restoration of the highlands-to-oceans continuum.

Significance to Coastal Biodiversity, Fisheries, Blue Carbon Dynamics, and Nutrient Fluxes

Even with the current state of its coastal areas, Manila Bay still supports: (1) A vast array of biota, including a highly diverse community of migratory birds in some stopover areas (e.g., in LPPCHEA, in Candaba and other coastal swamps in the Bulacan-Pampanga stretch, and in the Tortugas Wetland and Nature Park [Balanga, Bataan]), (2) marine mammals and turtles (for recent sighting records, see GMA

News (2009), Simeon 2018, Villanueva 2006; Fig. 7.6 and notes 1, 2, 3 therein), and (3) a substantial coastal fisheries production, yielding at least 250,000 tons annually (Fig. 7.2). Although at present this annual production is mainly derived from aquaculture (Fig. 7.2h, i), the contribution of the capture coastal fisheries is also huge (ca. 40,000 tons of fish landings annually; Fig. 7.2h, e, f), and likely excluded the catch from gleaners, fish corrals, and filter mesh nets (Fig. 7.2E; Fig. 7.6), and other subsistence fishers (Fig. 7.2g). In fact, particularly for fish corrals ("baklad"), this study counted a total of 1,593 units (Fig. 7.6), mostly probably yielding ca. 1,700 tons annually, assuming a conservative daily catch of at least three kilograms per fish corral. Filter mesh nets ("sanggab"; n = 167, Fig. 7.6) also appeared to be widespread in the northern Manila Bay area.

Other than the roles in biodiversity and fisheries production, the conservation and restoration of mangroves, particularly in the northern Manila Bay area, would likewise be critical to highlands-to-ocean nutrient fluxes as well as to the sequestration of atmospheric carbon and storage. As a result of the massive loss of mangroves in Manila Bay (> 70,000 ha, 98%; Table 7.2), we also lost the potential carbon burial rate of ca. 124,000 tC y^{-1} (sensu Alongi 2012; mean global burial rate = 174 g C m^{-2} y^{-1}) over the past decade now. Worse, by converting to fishponds, a huge amount of carbon stock stored in the mangrove soil (conservatively, ca. 25 million tC, at 350 MgC ha^{-1}; *sensu* Atwood et al. 2017) must have been released back to the atmosphere, in addition to the loss of aboveground biomass of about 7 million tC (*sensu* Alongi 2012; mean aboveground biomass = 95 tC ha^{-1}).

Bioshield Function Against Climate-related Threats (e.g., storm surge, sea-level rise, etc.)

The bioshield function of the mangrove areas in Manila Bay against storm surges also appeared to be huge (Fig. 7.7). Particularly in the northern section (Bulacan-Pampanga coastal areas), the projected extent of impact of storm surges could reach up to 20 km landward (Fig. 7.7, vis-à-vis Figs. 7.5 and 7.6), seriously threatening the respective coastal communities. With the expected rise in sea level due to climate change, such a projected impact of storm surges would become worse. Hence, a revisit and/or reorientation of the current efforts in managing Manila Bay and its environs should be an urgent matter. We know that the presence of mangroves may attenuate 13 to 66 percent of swell waves over 100 m of this coastal forest (McIvor et al. 2012). This means that a focus on restoring even just a kilometer belt (i.e., ca. 5% of the 20 km stretch) of mangroves along the northern coastline of the bay should result in a significant bio-shielding effect, attenuating waves and filtering disastrous debris that may also go with storm surges. In addition, mangrove systems are also efficient sediment traps (Van Santen et al. 2007). Restoring mangroves in these areas would therefore mitigate sea-level rise and maintain substrate elevation relative to the mean sea level, albeit rising.

If indeed restored, the sediment-trapping capacity of the mangroves in this area would also certainly help address some problems associated with land subsidence attributed to the excessive groundwater withdrawal in wide areas of Bulacan and Pampanga (Rodolfo and Siringan 2006; Rodolfo et al. 2003). Other areas along the

Manila Bay coastline (e.g., Cavite, Metro Manila, and Bataan) may also be revisited along parallel bioshield and climate change context.

Restoring the Northern Manila Bay Mangroves: Crucial to Sustaining the Highlands-to-oceans Continuum

As articulated in the foregoing sections, some deliberate and concrete efforts in conserving and enhancing the current state of the northern Manila Bay mangroves would indeed be crucial to sustaining the highlands-to-oceans continuum. The subsystems (i.e., terrestrial forests, rivers and lakes, wetlands and marshes, mangroves, estuaries, and the marine ecosystems), and the biophysical processes inherent in these systems linking such continuum should be revisited, understood more clearly, and managed more appropriately. Although the focus of such efforts may be most rational at the northern Manila Bay section (i.e., Bulacan-Pampanga areas), some parallel efforts may also be exerted at the Tagaytay Ridge—Bacoor Bay—Manila Bay connectivity (Figs. 7.1 and 7.3).

7.4 The Development-Conservation Battle of Manila Bay Coastal Areas

The Need for the Metropolis to Expand Laterally

With the population and infrastructures in Metro Manila getting unsustainably denser, lateral development inevitably is spilling over toward all directions: (1) southward—with the stretch up to Tagaytay Ridge now being sprawled with commercial and residential developments, (2) eastward—with development efforts encompassing even the slopes, valleys and plateaus of the Sierra Madre mountains, (3) westward—with more and more reclamation projects along the Manila Bay coastal areas, and (4) northward—coalescing with the existing cities in Bulacan, Pampanga, and Tarlac, as well as, exploring to develop "available" areas for flagship projects, e.g., the proposed new Manila International Airport in Bulacan (Cabuenas 2018), the main component of the 12,000-ha Aerotropolis development. Apparently linked with such lateral development effort, a 26,000-ha reclamation masterplan for Manila Bay has been proposed (see also Sect. 1.3; PRA 2011–2012), creating multi-purpose new dry lands (see Fig. 7.8 for spatially explicit details).

Implications to the Highlands-to-oceans Continuum and Coastal Functions of the Northern Manila Bay Mangroves

The conservation-development battle in Manila Bay appears to be incessant over the past centuries, with the development side even gaining stronger hold in more recent years. On the conservation side, Manila Bay received: (1) the rehabilitation continuing mandamus in 2008 by the Supreme Court (GR Nos. 171947–48), (2) the key biodiversity area (KBA) status for its northern section by DENR (Ambal et al. 2012), (3) several other conservation declarations, as well as, (4) more aggressive clean-up and rehabilitation initiatives accorded most recently by the DENR. On the development side however, there is the need for the metropolis to expand laterally, threatening to erase possibilities of reversion to the former mangrove habitat status of the northern Bay. If not carefully reconsidered, such masterplans would compromise the coastal functions of these areas, seriously threatening the ecological integrity of the highland-to-oceans continuum (vis-à-vis Figs. 7.1–7.8). If ever, these zones would be converted completely into mere water channels and a new metropolis, rather than conserved and enhanced in their roles as coastal wetlands and mangrove ecosystems.

7.5 Options for Manila Bay Coastal Systems in the Broader Context of Climate Emergency

From the foregoing and the real survival threat from climate projections (as reviewed most recently in Saintilan et al. 2020), the coastal habitats of Manila Bay (with emphasis on the northern section) play a number of critical functions (e.g., biodiversity and coastal fisheries, blue carbon and nutrient fluxes, bioshield against storm surges and sea-level rise, etc.) which should all be seriously considered in the climate emergency discourse, with concrete links to food security, coastal protection, and community resilience. Such insights applying these lenses may not have been clearly explicit until today, and hence, our gross under-valuation of these areas. We therefore advocate to reconsider Manila Bay development plans which may have high probabilities of further compromising the ecological integrity of the bay (e.g., reclamation masterplan, aerotropolis development, etc.). Rather, other strategic options to restore a significant amount of these former mangrove areas should be evaluated more seriously. Such ecosystem-based options may be pursued within the broader framework of the integrated coastal resource management (ICRM; see also White et al. 2006 for process details), which the Philippines has a long and institutionalized history, along with our ASEAN neighbor countries.

The spatially explicit details of such ecosystem-based options for the northern Manila Bay area should yet be worked out under the ICRM participatory process, but likely to include measures which promote protection, restoration, and/or enhancement of mangrove habitats, with the local governments units (LGUs: barangay, municipality, province) taking the crucial lead. For Manila Bay, this consensus-building would necessarily be an inter-LGU effort, which may consider the following as among the key decision handles: (1) the paleo-shoreline (see Figs. 7.1–7.8) as the landward boundary of the mangrove protection, restoration, and/or enhancement efforts, (2) the extent of storm surge level 4, and hazard 3 as the possible focus zone, if such intervention programs are to be designed to optimize the bioshield function of these systems (Fig. 7.7), and (3) practical, achievable, and consensus-based restoration targets (e.g., 10-, 20-, 50-, 90%; see Tables 7.3 and 7.4).

Table 7.3 Areal extent (in hectares) of fishponds (FP), mangroves (MGRV) outside ponds, mangroves within ponds: total extent for the northern Manila Bay Estuary (i.e., covering mainly Bulacan-Pampanga area, but may be extended to the northern coast of Bataan, as well as, Malabon-Navotas-Valenzuela area) versus the extent if restricted only within 1500 yBP shoreline limit (Soria 2009; see also Fig. 7.3)

Province/ Municipality	No. of Barangays*	Total for Northern Manila Bay Estuary (ha)**				Restricted within Paleo-shoreline at 1500 yBP (ha)**			
		FP	MGRV	MGRV within FP	Total	FP	MGRV	MGRV within FP	Total
Bataan	**63**	**2,895.68**	**6.52**	**25.72**	**2,902.20**	**2,895.17**	**6.52**	**25.72**	**2,901.70**
Abucay	7	462.51		4.52	462.51	462.51		4.52	462.51
Balanga City	6	275.23			275.23	275.23			275.23
Hermosa	6	987.09	4.48	0.14	991.57	987.09	4.48	0.14	991.57
Orani	11	421.67	2.05	0.68	423.72	421.67	2.05	0.68	423.72
Orion	13	171.54		7.85	171.54	171.03		7.85	171.03
Pilar	9	257.03		9.48	257.03	257.03		9.48	257.03
Samal	11	320.6		3.04	320.6	320.6		3.04	320.6
Bulacan	**100 (61)**	**20,345.75**	**151.93**	**262.88**	**20,497.68**	**15,850.35**	**151.93**	**157.87**	**16,002.28**
Balagtas	2 (0)	198.72		0.72	198.72				
Bocaue	3 (0)	195.01		17.43	195.01				
Bulacan	13 (6)	5,729.72	27.46	71.78	5,757.18	4,302.73	27.46	67.44	4,330.19
Calumpit	5 (1)	265.2			265.2	31.83			31.83
Guiguinto	2 (0)	29.06			29.06				
Hagonoy	26 (23)	6,311.96	49.37	11.82	6,361.33	5,925.68	49.37	11.16	5,975.05
Malolos City	18 (8)	1,533.99	9.87	1.01	1,543.86	1,162.61	9.87	1.01	1,172.48
Marilao	3 (0)	78.24		7.42	78.24				
Meycauayan City	4 (1)	1,033.36	0.11	70.02	1,033.47	86.83	0.11	1.91	86.95

(continued)

Table 7.3 (continued)

Province/ Municipality	No. of Barangays*	Total for Northern Manila Bay Estuary (ha)**				Restricted within Paleo-shoreline at 1500 yBP (ha)**			
		FP	MGRV	MGRV within FP	Total	FP	MGRV	MGRV within FP	Total
Obando	10	1,513.85	53.31	9.1	1,567.16	1,462.10	53.31	8.7	1,515.41
Paombong	14 (12)	3,456.63	11.8	73.58	3,468.43	2,878.57	11.8	67.65	2,890.38
Metropolitan Manila	**21 (19)**	**748.06**	**10.93**	**8.42**	**758.99**	**508.69**	**10.93**	**0.59**	**519.62**
Malabon	10	246.27	1.68	0.35	247.95	246.27	1.68	0.35	247.95
Navotas	1	219.79	9.25		229.04	219.79	9.25		229.04
Valenzuela	10 (8)	282		8.07	282	42.63		0.24	42.63
Pampanga	**186 (125)**	**30,008.93**	**125.75**	**86.13**	**30,134.68**	**25,560.02**	**125.75**	**86.13**	**25,685.78**
Apalit	1 (0)	44.98			44.98				
Bacolor	4 (1)	272.57			272.57	93.09			93.09
Floridablanca	4 (0)	36.7			36.7				
Guagua	31 (27)	2,608.39			2,608.39	2,364.17			2,364.17
Lubao	36 (35)	8,152.85	59.87	83.71	8,212.72	8,125.56	59.87	83.71	8,185.43
Macabebe	24 (13)	2,469.32	0.55		2,469.87	1,594.92	0.55		1,595.47
Masantol	26	6,943.19	26.54	1.55	6,969.73	6,928.69	26.54	1.55	6,955.24
Mexico	2 (0)	1.2			1.2				
Minalin	14 (9)	2,741.10			2,741.10	1,425.16			1,425.16
Porac	3 (0)	50.85			50.85				

(continued)

Table 7.3 (continued)

Province/ Municipality	No. of Barangays*	Total for Northern Manila Bay Estuary (ha)**				Restricted within Paleo-shoreline at 1500 yBP (ha)**			
		FP	MGRV	MGRV within FP	Total	FP	MGRV	MGRV within FP	Total
San Fernando City	3 (0)	194.32			194.32				
San Luis	3 (0)	19.81			19.81				
Santa Ana	8 (0)	749.56			749.56				
Santa Rita	9 (2)	304.37			304.37	14.54			14.54
Santo Tomas	6 (0)	405.83			405.83				
Sasmuan	12	5,013.90	38.79	0.87	5,052.69	5,013.90	38.79	0.87	5,052.69
Grand Total	**370 (286)**	**53,998.41**	**295.14**	**383.16**	**54,293.55**	**44,814.23**	**295.14**	**270.31**	**45,109.37**

Legend *, numbers in parenthesis indicate the number of barangays within the 1500 yBP paleo-shoreline (Soria 2009; see also Fig. 7.3) if less than the total for northern Manila Bay estuary; **, corresponding set of information at the barangay level will be made available as *Appendix A/Supplemental Data*

Table 7.4 Focusing on the northern Manila Bay Estuary: Areal extent (in hectares) of fishponds (FP), mangroves (MGRV) outside ponds, mangroves within ponds, within zones exposed to storm surge level 4—hazard 3 (see also Fig. 7.7) vis-à-vis sample options for reforestation targets reverting some fishponds back to mangroves

Province/ Municipality	No. of Barangays	FP and MGRV within zones exposed to Storm Surge Level 4-Hazard 3*				Pursuing mangrove restoration: Target options for FP to be reverted back to mangrove forests*,**				
		FP	MGRV outside FP	MGRV within FP	Total Area	10%	20%	50%	70%	90%
Bataan	**63**	**2,895.68**	6.52	**25.72**	**2,902.20**	**283.7**	**573.92**	**1,444.58**	**2,025.02**	**2,605.46**
Abucay	7	462.51		4.52	462.51	46.25	92.5	231.26	323.76	416.26
Balanga City	6	275.23			275.23	27.52	55.05	137.62	192.66	247.71
Hermosa	6	987.09	4.48	0.14	991.57	94.68	193.84	491.31	689.62	887.94
Orani	11	421.67	2.05	0.68	423.72	40.33	82.7	209.81	294.56	379.3
Orion	13	171.54		7.85	171.54	17.15	34.31	85.77	120.08	154.38
Pilar	9	257.03		9.48	257.03	25.7	51.41	128.52	179.92	231.33
Samal	11	320.6		3.04	320.6	32.06	64.12	160.3	224.42	288.54
Bulacan	**54**	**16,300.16**	**150.52**	**222.9**	**16,450.68**	**1,494.55**	**3,139.62**	**8,074.82**	**11,364.96**	**14,655.09**
Bocaue	1	7.93			7.93	439.22	926.40	2,387.92	3,362.27	4,336.63
Bulacan	8	5,001.82	27.46	70.99	5,029.28	120.81	251.49	643.52	904.87	1,166.23
Hagonoy	13	4,823.80	47.96	10.38	4,871.76	7.54	15.07	37.68	52.75	67.82
Malolos City	8	1,296.91	9.87	1.01	1,306.77	97.69	195.49	488.9	684.5	880.11
Marilao	2	75.35		7.23	75.35	103.4	260.12	730.27	1,043.70	1,357.13
Meycauayan City	3	977.91	0.11	66.43	978.03	249.63	511.07	1,295.39	9 1,818.27	2,341.15
Obando	10	1,513.85	53.31	9.1	1,567.16	439.22	926.40	2,387.92	3,362.27	4,336.63
Paombong	9	2,602.59	11.8	57.76	2,614.39	120.81	251.49	643.52	904.87	1,166.23

(continued)

Table 7.4 (continued)

Province/ Municipality	No. of Barangays	FP and MGRV within zones exposed to Storm Surge Level 4-Hazard 3*				Pursuing mangrove restoration: Target options for FP to be reverted back to mangrove forests*,**				
		FP	MGRV outside FP	MGRV within FP	Total Area	10%	20%	50%	70%	90%
Metropolitan Manila	**21**	**748.06**	**10.93**	**8.42**	**758.99**	**64.97**	**140.87**	**368.56**	**520.36**	**672.16**
Malabon	10	246.27	1.68	0.35	247.95	23.12	47.91	122.3	171.88	221.47
Navotas	1	219.79	9.25		229.04	13.65	36.56	105.27	151.08	196.89
Valenzuela	10	282		8.07	282	28.2	56.4	141	197.4	253.8
Pampanga	**38**	**11,085.63**	**112.91**	**78.95**	**11,198.54**	**1,006.94**	**2,126.80**	**5,486.36**	**7,726.07**	**9,965.78**
Lubao	13	5,188.23	59.49	77.61	5,247.72	465.28	990.05	2,564.37	3,613.91	4,663.46
Masantol	19	3,914.74	19.17	0.48	3,933.91	374.22	767.61	1,947.78	2,734.57	3,521.35
Sasmuan	6	1,982.66	34.25	0.87	2,016.91	167.44	369.13	974.21	1,377.59	1,780.97
Grand Total	**180**	**31,029.52**	**280.88**	**335.99**	**31,310.41**	**2,850.16**	**5,981.20**	**15,374.32**	**21,636.40**	**27,898.48**

Legend *, corresponding set of information at the barangay level will be made available as *Appendix B/Supplemental Data*; **, calculations of respective hectarage (y) relative to restoration targets (***RT%***) were implemented following ***y*** = *(Total Area x* ***RT%****)—(MGRV outside FP)*

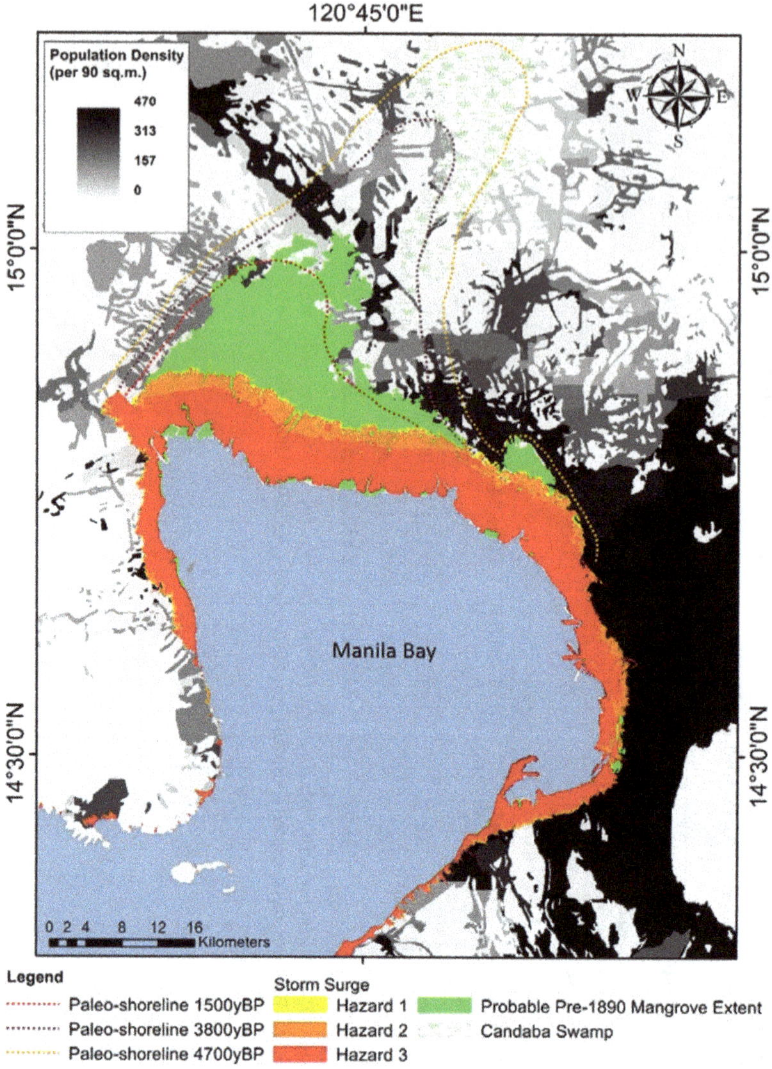

Fig. 7.7 Bioshield function: storm surge hazard exposure of Manila Bay coastal areas (data source: Project NOAH downloadable datasets; see also Lapidez et al. 2015; Lagmay and Kerle 2015; Tablazon et al. 2015; Lagmay et al. 2015), highlighting the importance of restoring mangrove forests, particularly in the northern section (i.e., Bulacan-Pampanga stretch) of the bay. The 2015 population density information is also superimposed

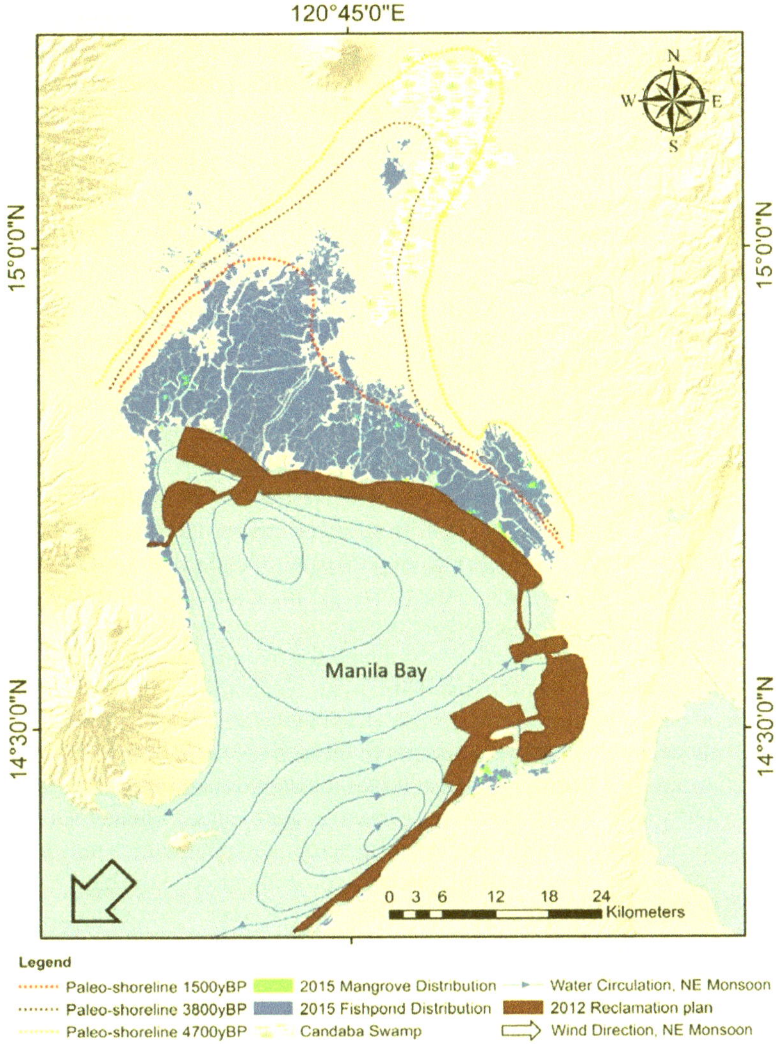

Fig. 7.8 The need for lateral expansion of the metropolis: the 2012 reclamation masterplan and its potential disruption to the functional integrity of the mangrove areas (especially the northern section of Manila Bay), and the highlands-to-oceans connectivity of the various subsystems within the broader Manila Bay complex

More specifically, about 83 percent of the ca. 54,000-ha total extent of historical mangrove area in the northern Manila Bay is within the 1500 yBP paleo-shoreline (*sensu* Soria 2009), encompassing 286 of the total 370 barangays (Table 7.3). If restoration efforts will be focused on areas exposed to storm surge hazards (Fig. 7.7) only, the affected area would be further reduced to ca. 31,000 hectares (i.e., 57% of the historical potential), involving some 180 barangays (Table 7.4). This means that full mangrove reversion of areas within the 1500 yBP shoreline and storm surge zone would require mangrove reforestation of some 54,000 and 31,000 hectares, respectively. If 10 percent would be set as the minimum practical, achievable, and consensus-based goal, then we are still looking at some 3,100–5,400 hectares of spatially explicit reforestation targets (Table 7.4).

7.6 Conclusion

This chapter revisits the changes in mangrove occurrence and distribution relative to the spatio-temporal changes elsewhere in the 200 km coastline of the entire Manila Bay. Using topographic maps, brackish water ponds, and toponyms, we reconstructed the Bay's historical mangrove extent. About 75,000–90,000 hectares of mangroves apparently existed in the pre-1900s, comprising some 7–9 percent of the Pampanga River basin, which drains 50 percent of the total freshwater volume into Manila Bay. Such reconstructed mangrove extent was strongly supported by data from recent field surveys (distribution of mangrove remnant patches and salinity measurements), as well as previous work on paleo-shorelines. At present, only barely one percent (<1,000 ha) is left, reflecting massive deforestation. Conversion into brackish water fishponds, salt pans, and rice paddies accounted for the majority of these mangrove losses, as well as development into residential-, and industrial settlements. Realizing such historical natural proportion (1:2; historical mangroves: Manila Bay), we underline the significance of mangrove reversion in these areas to the coastal biodiversity, fisheries, blue carbon dynamics, nutrient fluxes, bioshield function, and overall environmental vitality of the bay, considering present and projected climate scenarios. We therefore advocate to reconsider Manila Bay development plans which may have high probabilities of further compromising the ecological integrity of the bay (e.g., reclamation masterplan, aerotropolis development, etc.). Rather, other strategic options to restore a significant amount of these former mangrove areas should be evaluated more seriously. With this Manila Bay case as a microcosm, we stress the need for more environmental scientists looking through broader spatio-temporal lenses in our work toward conserving the Philippine coastal ecosystems.

Acknowledgements We thank the two research programs whose collective effort made this contribution possible: (1) the CHED-PCARI project entitled "Developing Information Infrastructure for Managing Antibiotics and Endocrine Disrupting Substances in Pampanga River Basin and its Coastal Environs: Maps, Transport Models and Bioindicators of Ecological and Public Health Risks," and (2) the DOST-PCIEERD JICA-JST collaborative program on the "Integrated Assessment and Monitoring of Blue Carbon Ecosystems for Conservation and Adaptive Management" (IAMBlueCECAM) Program and the "Comprehensive Assessment and Conservation of Blue Carbon Ecosystems and their Services in the Coral Triangle" (BlueCARES) Project. Specifically, we thank IAMBlueCECAM Project 6 entitled "Suitability mapping for guiding mangrove replanting (SuitMaPs)." We also thank Dr. Cesar L. Villanoy for allowing us to use his laboratory's CastAway-CTD, which was instrumental in generating Fig. 7.4f.

We also acknowledge the UP Resilience Institute (UPRI) for the Research and Creative Writing (RCW) Grant.

References

Alongi DM (2012) Carbon sequestration in mangrove forests. Carbon Manag 3(3):313–322

Ambal RGR, Duya MV, Cruz MA, Coroza OG, Vergara SG, de Silva N, Molinyawe N, Tabaranza B (2012) Key biodiversity areas in the philippines: priorities for conservation. J Threat Taxa 4(8):2788–2796

Armada NB (1994) Capture fisheries of manila bay. Final report on capture fisheries component of the resource and ecological assessment of manila bay. Fisheries sector program, department of agriculture, Quezon City

Atwood TB, Connolly RM, Almahasheer H, Carnell PE, Duarte CM, Ewers Lewis CJ, Irigoien X, Kelleway JJ, Lavery PS, Macreadie PI, Serrano O, Sanders CJ, Santos I, Steven ADL, Lovelock CE (2017) Global patterns in mangrove soil carbon stocks and losses. Nat Clim Chang 7:523–528

Cabuenas JV (2018) NEDA Board approves negotiation report on P735-B Bulacan airport. GMA News Online, 24 Dec 2019. https://www.gmanetwork.com/news/money/economy/679255/neda-board-approves-negotiation-report-on-p735-b-bulacan-airport/story/

Carvalho FP, Villeneuve JP, Cattini C, Tolosa I, Bajet CM, Navarro-Calingacion MF (2009) Organic contaminants in the marine environment of manila bay, Philippines. Arch Environ Contam Toxicol 57:348–358

de las Alas JG, Sodusta JA (1985) A model for the wind-driven circulation of Manila Bay. Nat Appl Sci Bull 37:159–170

DENR-NCR. n.d. Las Piñas-Parañaque critical habitat and ecotourism area. http://old-ncr.denr.gov.ph/index.php/89-webpage/338-brochure. Accessed 22 Jun 2020

Dorenbosch M (2006) Connectivity between fish assemblages of seagrass beds, mangroves and coral reefs: Evidence from the Caribbean and the western Indian Ocean. Ph.D Thesis, Faculty of Science, Radboud University Nijmegen, The Netherlands

GMA News (2009) 100 melon-headed whales stranded off Bataan, 3 reported dead. GMA News Online, 10 Feb 2009. https://www.gmanetwork.com/news/news/regions/148054/100-melon-headed-whales-stranded-off-bataan-3-reported-dead/story/

English S, Wilkinson C, Baker V (eds) (1994) Survey manual for tropical marine resources. Townsv: Aust Inst Mar Sci

Ewers Lewis CJ, Irigoien X, Kelleway JJ, Lavery PS, Macreadie PI, Serrano O, Sanders CJ, Santos I, Steven ADL, Lovelock CE (2017) Global patterns in mangrove soil carbon stocks and losses. Nat Clim Chang 7:523–528

FMB (2018) Philippines forestry statistics 2018. Forest management bureau, department of environment and natural resources. https://forestry.denr.gov.ph/index.php/statistics/philippines-forestry-statistics. Accessed 24 Jan 2020

FSP (2004–2006) Fisheries statistics of the Philippines 2004–2006: 15. Bureau of agricultural statistics, department of agriculture, Quezon City. https://psa.gov.ph/content/fisheries-statistics-philippines. Accessed 29 Nov 2019

FSP (2007–2009) Fisheries Statistics of the Philippines 2007–2009: 18. Bureau of agricultural statistics, department of agriculture, Quezon City. https://psa.gov.ph/content/fisheries-statistics-philippines. Accessed 29 Nov 2019

FSP (2010–2012) Fisheries Statistics of the Philippines 2010–2012: 21. Bureau of agricultural statistics, department of agriculture, Quezon City. https://psa.gov.ph/content/fisheries-statistics-philippines. Accessed 29 Nov 2019

FSP (2013–2015) Fisheries statistics of the Philippines 2013–2015: 24. Philippine statistics authority, Quezon City. https://psa.gov.ph/content/fisheries-statistics-philippines. Accessed 29 Nov 2019

FSP (2016–2018) Fisheries Statistics of the Philippines 2016–2018: 27. Philippine statistics authority, quezon City. https://psa.gov.ph/content/fisheries-statistics-philippines. Accessed 29 Nov 2019

Jacinto GS, Sotto LPA, Senal MIS, San Diego-McGlone ML, Escobar MTL, Amano A, Miller TW (2011) Hypoxia in Manila Bay, Philippines during the northeast monsoon. Mar Pollut Bull 63:243–248

Lagmay AMF, Kerle N (2015) Typhoons: Storm-surge models helped for Hagupit. Nature 519:414. https://doi.org/10.1038/519414b

Lagmay AMF, Agaton RP, Bahala MC, Briones JT, Cabacaba KMC, Caro CVC, Dasallas LL, Gonzalo LIL, Ladiero CN, Lapidez JP, Mungcal MTF, Puno JVR, Ramos MC, Santiago J, Suarez JK, Tablazon JP (2015) Devastating storm surges of Typhoon Haiyan. Int J Disaster Risk Reduct 11:1–12

Lapidez JP, Tablazon J, Dasallas L, Gonzalo LA, Cabacaba KM, Ramos MMA, Suarez JK, Santiago J, Lagmay AMF, Malano V (2015) Identification of storm surge vulnerable areas in the Philippines through the simulation of Typhoon Haiyan—induced storm surge levels over historical storm tracks. Natl Hazards Earth Syst Sci 15:1473–1481. https://doi.org/10.5194/nhess-15-1473-2015

Long JB, Giri C (2011) Mapping the Philippines' mangrove forest using landsat imagery. Sensors 11(3):2972–2981

McIvor AL, Möller I, Spencer T, Spalding M (2012) Reduction of wind and swell waves by mangroves. natural coastal protection series: Report 1. Cambridge Coastal Research Unit Working Paper 40. Publ Nat Conserv Wetl Int. p 27

MBSDMP (2019) Manila bay sustainable development master plan. http://mbsdmp.com/resources. Accessed 30 Nov 2019

NEDA (2017) Terms of Reference for the formulation of the Manila Bay Sustainable Development Master Plan (MBSDMP). National Economic Development Authority (NEDA), 23 pages. http://www.neda.gov.ph/terms-of-reference-tor-for-the-formulation-of-the-manila-bay-sustainable-development-master-plan-mbsdmp/. Accessed 30 Nov 2019

OPMBCS (2018) Operational Plan for the Manila Bay Coastal Strategy 2017–2022. Manila Bay Mandamus Agencies

PEMSEA and MBEMP TWG-RRA (2004) Manila Bay: Refined Risk Assessment. PEMSEA Technical Report No. 9, p 169 Global Environment Facility/United Nations Development Programme/International Maritime Organization Regional Programme on Building Partnerships in Environmental Management for the Seas of East Asia (PEMSEA), and Manila Bay Environmental Management Project (MBEMP), Technical Working Group for Refined Risk Assessment (TWG-RRA) Quezon City, Philippines

PSA (2018) 2018 Philippine statistical yearbook. Philippine statistics authority, Quezon City, Philippines

PRA. 2011–2012. Mapping new economic terrain: Progress through land reclamation. Philippine Reclamation Authority. 2011–2012 Corporate Report, p 32. https://www.pea.gov.ph/images/pra_images/pdf/AnnualReport2012_flattened.pdf

Rodolfo K (2013a) Why reclamation of nearshore Manila Bay is a very bad idea. Philippine Star, 03 October 2013. https://www.philstar.com/business/science-and-environment/2013/10/03/1240670/why-reclamation-nearshore-manila-bay-very-bad-idea

Rodolfo K (2013b) Statement on the geological hazards of Manila Bay in reference to "An Environmental Impact Statement (EIS) Report for the proposed 148-hectare manila Goldcoast Reclamation Project along the Coast of the City" by Technotrix Consultancy Services. Submitted in support of the oral testimony at DENR Public Hearing, 18 November 2013, p 21

Rodolfo K (2014) On the geological hazards that threaten existing and proposed reclamations of Manila Bay. Philipp Sci Lett 7(1):228–240

Rodolfo K, Siringan FS (2006) Global sea-level rise is recognised, but flooding from anthropogenic land subsidence is ignored around northern Manila Bay, Philippines. Disasters 30(1):118–139

Rodolfo K, Siringan FS, Remotique CT, Lamug CB (2003) Worsening floods around northern Manila Bay, Philippines: Research-based analysis from physical and social science perspectives. Philipp Sociol Rev 51:17–40

Saintilan N, Khan NS, Ashe E, Kelleway JJ, Rogers K, Woodroffe CD, Horton BP (2020) Thresholds of mangrove survival under rapid sea level rise. Science 368(6495):1118–1121
Samson MS, Rollon RN (2008) Growth performance of planted mangroves in the Philippines: revisiting forest management strategies. Ambio 37(4):234–240. https://doi.org/10.1579/0044-7447(2008)37[234:GPOPMI]2.0.CO;2
Samson MS, Rollon RN (2011) Mangrove revegetation potentials of Brackish-Water pond areas in the Philippines. In aquaculture and the Environment—A Shared Destiny, edited by Dr. Barbara Sladonja, 31–50. InTech. http://www.intechopen.com/books/aquaculture-and-theenvironment-a-shared-destiny/mangrove-revegetation-potentials-of-brackish-water-pond-areas-in-thephilippines
Santiago E (1997) The levels and distribution of Polycyclic Aromatic Hydrocarbons (PAH) contamination in bottom sediments in Manila Bay. Science Diliman 9(1 & 2):16–28
Simeon LM (2018) 12 dolphins rescued in Manila Bay. The Philippine Star, 2 Mar 2018. https://www.philstar.com/headlines/2018/03/02/1792770/12-dolphins-rescued-manila-bay
Siringan FP, Azanza RV, Macalalad NJH, Zamora PB, Sta MYY, Maria. (2008) Temporal changes in the cyst densities of Pyrodinium bahamense var. compressum and other dinoflagellates in Manila Bay, Philippines. Harmful Algae 7:523–531
Soria JLA (2009) Paleoenvironmental and paleo-sea level reconstructions, and natural compaction in the delta complex bordering north of Manila Bay. Master's thesis, College of Science, University of the Philippines.
Sta Maria EJ, Siringan FP, Bulos AdM, Sombrito EZ (2009) Estimating sediment accumulation rates in Manila Bay, a marine pollution hot spot in the Seas of East Asia. Mar Pollut Bull 59:164–174
Tablazon J, Caro CV, Lagmay AMF, Briones JBL, Dasallas L, Lapidez JP, Santiago J, Suarez JK, Ladiero C, Gonzalo LA, Mungcal MTF, Malano V (2015) Probabilistic storm surge inundation maps for Metro Manila based on Philippine public storm warning signals. Natl Hazards Earth Syst Sci 15:557–570. https://doi.org/10.5194/nhess-15-557-2015
Van Santen P, Augustinus PGEF, Janssen-Stedler BM, Quartel S, Tri NH (2007) Sedimentation in an estuarine mangrove system. J Asian Earth Sci 29:566–575
Villanueva R (2006) Whale found in Manila Bay. The Philippine Star, 17 May 2006. https://www.philstar.com/headlines/2006/05/17/337048/whale-found-manila-bay
Villeneuve JP, Cattini C, Bajet CM, Navarro-Calingacion MF, Carvalho FP (2010) PCBs in sediments and oysters of Manila Bay, the Philippines. Int J Environ Health Res 20(4):259–269
Wang SF, Tang DL, He FL, Fukuyo Y, Azanza RV (2008) Occurrences of harmful algal blooms (HABs) associated with ocean environments in the South China Sea. Hydrobiologia 596:79–93
White A, Deguit E, Jatulan W, Eisma-Osorio L (2006) Integrated coastal management in philippine local governance: evolution and benefits. Coast Manag 34:287–302

Rene N. Rollon Ph.D. is a Professor at the Institute of Environmental Science & Meteorology (IESM), College of Science (CS), University of the Philippines Diliman. Heading the Aquatic Ecology Laboratory at IESM, Dr. Rollon leads research projects looking at various fields of interest including aquatic ecosystems, coastal marine biology and ecology, mangrove and seagrass ecology, coastal resource management, environmental impact assessment, and general environmental sciences. He served as the Director of IESM for two terms (2009-2015). He also served (2019-2023) as the Director of the Science and Society Program of the College of Science.

Giannina Marie G. Albano is a Senior Science Research Specialist of the IAMBlueCECAM Project 6 titled, "Suitability mapping for guiding mangrove replanting" (SuitMaPs). With an MS degree in Marine Science, Ms. Albano worked on various research projects dealing with coastal sediment integrity, mangrove recolonization in abandoned, unutilized and underdeveloped fishpond areas, nationwide assessment of coral reef environments (NACRE-associated habitats: mangroves and seagrasses), and coral reef visualization and assessment (CoRVA).

Green Ann A. Cruz is a Science Research Specialist for "Suitability mapping for guiding mangrove replanting" (SuitMaPs). While pursuing an MS in Environmental Science, she studies carbon stock in mangrove forests, both in the aboveground and belowground reservoirs. She is part of the nationwide assessment of coral reef environments (NACRE-associated habitats: mangroves and seagrasses), and coral reef visualization and assessment (CoRVA) research projects.

Maia Patricia D. Garcia is a Research Fellow of the Commission on Higher Education Philippine-California Advanced Research Institutes Project (PCARI) looking at impacts of selected micropollutants on the aquatic ecosystems, including mangroves in the coastal floodplains, associated with the Pampanga River Basin. Ms. Garcia focuses on the fish ecology and fisheries aspects of the project while currently pursuing her MS degree in Environmental Science.

Kristine Camille V. Buenafe works as a Research Fellow of the Commission on Higher Education Philippine-California Advanced Research Institutes Project (PCARI), focusing on the macrobenthic community aspect. Ms. Buenafe is currently pursuing her MS degree under the Erasmus-Mundus Joint Masters Programme in Tropical Biodiversity and Ecosystems (Tropimundo).

Anisette Monique G. Albano works as a Project Assistant under "Suitability mapping for guiding mangrove replanting" (SuitMaPs), significantly contributing to the planning, field data collection, analysis, and writing of this chapter, "The Need for a Clear Rehabilitation Roadmap for Manila Bay Coastal Ecosystems: A Microcosm of Climate Emergency in the Philippine Context".

Betty May R. Villamayor worked at the Aquatic Ecology Laboratory in IESM as a Senior Research Associate under various research projects on mangrove and seagrass ecology, culminating to her MS Environmental Science thesis on the impact of super typhoon Haiyan on mangroves. She contributed significantly to the planning, field data collection, analysis, and report writing of research projects, SuitMaps, NACRE, CoRVA, and CHED-PCARI. She is pursuing her PhD at the National University or Ireland (NUI Galway).

Chapter 8
Flood Hazard Mapping with Nonstationary Extreme Values under Changing Climate for Cagayan De Oro River Basin

Guillermo Q. Tabios III

Abstract Flood hazard mapping is vital for forecasting, early warning, and planning for flood disasters. Flood inundation modeling and simulation is conducted to describe the spatial extent and distribution of flood waters over time, for a given hydrologic design flood. Engineers use the 'return period' or probability of occurrence as the basis for designing flood management schemes or control structures to protect vulnerable communities. However, with the changing climate and urbanization, the flood frequency analysis should be performed in the framework of nonstationary extreme value analysis rather than treating the river flow process as stationary in time which is the usual practice. This chapter applies flood hazard mapping with nonstationary extreme values under a changing climate for the Cagayan de Oro River Basin which was devastated during Typhoon Sendong in December 2011. This chapter presents several policy implications: first, for the engineering practice of DPWH to utilize nonstationary extreme value analysis over stationary analysis. Second, policies and decisions should be based on actual trends or shifts from observed data rather than just relying on climate change design parameters and projections that depend on scenarios, storylines, and global climate models that may not be appropriate for Philippine climate conditions. A strategic policy that should be adopted is piecewise engineering or optimal project sequencing and staging that considers the evolutionary nature and uncertainty of climate change.

Keywords Storm rainfall · Nonstationary extreme value analysis · Design return period

Supplementary Information The online version contains supplementary material available at https://doi.org/10.1007/978-981-99-7804-5_8.

G. Q. Tabios III (✉)
University of the Philippines Diliman, Quezon, Philippines
e-mail: gtabios@up.edu.ph

K. B. Berse et al. (eds.), *Climate Emergency in the Philippines*, Disaster Risk Reduction, https://doi.org/10.1007/978-981-99-7804-5_8

Key Points

- In flood hazard mapping, engineers and experts conduct flood inundation modeling and simulation for the purpose of forecasting, early warning, planning, and preparing for flood disasters.
- The climate emergency compels us to rethink and redesign our methods so that we can better prepare for climate change impacts and disasters in the future.
- The method of flood hazard mapping utilizing the framework of nonstationary extreme value analysis, rather than stationary analysis, was applied in the case of the Cagayan de Oro River Basin after Typhoon Sendong in 2011.
- Piecewise engineering or optimal project sequencing and staging should be adopted, as it considers the evolutionary nature and uncertainty of climate change.

Summary For Policymakers

In flood hazard mapping, flood inundation modeling and simulation is conducted to describe the temporal distribution and spatial extent of flooding for a given hydrologic design flood which has an associated return period or probability of occurrence.

- The return period has been used in engineering practice as the basis of the design of flood control structures or nonstructural flood management schemes to provide a certain level of flood protection (i.e., or conversely flood risk aversion) to the community.
- Flood frequency or extreme value probability analysis is performed to define the design flood magnitude and the associated return period with the assumption that the underlying probability distribution of the river flow process (implying also its generating mechanism) is stationary in time.

However, with changing climate, including land use changes and/or urbanization, the river flow process should be treated as nonstationary. In this case, the flood frequency analysis must be performed in the framework of nonstationary extreme value analysis.

This is illustrated here in the case of Cagayan de Oro City, which was devastated during Typhoon Sendong in December 2011. A two- dimensional (2d) flood inundation model is used with inflows (2d model input) generated from a watershed model given the historical design storm rainfall assumed to be stationary in time. For comparison, the design storm rainfall and associated return period were evaluated based on nonstationary extreme value analysis to account for changing climate.

In this case, the return period of the design flood discharge is associated to the return period of the storm rainfall. It may be noted that while the design flood discharge and return period of Cagayan de Oro River could change also due to land use change and urbanization and thus nonstationary as far as the flood discharge probability characteristics. However, only nonstationary extreme value analysis of storm rainfall is conducted here thus calculated design flood discharge only reflects changing climate through the nonstationarity of rainfall regimes.

Since the engineering practice of DPWH is to design flood control infrastructures based on return period, the analysis of design flood for the corresponding return period should utilize nonstationary extreme value analysis rather than stationary analysis with changing climate including land use change and urbanization.

Several policy implications can be drawn in this study as follows.

- Since the engineering practice of the Department of Public Works and Highways (DPWH) is to design flood control infrastructures based on return period (e.g., 25-year return period for local drainage culverts or 50-year return period for national drainage system which are revised guidelines with climate change), the analysis of design flood for the corresponding return period should utilize nonstationary extreme value analysis rather than stationary analysis with changing climate including land use change and urbanization.
- Considering climate change, the approach here is to define future design flood and associated return period based on actual trends or shifts from observed data so this can be better appreciated by decision-makers rather than deriving climate change design parameters from results of the climate projections (i.e., IPCC) which depends on the scenarios or storylines (i.e., four to five of them) and the specific global climate model used (i.e., about 25 models are available and five are known to be appropriate to Philippine climatic conditions).
- Since nonstationary, probability analysis is employed here, the flood risk at different design life (time windows) of the projects can be easily calculated which is useful in optimal sequencing and staging flood control projects or components thereof to properly time investments and financial resources.
- Optimal project sequencing and staging, also referred to as piecewise engineering is a strategic policy to adopt in view of the evolutionary nature and uncertainty of climate change (due to advancing climate science and modeling with changing storylines and projections over time and depending on actual climate change mitigation measures implemented). Land use change and urbanization can also be evolutionary in nature in view of population changes, socio-economic changes as well as political changes reflected in changing objectives, priorities, and uncertainties in government policies and administration.

Piecewise engineering or optimal project sequencing and staging should be adopted, as it considers the evolutionary nature and uncertainty of climate change.

8.1 Introduction

In flood hazard mapping, flood inundation modeling and simulation is conducted to describe the temporal distribution and spatial extent of flooding for a given hydrologic design flood which has an associated return period or probability of occurrence. The return period has been used in engineering practice as the basis of the design of flood control structures or nonstructural flood management schemes to provide a certain level of flood protection (i.e., or conversely flood risk aversion) to the community.

Flood frequency or extreme value probability analysis is performed to define the design flood magnitude and the associated return period with the assumption that the underlying probability distribution of the river flow process (implying also its generating mechanism) is stationary in time. However, with changing climate, including land use changes and/or urbanization, the river flow process should be treated as nonstationary. In this case, the flood frequency analysis should be performed in the framework of nonstationary extreme value analysis and it is illustrated here in the case of the Cagayan de Oro River Basin.

The Cagayan de Oro River (CDOR) Basin is situated in Region X in the Northern Mindanao Region with a total catchment area of 1,364 km^2. Its headwaters are in Mt. Kitanglad at an elevation of about 2,900 m and the river runs for 97 km in which the first 75 km has a steep bed slope, then becomes a mild and gentle slope as it approaches the relatively flat lowland area of the highly urbanized city of Cagayan de Oro with elevations around 15 to 20 m above MSL (mean sea level), and finally the river ends in Macajalar Bay at MSL. Past studies of the flow capacities of the existing river channel around Cagayan de Oro City range from 700 to 1,500 m^3/s corresponding to the 2–5-year flood recurrence interval. Major flood events in CDOR Basin recorded occurred in 1916, 1957, 1982, 1998, and 2009. Then the recent Tropical Storm (TS) Sendong in December 2011 and TS Pablo the following year, both producing widespread inundation and extensive loss of life, damage to infrastructure, and disruption in socio-economic activities in Cagayan de Oro City (JICA 2014). TS Sendong is considered to be the maximum historic flood experienced by the city and the most devastating which estimated to be a 50-year return period flood. At around 1 AM on December 17, 2011, the floodwaters rapidly rose toward the populated area within floodplains of the Cagayan de Oro River which killed 1,260 people especially those living along the river banks and its flood plains. The flood inundation in the urban area was less than one day and the estimated flood economic damage was 1.7 billion pesos.

A flood hazard mapping of CDOR was conducted by NHRC (2016) and also reported by Tabios (2017) for alternative flood mitigation plans for the river around Cagayan de Oro City. The NHRC (2016) study was essentially to revise the original flood risk management plan of CDOR of June 2011 in response to the flood devastation brought by TS Sendong in December 2011. A two-dimensional (2d) flood inundation model was used with watershed inflows based on historical design storm rainfall assumed to be stationary in time (NHRC 2016). In this chapter, the same flood model and its accompanying watershed model were used; however, the design storm rainfall and associated return period are based on nonstationary extreme value analysis to account for changing climate. Note that while the design flood discharge and return period of CDOR could change due to land use change and urbanization in addition to changing climate, only the nonstationary extreme value analysis of storm rainfall is considered in this study, thus, the calculated design flood discharge only reflects changing climate through nonstationarity of rainfall regimes.

8.2 Nonstationary Hydrologic Design Flood

Stationary versus Nonstationary Extreme Value Analysis

In flood hazard mapping, the hydrologic design flood is specified which has an associated return period. In particular, the return period also referred to as the average recurrence interval is statistically defined as the expected waiting time for the occurrence of a flood exceeding a certain design flood magnitude (Salas et al. 2018). If historical data of flood events such as flow discharge or water stages are available, one can perform flood frequency or probability analysis of say the flood flows to determine the flood design magnitude and corresponding probability of occurrence or return period. One can also calculate the design flood magnitude using a rainfall-runoff model of the watershed with storm rainfall hyetograph as input. In this case, the associated probability of occurrence or return period of the storm rainfall event is assumed to be the return period of the design flood magnitude. To account for changing climate, nonstationary extreme value analysis should be employed instead of stationary analysis. Details of the nonstationary extreme value analysis are presented in the Annex section at the end of this chapter.

Nonstationary Hydrologic Design Parameter for CDOR based on Storm Rainfall

As mentioned earlier, the flood hazard mapping conducted here is based on a flood hydrograph generated from storm rainfall using a watershed model. The 24-h annual maxima basin rainfall is utilized in this study. The time series data is shown in Fig. 8.1. The data is concatenated from the 24-h annual maxima basin rainfall from 1985 to 2012 taken from JICA (2014) study and point rainfall data from Cagayan de Oro City station from 1949 to 1984 which was properly converted to basin rainfall using an areal reduction factor of 0.6574 (Kok et al. 2017; Pietersen et al. 2015).

The annual maxima basin rainfall clearly exhibits an increasing trend definitely suggesting nonstationarity and can be fairly attributed to changing climate regimes. For purposes of fitting the general extreme value (GEV) distribution, it is assumed that only the location parameter accounts for the nonstationarity while the scale and shape parameters are assumed constant. This is given by Eq. (3a) in the Annex section at the end of this chapter.

The parameters of the fitted GEV distribution (assuming a stationary probability distribution) estimated by the method of moments are $\boldsymbol{x_0} = 70.055$, $\boldsymbol{a} = 26.525$ and $\boldsymbol{b} = -0.0533$ is the location, scale and shape parameter, respectively. To parameterize nonstationarity, the location parameter is replaced by a linear trend function given by

$$x_0(t) = -880.11 + 0.48682t + e \tag{8.1}$$

With the fitted nonstationary GEV probability distribution function, several nonstationary parameters or indices can be computed and compared to their stationary counterparts. Figure 8.2 shows the relationship between the nonstationary return

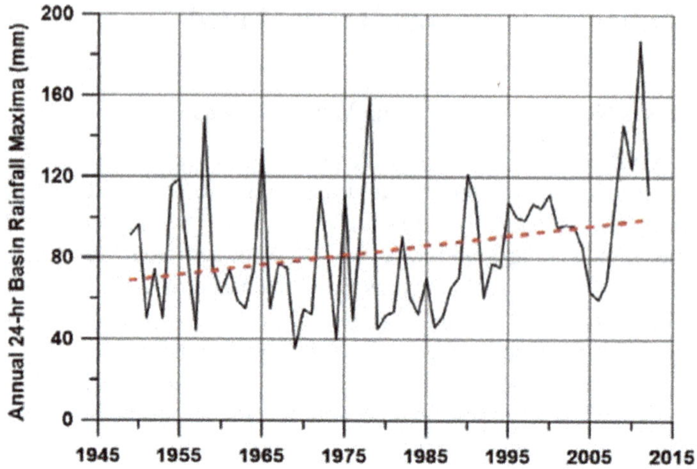

Fig. 8.1 Time series plot of 24-h annual maxima basin rainfall of Cagayan de Oro River Basin

period and stationary return period. Note that this particular curve is the result of the fitted GEV distribution with its accompanying fitted time-varying location parameter. It is shown here that if a flood control project is to provide a 50-year (nonstationary T in Fig. 8.2) level of protection over the life of the project, then perhaps it should be designed at 120-year return period (stationary T or initial return period).

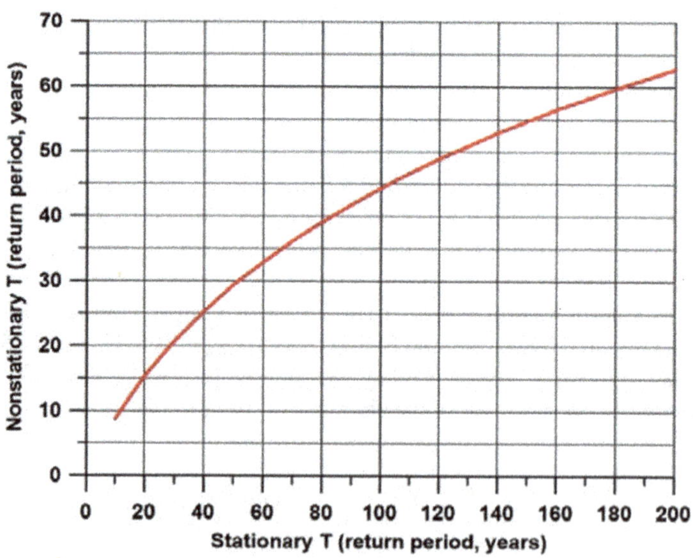

Fig. 8.2 Nonstationary return period T as a function of the stationary T

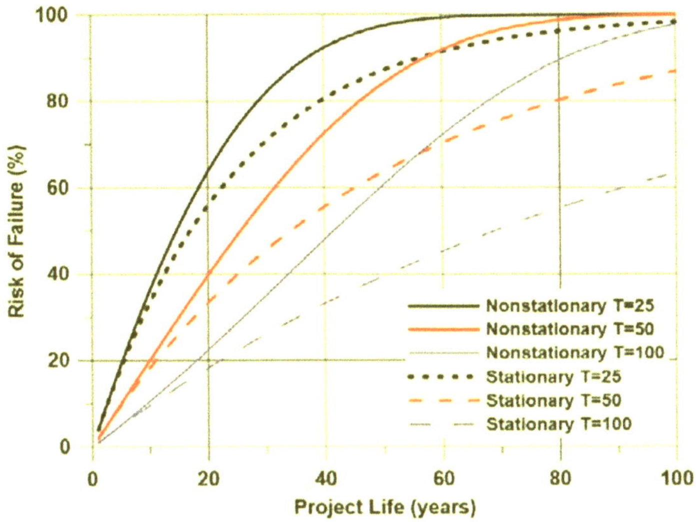

Fig. 8.3 Stationary and nonstationary risk of failure as a function of project life at selected return periods T = 25, 50 and 100 years

In calculations of risk of failure in stationary versus nonstationary extreme value analysis, Fig. 8.3 shows the risk probabilities (in percent) plotted as a function of project life at selected return periods T = 25, 50, and 100 years under stationary and nonstationary analyses. As shown here, the risk of failure for the nonstationary cases is higher compared to the stationary cases as the project life increases. In relation to the stationary and nonstationary return period, to ensure that a project is able to provide a 10 percent risk aversion over the project life of 60 years, one has to design the infrastructure at 50-year return period (nonstationary case in Fig. 8.3) rather than at 25-year return period (stationary case).

Finally, Fig. 8.3 shows that the 24-h annual basin rainfall of the Cagayan de Oro River basin was plotted as a function of the return period for stationary and nonstationary cases. For instance, the basin rainfall during Typhoon Sendong is around 187 mm (see Fig. 8.1) which has an associated return period of around 58-year based on stationary case while the estimates will only be 30-year and 11-year return periods when the years of reckoning are in 2011 and 2061, respectively (i.e., when analysis are conducted under nonstationary conditions). If one designs a flood infrastructure with a design flood associated to a 58-year return period, then the design 24-h annual rainfall maxima should be 243 mm to ensure a decent level of protection over 50 years (i.e., until 2061).

Flood Inundation Modeling

In this study, the two-dimensional (2d) flood hydraulic model used in the same flood inundation model used by NHRC (2016) and also in Tabios (2017) for the Cagayan de Oro River Basin. Details of this 2d flood hydraulic model are given in the Annex section at the end of this chapter.

The Cagayan de Oro River basin model is shown in Fig. 8.5 with a drainage area of 1,364 km^2. The 2d hydraulic modeled area is shown in the finite volume (element) mesh in Fig. 8.6 which is located in the north portion of the basin (indicated in a small inset of the boxed figure of the entire watershed) where the river drains into Macajalar Bay.

8.3 Flood Inundation Model Simulations

Flood Mitigation Diking Configuration

In the NHRC (2016) study and also described in Tabios (2017), several meetings with DPWH engineers and a series of public consultations held in Cagayan de Oro City were conducted to arrive at seven (7) alternative flood mitigation plans evaluated in that study. The recommended flood mitigation plan currently being implemented is the case with the diking system realigned so that the Paseo del Rio and Torre de Oro developments are on the land side of the dike rather than being on the riverside as originally proposed by the JICA (2014) study. For purposes of this study, two flood mitigation diking configurations are subjected to flood inundation model simulations which are the existing diking system (Case 2 in Tabios 2017) and the new flood mitigation diking system (Case 4 in Tabios 2017).

Inflow Flood Hydrograph Computations with Watershed Model

Three (3) 24-h storm rainfall totals are considered here to calculate the inflow flood hydrographs in the 2d model as follows: (a) 127 mm; (b) 187 mm; (c) 243 mm. The 127 mm 24-h rainfall total was the basin rainfall used in the JICA (2014) study and required by DPWH in the review and value engineering study of the flood mitigation plan of CDOR conducted by NHRC (2016). The 187 mm rainfall total is the computed basin rainfall associated to the 58-year return period computed by JICA (2014) but was not used in their flood simulation study. The 243 mm is the 24-h rainfall total associated to the 58-year return period based on nonstationary fitted GEV distribution in 2061 as shown in Fig. 8.5.

For this 24-h rainfall totals, the hourly rainfall hyetograph was distributed using the same time pattern or distribution as Typhoon Sendong which occurred in December 2011. With these rainfall hyetographs, the corresponding flood hydrographs were calculated from the watershed model given in Table 8.1. Note that in the 2d modeled area, the major inflow points are upstream of the modeled area at Pelaez Bridge referred to as the main inflow and another one in the middle, eastern point of the modeled area referred to as Sub-basin 67 inflow. It may be noted that the peak flow of the Cagayan de Oro River at Pelaez Bridge from the JICA (2014) study is 4,924 m^3/s with a 24-h rainfall of 127 mm in contrast to the peak flows of 9,358 and 12,358 m^3/s for the 24-h rainfall totals of 187 and 243 mm, respectively.

Table 8.1 Storm rainfall hyetographs and corresponding flood hydrographs (CMS is m^3/s)

Hour	127 mm Storm Rainfall			187 mm Storm Rainfall			243 mm Storm Rainfall		
	Rain (mm)	Main Inflow (CMS)	Sub-Basin 67 Inflow (CMS)	Rain (mm)	Main Inflow (CMS)	Sub-Basin 67 Inflow (CMS)	Rain (mm)	Main Inflow (CMS)	Sub-Basin 67 Inflow (CMS)
1	0.39	131.74	0.64	0.57	185.21	1.58	0.75	243.19	1.75
2	0.39	131.75	0.64	0.57	237.47	1.60	0.75	291.33	1.76
3	0.82	131.82	0.64	1.21	281.37	1.60	1.57	326.22	1.73
4	0.82	132.04	0.65	1.21	311.21	1.57	1.57	350.99	1.90
5	1.18	134.03	0.65	1.74	329.78	1.67	2.26	371.39	2.12
6	1.18	155.71	0.76	1.74	342.34	1.82	2.26	395.48	2.47
7	1.44	194.28	0.95	2.12	355.52	2.06	2.76	428.84	2.75
8	2.08	301.58	1.47	3.06	374.18	2.26	3.98	474.92	2.95
9	3.02	599.73	2.93	4.45	401.54	2.47	5.78	541.96	3.68
10	5.73	1060.63	5.18	8.44	442.97	2.88	10.96	652.50	5.04
11	12.58	1881.88	9.20	18.52	512.96	3.96	24.07	903.17	8.49
12	52.00	3121.92	15.26	76.56	668.14	6.52	99.49	1737.61	18.34
13	22.36	4326.58	21.14	32.92	1184.73	13.91	42.78	8288.83	75.65
14	8.01	4924.51	24.07	11.79	5430.71	57.12	15.32	12,772.81	49.15
15	4.19	4556.73	22.27	6.17	9369.01	38.60	8.02	10,358.62	22.57
16	2.85	3787.47	18.51	4.20	8156.07	18.02	5.45	6706.05	11.88
17	1.76	3063.28	14.97	2.59	5440.38	9.45	3.37	4276.47	7.93
18	1.45	2473.08	12.09	2.13	3533.64	6.26	2.77	2911.06	5.83
19	1.21	2010.56	9.83	1.78	2433.43	4.63	2.31	2141.56	4.57
20	0.90	1653.61	8.08	1.33	1807.70	3.62	1.72	1686.15	3.96
21	0.90	1378.36	6.74	1.33	1430.44	3.08	1.72	1394.22	3.36
22	0.90	1164.82	5.69	1.33	1183.70	2.70	1.72	1194.01	3.05
23	0.46	997.60	4.88	0.68	1011.73	2.46	0.88	1051.67	2.92
24	0.46	865.23	4.23	0.66	887.63	2.35	0.85	941.56	2.58

Referring to Fig. 8.4, the 24-h rainfall total of 127 mm based on the fitted stationary GEV distribution has an associated return period of about 10 years with a peak flow of 4,924 m^3/s; that is about half of the computed peak flood flow of 9,358 m^3/s but with an associated return period of 58 years. Should one design the flood infrastructure based on the 24-h total rainfall of 187 mm, the cost may be twice or three times compared to designing at 127 mm but its associated return period provides almost six (6) times higher level of protection. There are other factors to be considered in constructing major flood infrastructures but long-term costs may be one of the priority parameters to decide what will be actually constructed.

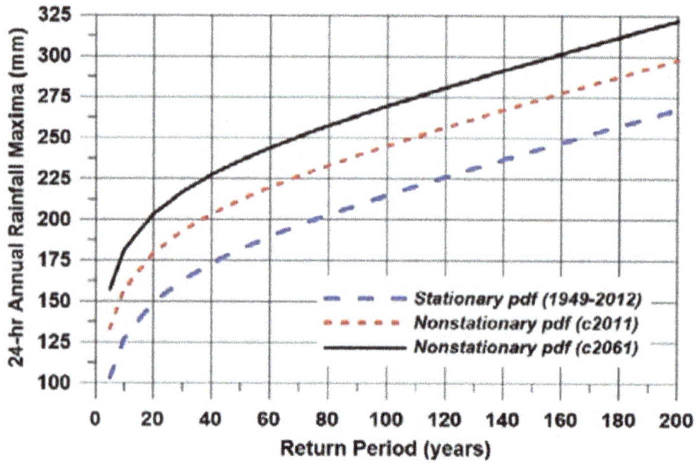

Fig. 8.4 24-h annual maxima basin rainfall of Cagayan de Oro River basin as a function of return with GEV probability density function (pdf) for cases of stationary analysis and nonstationary analysis at pdf in 2011 and 2061

8.4 Flood Hazard Mapping

For purposes of flood hazard mapping which is generally required in flood risk management, the flood inundation model simulations were conducted for the two (2) flood mitigation configurations, namely, (a) existing diking system and (b) new flood mitigation diking system which was recommended in the NHRC (2016) study. Thus, for the two physical configurations and three 24-h storm rainfall totals, the model simulations were conducted for a period of 33 h and the results are shown here in terms of maximum water depths for the two diking configurations (for three storm rainfalls) as shown in Fig. 8.7. The water depths are plotted by class posting at every finite element cell whereby different colors represent different ranges of depths as indicated in the legend. Essentially, these two figures provide flood hazard maps where the information can be used to determine flood prone areas, to guide flood zoning for build or no-build zones, as the basis for putting flood markers (historical and simulated floods), and to identify staging and escape routes or pathways to safe and evacuation areas. Other ways to display or present the flood hazard maps may be to plot the water surface elevations, vector of flow velocities, or flooding durations at a particular depth. The depth-duration maps in particular may be useful to assess property damages while the velocity vectors (magnitude and directions) may be useful to ensure safe human and land vehicle movement in flooded areas.

For a brief discussion of the flood hazard maps in Fig. 8.7, the simulations with 127 mm 24-h rainfall total was the design storm rainfall with its corresponding design flood discharge of 4,924 m^3/s. In these figures, the resulting flood inundated areas are fairly contained within the diking system and in fact this was the basis for recommending the new or realigned flood diking configuration in CDOR. On the other hand,

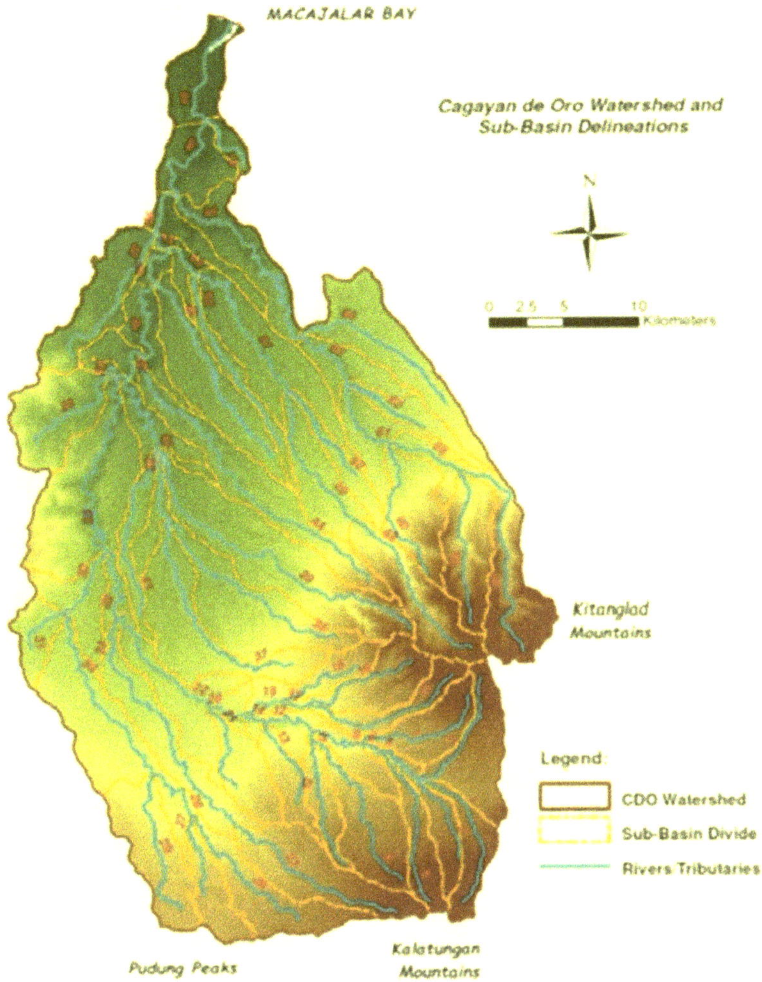

Fig. 8.5 Cagayan de Oro River Basin watershed area with drainage area of 1,370 km^2

the storm rainfall totals of 187 and 243 mm with associated peak flood discharges of 9,369 and 10,358 m^3/s, respectively, would result in flood waters spilling over the dikes which are actually associated to 58-year return period, level of protection (i.e., respectively, in 2011 and 2061). Note that as shown in Fig. 8.3 earlier, the risk of failure of the 58-year return period flood over 50 years period is almost 70 percent (70%) and 90 percent (90%), based on stationary and nonstationary extreme value analysis. Admittedly, these risk of failure probabilities are very high and definitely it implies that the design flood of the 58-year return period provides a very low level of flood protection to the communities under changing climate.

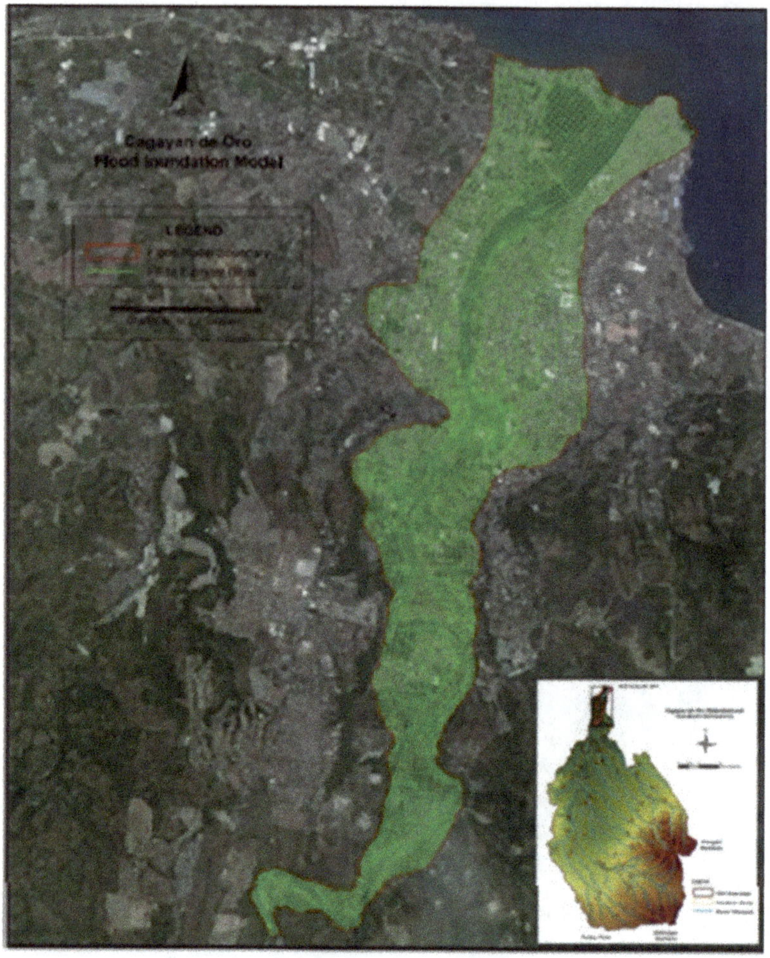

Fig. 8.6 Cagayan de Oro River (2d finite volume mesh) covered in the 2d flood inundation model. The inset figure shows the location of the mesh area relative to the entire Cagayan de Oro River Basin

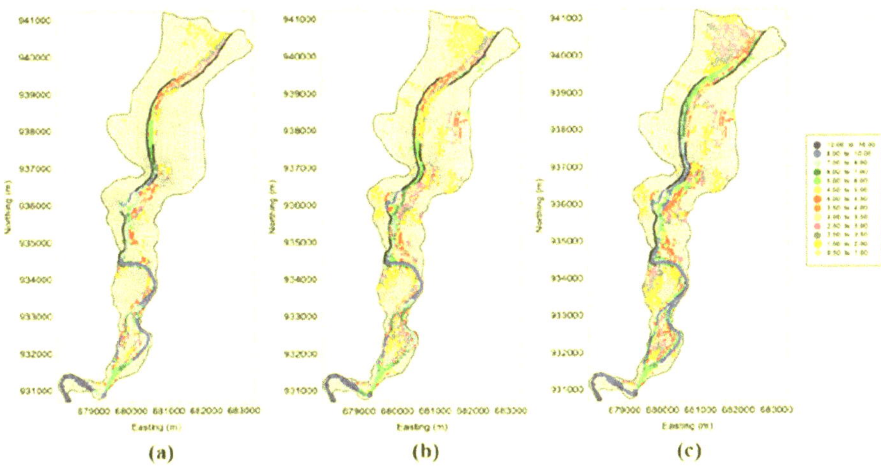

Fig. 8.7 Flood hazard map of Cagayan de Oro City in terms of water depths (m) with the new (ongoing project) flood control river diking system (indicated by black curve/line) at 24-h rainfall of **a** 127 mm, **b** 187 mm, and **c** 243 mm

8.5 Conclusion

This chapter presented flood hazard mapping through flood inundation modeling and simulation for a given the hydrologic design flood. In practice, the design flood magnitude and the associated return period or probability of occurrence are commonly defined under the assumption that the underlying probability distribution of the river flow process is stationary in time. However, with changing climate, including land use changes and/or urbanization, the river flow process should be treated as nonstationary. Based on 24-h annual maxima basin rainfall data of the Cagayan de Oro River basin, a nonstationary general extreme value distribution was fitted to account for the increasing trend exhibited in the basin rainfall time series plot from 1949 to 2012. The increasing rainfall trend may be attributed to changing climate, and the use of rainfall in particular can be considered isolated or free from the effects of land use change and urbanization on local scales. With the derived storm rainfall totals estimated by nonstationary extreme value analysis, flood inundation modeling was conducted using the 2d hydraulic model of the urban area of Cagayan de Oro City with flood inflow hydrographs generated using the watershed (rainfall-runoff) model for the various design storm rainfalls. The flood hazard maps produced from flood inundation modeling can be utilized for flood mitigation planning or flood risk management such as to determine flood prone areas, for land zoning of build and no-build zones as well as for early warning and flood evacuation staging and escape routing procedures. Finally, the importance of providing adequate flood level of protection to the community and people cannot be overemphasized as seen here that observed storm rainfall total of 187 mm with associated peak flood discharge of 9,369 that would result in floodwaters spilling over the dikes is only a 58-year return period event. At this return period, the risk of failure for a project life of 50 years period is almost 70 percent (70%) and 90 percent (90%), based on stationary and nonstationary extreme value analysis. These risk probabilities are very high and definitely it implies that the design flood at a 58-year return period provides a very low level of flood protection to the communities especially considering changing climate or climate change.

Several policy implications can be drawn in this study as follows. First, since the engineering practice of DPWH is to design flood control infrastructures based on return period (e.g., 25-year return period for local drainage culverts or a 50-year return period for national drainage systems which are revised guidelines with climate change), the analysis of design flood for the corresponding return period should utilize nonstationary extreme value analysis rather than stationary analysis with changing climate including land use change and urbanization. Second, considering climate change, the approach here is to define future design flood and associated return period based on actual trends or shifts from observed data so this can be better appreciated by decision-makers rather than deriving climate change design parameters from results of the climate projections (i.e., IPCC) which depends on the scenarios or storylines (i.e., 4 to 5 of them) and the specific global climate model used (i.e., about 25 models are available and 5 are known to be appropriate to Philippine climatic conditions). Third, since nonstationary, probability analysis is employed here, the flood risk at different design life (time windows) of the projects can be easily calculated which is useful in optimal sequencing and staging flood control projects or components thereof to properly time investments and financial resources. Optimal project sequencing and staging, also referred to as piecewise engineering is a strategic policy to adopt in view of the evolutionary nature and uncertainty of climate change (due to advancing climate science and modeling with changing storylines and projections over time and depending on actual climate change mitigation measures implemented). Land use change and urbanization can also be evolutionary in nature in view of population changes, socio-economic changes as well as political changes reflected in changing objectives, priorities, and uncertainties in government policies and administration.

References

Burnash RJC, Ferral, RL, McGuire RA (1973) A generalized streamflow simulation system: Conceptual modeling for digital computers, National Weather Service, California, Dept. of Water Resources, Sacramento, CA

Cheng L, Aghakouchak A, Gillelad E (2014) Non-stationary extreme value analysis in a changing climate. Clim Change 127:353–369

JICA (Japan International Cooperation Agency) (2014) Preparatory survey for flood risk management project for cagayan de oro river (FRIMP-CDOR). Rep Submitt Dep Public Work Highw, Manila Office, Nippon Koei, Company Ltd, Tokyo

Kok K, Yu W, Sikek LM, Jung K (2017) Derivation of areal reduction factor for design application in Malaysia using grid- based approaches, In Proc of 37th IAHR World Congress, Kuala Lumpur, August 13–18, p 10

Kuester JL, Mize JH (1973) Optimization Techniques with FORTRAN. McGraw Hill Book Co, New York

NHRC (National Hydraulic Research Center) (2016) Review and value engineering study of the flood risk management project for Cagayan de Oro River, Final Report to Department of Public Works and Highways (DPWH), Manila, August

Pietersen J, Gericke J, Smithers J, Woyessa Y (2015) Review of current methods for estimating areal reduction factors applied to South African design point rainfall and preliminary identification of new methods, Jour of the South African Inst of Civil Engg, 57:1, pp 16–30

Salas JD, Obeysekera JTB (2014) Revisiting the concepts of return period and risk for nonstationary hydrologic extreme events. Jour Hydraul Eng, ASCE 19:554–568

Salas JD, Obeysekera JTB, Vogel RM (2018) Techniques for assessing water infrastructure for nonstationary extreme events: a review. Hydrol Sci J 63(3):325–352

Sraj MA, Viglione JP, Bloschl G (2016) The influence of non-stationary in extreme hydrological events flood frequency estimation. Jour Hydrol Hydromechs 64(4):426–437

Tabios GQ III (2008) Two-dimensional hydraulic model with mobile bed for dam-break problems, Jour Hydrol Environ, IHES, 4(1), pp 17–25

Tabios GQ III (2017) Alternative flood mitigation plans for cagayan De Oro River, In Proc. UNESCO-JASTIP Joint Symposium on Intra-Regional Water Security and Disaster Management (eds By Soriano RS, Tabios GQ III), Quezon City, Metro Manila, November 13–16, pp 52–53

Tabios GQ III, Obeysekera JTB, Salas JD (1986) National Weather Service model—PC version. Colorado State University, Fort Collins, Colorado, August, Dept. of Civil Engineering

USACE (U.S. Army Corps of Engineers). (1985). HEC-1 flood hydrograph package user's manual. Hydrologic Engineering Center, Davis, California, January

Guillermo Q. Tabios III Ph.D. is a Professor Emeritus and Research Fellow of the Institute of Civil Engineering and the National Hydraulic Research Center, respectively, in the University of the Philippines Diliman, and an Academician of the National Academy of Science and Technology of the Philippines.

Chapter 9
Climate Change Strategies in the Davao River Basin

Ruth U. Gamboa, Ludivina M. Porticos, and Joseph E. Acosta

Abstract For the past 20 years, the frequency and magnitude of the flash floods and landslides in the Davao River Basin (DRB) have wrought destruction in lives, properties, public and private investments in development, and the physical and mental health of surviving families in affected *barangays (villages)*. The past decade experienced an average of two floods/year. Twenty-one of the 36 vulnerable barangays in the Basin are prone to 100-year return flooding. Four (*Brgys* 8-A, Ma-a, Mandug, Tigatto) are most vulnerable to recurring floods and landslides due to extreme weather events associated with climate change. This chapter discusses climate change strategies to help address the climate emergency in the DRB. One recommendation is to include nature-based solutions (NbS) aside from hard engineering alone. NbS considers nonstructural measures such as the indigenous knowledge of the four tribes (Obu-Manuvu, Matigsalug, Ata, and Bagobo-Klata) in nature conservation. Another is to create a local version of the Climate Change Commission's database-sharing system. The system has to be continuously populated, open-access, interactive, and staffed by a technical unit with plantilla items. Disaster risk and climate information need sound evidence gathered not only by the "experts" but in partnership with the local communities and civil society.

Keywords Floods · Information sharing system · Indigenous peoples · Community science

Key Points

- Fifty-eight percent (58%) of the flood-prone *barangays* (villages) located within the Davao River Basin are vulnerable to a 100-year return flood cycle.

R. U. Gamboa (✉) · J. E. Acosta
University of the Philippines Mindanao, Davao, Philippines
e-mail: rugamboa@up.edu.ph

L. M. Porticos
University of Immaculate Concepcion, Davao, Philippines

K. B. Berse et al. (eds.), *Climate Emergency in the Philippines*, Disaster Risk Reduction, https://doi.org/10.1007/978-981-99-7804-5_9

- Of the 22 floods that swelled the river in the past 20 years, 20 occurred in the past decade, or an average of two floods/year. Those frequent floods manifest the climate change emergency experienced in the DRB. They are destructive and repeatedly set back development efforts.
- Policymakers can invest in an information-sharing system that, is continuously populated, open-access, and interactive as a tool for climate change strategy.
- Local government must mainstream community science where local communities drive and implement scientific research and monitoring characterized by place-based knowledge, social learning, collective action, and empowerment.

Summary for Policymakers
The vulnerability to flooding of the agricultural and built-up areas in the Davao River Basin was once ascribed to its topography, slope, and vegetation. Today, the Basin faces a climate change emergency.

- The frequency and magnitude of the flash floods, with landslides in some, in the past 20 years have wrought destruction in lives, properties, public and private investments, as well as affected the mental health of victim families.
- Big-time flood management projects are counterproductive if they marginalize local communities at the decision table. The government will still bear disaster relief costs, encompassing financial, political, socio-economic, and even moral (e.g., slow to respond) accountabilities.

Time and money spent on rebuilding and reconstruction can be considered lost opportunities. Nevertheless, what if some of the costs pay for levelling up the state of science and technology for planning, monitoring, evaluation, and learning? The levelling up includes: a) adopting nature-based solutions which promote green infrastructure and nonstructural intervetions, and b) establishing a user-friendly, open-access, and interactive state-of-the-art information and communication sharing system.

What if hard core science also embrace community science where technical experts and trained researchers partner with an educated and empowered community in monitoring and ground conservation, among others? What if there is some matching of sophisticated and costly technological solutions with indigenous knowledge and nature-based solutions?

The frequency of flash floods due to changing weather patterns makes *barangays* and *puroks* (house clusters) in flood-prone areas most vulnerable.

- An average of two floods/year in the last decade speaks of emergency and urgency (DCDRRMC Data).
- Four DRB *Barangays* are most vulnerable in terms of flood extent, population, exposure to a 100-year return flood, and vulnerability of water district infrastructure, namely, *Brgys* 8-A, Ma-a, Mandug, and Tigatto (DREAM-LIDAR Flood Hazard Study).
- The July–September average monthly rainfall of about 150 mm could swell between 180 and 280 mm. In 1/50 events of extreme rainfall, that is higher than

the monthly average of 150 mm, one will be about 200 mm. For appreciation, the flash flood in Matina Pangi in June 2011 resulted from some 200 mm of rainfall that fell in about five days, 60 mm of which was in the last three hours before the flash flood (ICHARM UNESCO Project).

- The three big floods that affected the City in the past 20 years (Jan 2002, Jun 2011, and Dec 2017) show that the early two were flash floods from thunderstorms accompanied by heavy downpours. The third and most destructive in terms of affected families and damaged properties was due to a super typhoon, *Vinta*, that did not directly hit the City but passed by its northern tip, causing the headwater of the Davao River to swell, bringing flash flood downstream (DCDRRMC Data).

The Climate Change Commission can assist the City in creating LCCDIES (Local Climate Change Database And Information Exchange System), a local version of NICCDIES (National Integrated Climate Change Database And Information Exchange System), which must be continuously populated, open-access, and interactive.

To help address this challenge, the Climate Change Commission, as part of its NICCDIES (National Integrated Climate Change Database And Information Exchange System) initiative (CCC Phil, 2024) can assist the City in creating LCCDIES (Local Climate Change Database and Information Exchange System), a local version of NICCDIES, which is continuously populated, open-access, and interactive.

- The City will have its own LCCDIES that is linked to the NICCDIES.
- Enhance the competence and retool the technical team who will be incharge of the LCCDIES. Competence will include the ability to analyze, make connections, forecast trends, and communicate the complex science of climate change in layman's language.
- Provide plantilla items for this technical team to protect them from rapid turnover when political administrations change as a result of elections.
- The LCCDIES should come up with a standard data monitoring template that the local DRRMC (Disaster Risk Reduction and Management Council) can use.

Policymakers at the local level must not only latch in with NICCDIES via LCCDIES but also mainstream community science where "scientific research and monitoring are driven and co-implemented by local communities and characterized by place-based knowledge, social learning, collective action, and empowerment" (Anthony et al. 2020).

- Engage the four Ancestral groups (Obu-Manuvu, Matigsalug, Ata, and Bagobo-Klata) as co-managers of the uplands and value their indigenous knowledge.

Respect their indigenous ways side-by-side with Western science and acknowledge their *Pusaka* paradigm of respecting and living with nature.

- The city and *barangay* government units need to recognize that hard (gray) engineering solutions need to be integrated with green, nature-based solutions (NbS). The latter includes partnering with "green" civil groups who are sources of ground information and place-based volunteers. These groups are not only articulate in the science of their respective advocacies, but they are also effective in harnessing human and logistical resources.
- Davao City with two State Universities and about ten private universities and colleges, must tap the expertise of the academe as objective facilitators in realizing a community of local scientists where local knowledge is acknowledged, community efforts recognized, and social learning with local and external experts solicited.

> Engage the four ancestral groups (Obu-Manuvu, Matigsalug, Ata, and Bagobo-Klata) as co-managers of the uplands and value their indigenous knowledge. Acknowledge their *Pusaka* paradigm of respecting and living with nature.

9.1 Introduction

Davao City "is outside the typhoon belt and lacks major seasonal variations". (Davao City Comprehensive Development Plan 2012–2021). This claim used to be a selling point to attract business investments. But today, the same claim faces a "wet" reality that in the last decade, four typhoons and 16 monsoon rains caused floods (Table 9.1) in different low-lying areas of the Davao River Basin. The most damaging was a super typhoon, *Vinta*, in December 2017. It did not directly hit the City but flooded low-lying areas. Several of those floods were the "first flash flood" in certain residential areas, according to narratives heard on television news. During the same decade, the City also experienced floods in other rivers, especially the Talomo River.

Table 9.1 shows some clustering of floods during the years of protracted La Nina. Twenty of the 22 floods happened in the last 10 years (of the data). The El Nino-La Nina cycle (scientifically called, ENSO or El Nino/Southern Oscillation) is among the drivers of global climate patterns. A 20-year data is short in the study of climate change and may not yield a more quantitative correlation between ENSO and flooding in the river basin. This is where a faithful and accurate monitoring and recording system needs to be in place. The flood events not only alter the meander of the DRB and upset its natural services, but they take away lives and livelihoods, flatten houses and commercial structures, compromise people's water supplies, affect food supply and demand, and threaten the community's physical and mental health and safety as well (Public Health England 2020).

Table 9.1 Davao River floods and ENSO events (1996–2019). Expanded from the DCDRRMC (Davao City Disaster Risk Reduction Management Council) data

Year	El Nino	La Lina	# Floods in Davao River	Estd damage (*in Millions of PhP*)
2000		Strong	1	0.22
2001		Moderate		
2002	Moderate		1	74.7
2003				
2004	Weak			
2005		Moderate		
2006	Weak			
2007		Strong		
2008				
2009	Moderate			
2010		Protracted	3	0.565
2011			7	3.142
2012			2	No estimate
2013			2*	0.19
2014	Strong		1*	No estimate
2015				
2016		Protracted	2	0.015
2017			2*	163.439
2018	Weak			
2019			1	No estimate
Total			22	

* with typhoon: 2013, *Zoraida*; 2014: *Basyang*; 2017: *Auring* and *Vinta*

9.2 The Davao River Basin (DRB)

The Davao River Basin (DRB) covers a total land area of 175,960 has, with a river extending 86 km (Map 9.1). It is the largest of the City's eight watersheds, the third largest river basin in Mindanao, and one of the 18 major river systems in the country. The Basin extends from parts of three neighboring provinces, namely, Kitaotao, Quezon, San Fernando (Fig. 9.1), and New Valencia of Bukidnon; Arakan and Magpet of North Cotabato; Talaingod and a small part of Panabo of Davao del Norte. It meanders through the floodplains of Davao City and drains into Davao Gulf. It is a primary aquifer with surface water primarily for domestic and agri-industrial uses (University of Philippines Training Center for Applied Geodesy and Photogrammetry 2015). Using some state-of-the-art water harvesting technology, the surface water at the Panigan-Tamugan junction will soon be a bulk supply for the Davao City Water District.

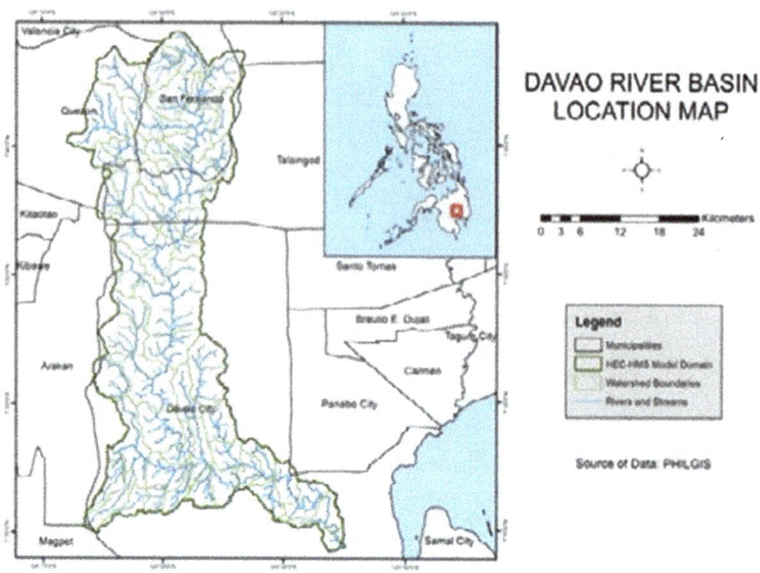

Map 9.1 Davao River Basin (Lifted from UP TCAGP, 2015)

The DRB is 77% forestland. It includes 137 *barangays,* 113 (82%) of which are within the political jurisdiction of Davao City. As of the 2015 Census, ninety-five (95%) of the Basin's residents are *taga cuidad* (city/Christian folks). The Basin hosts some 55% of the business establishments, which contribute about 63% to the city's capital investments (DRBMDP, Davao River Basin Management Development Plan Master Plan 2014).

Historically, the Basin is home to four Ancestral Tribes (Map 9.2): Obu-Manuvu, Matigsalug, Ata, and Bagobo-Klata (Fig. 9.2). Today, the Basin covers four CADTs (Certificate of Ancestral Domain Titles), one for each tribe (Table 9.2). Collectively, the City has 11 tribes spread across the eight watersheds. Excluding the Tinananon Tribe of North Cotabato, the four tribes occupy about 80% of the Basin. It is logical, therefore, that the tribes should bear a larger share in the stewardship of the uplands, taking an active role in tempering the negative impacts of climate change, such as flooding and landslides.

9.3 Why is There a Climate Change Emergency?

In 2016, UNESCO-Jakarta funded HELP-Davao (Hydrology for Environment Life and Policy) to study the resilience of urban water systems in Davao City during extreme floods. The study (Sales et al. 2016) looked at four of the eight river systems, namely: Davao, Lasang, Lipadas, and Talomo. It covered 67 barangays, 36 of which

Fig. 9.1 Davao River Basin headwater at the Salug-Mitimbus Junction, Barangay San Jose, San Fernando, Bukidnon. (*Source* DENR XI with permission)

are in the DRB. It made intensive use of the Davao River Basin maps generated by the 2014 DREAM (Nationwide Disaster Risk and Exposure Assessment for Mitigation) Program.

The highlights of the study are as follows:

1. Twenty one (58%) of the 36 *barangays* are prone to 100-year return floods.
2. Nine of the 10 most flood-prone *barangays* are within the political jurisdiction of Davao City (Table 9.3).
3. Nine of the 10 *barangays* with the highest exposure to 100-year return floods are in the DRB (Table 9.4). *Brgy.* Calinan is part of the Talomo Watershed.
4. Seven of the 10 most populated flood-prone *barangays* are in the DRB (Table 9.5).
5. Six of the 10 *barangays* where the Water District pipes and wells are most vulnerable to 100-year return floods are in the DRB (Table 9.6).
6. The river is brownish throughout the year as a result of soil run-off during heavy rainfall (2017 Water Quality Monitoring Area-DENR XI).

Five *Barangays* are common in Tables 9.3–9.6, namely, 8-A, Ma-a, Mandug, Tigatto, and Calinan. The first four are in the DRB, while Calinan is part of the Talomo Watershed. The four *Barangays* are most exposed to floods vis-a-vis other *barangays*. In a 100-year return flood, they will likely suffer the most impacts in terms of population and domestic water shortage (A 100-year flood does not come once every 100 years; it simply means that a flood of a certain severe magnitude has a 1% chance of hitting in any given year. A 1/100 event could happen in any year, multiple years in a row, or not within 100 years). The HELP-Davao study was disseminated in 2017. It would be interesting to look at the four *Barangays*' flood and disaster adaptations since then. They may become model cases or sources of learned lessons. In the race to economic development, the City government needs to bear in mind that important changes in the Basin's ecology, such as local land use, river engineering, and

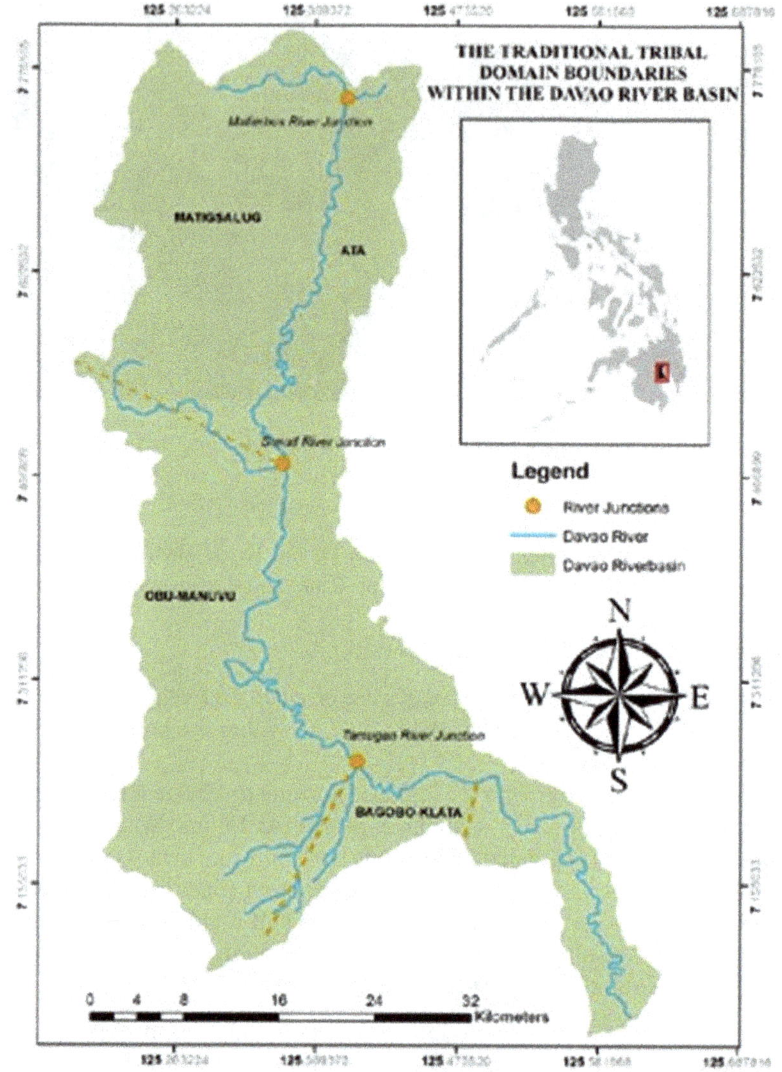

Map 9.2 Traditional physical boundaries for the four Ancestral Tribes in the Davao River Basin. (*Source* IPO MUADTELLI, Feb 2020)

the changing climate patterns, "can affect at what point a '100-year flood' is designated" (USGS, accessed January 2024). More recently, heavier rainfall and a more considerable magnitude of flash floods have led hydrologists to consider 500-year floods. The brownish water is an early warning of river's quality. While that color is directly related to the soil and sediment particles that are eroded with the running water, the run-off that sweeps those particles likewise brings with it other pollutants

Fig. 9.2 Bronze representation of the four tribes of the Davao River Basin at the Dabaw Museo, a public facility. Photos by RUGamboa

Table 9.2 The four Certificates of Ancestral Domain Titles (CADT) in the Davao River Basin. (*Source* IPO MUADTELLI, Feb 2020. The information here is more current than what is in the 2014 DRBMDP and the two have to be harmonized)

CADT no	Hectares	Tribes
CADT-R10-KIT-0703-0011 (102,324.8186 has and collectively owned by 7 tribes of Matigsalug-Manobo)	89,102.82	1. Tigwahanon Tribe of San Fernando, Bukidnon 2. Matigsalug Tribe of Kitaotao, Bukidnon 3. Pulangihon Tribe of Quezon, Bukidnon 4. Kulamanon Tribe of Kibawi, Bukidnon 5. Tinananon Tribe, Arakan, North Cotabato
	6,000	6. Obu-Manuvu, Portions of Brgys. Baganihan, Marilog District
	7,222	7. Ata Tribe, Paquibato District, Davao City (but Kulamanon and Ata, now called Matigsalug, are the ones currently managing
CADT-R11-DAV-1108-091 (30,309.6853 has)	26,209.69	Obu-Manuvu
	4,100	UP Mindanao Land Grant
CADT-R11-DAV-0915-185 (9,087.5002 has)	6,843.43	Bagobo-Klata
	2,244.07	Bagobo Tagabawa
CADT-R11-DAV-0216-160	87,664.87	Tinananon Tribe of Arakan, North Cotabato

such as harmful bacteria, toxic residues, industrial effluents, and also domestic and solid wastes from non-point sources like agri-industrial farms and settlements.

Table 9.3 Top ten flood-prone *barangays* in Davao City (Lifted from "Sustainability Science Demonstration Site on Enhancing Resilience to Disasters of Urban Water Systems of Mindanao in the Philippines" Final Report. 2016. UNESCO-Jakarta). New Valencia belongs to Bukidnon province but forms part of the headwaters of the DRB

River system	Barangay	Area prone to flooding (%)	Total area (has.)	Flood-prone area (has.)
Davao River Basin	1-A	90.98	11.94	10.86
Talomo Watershed	Calinan	78.84	863.47	680.76
Davao River Basin	8-A	67.38	159.13	107.22
Davao River Basin	5-A	64.54	36.27	23.41
Davao River Basin	Tigatto	64.41	770.02	495.97
Davao River Basin	Waan	51.06	515	262.96
Davao River Basin	Ma-a	46.22	1,014.91	469.09
Davao River Basin	40-D	46.07	9.12	4.2
Davao River Basin	Mandug	43.61	898.71	391.93
Davao River Basin	New Valencia	43.52	975.96	424.74

Table 9.4 Top 10 *barangays* with the highest exposure to 100-year return floods. (Lifted from "Sustainability Science Demonstration Site on Enhancing Resilience to Disasters of Urban Water Systems of Mindanao in the Philippines" Final Report. 2016. UNESCO-Jakarta)

River system	Flood-prone barangay	Exposure Index (EI)	Ranking
Talomo Watershed	Calinan	0.10108	1
Davao River Basin	Tigatto	0.09222	2
Davao River Basin	Ma-a	0.08901	3
Davao River Basin	Mandug	0.0509	4
Davao River Basin	8-A	0.04668	5
Davao River Basin	New Valencia	0.04503	6
Davao River Basin	1-A	0.04431	7.00
Davao River Basin	Waan	0.04065	8
Davao River Basin	5-A	0.04038	9
Davao River Basin	Callawa	0.03997	10

9.4 What Are the Impacts and Threats of this Climate Change Emergency?

According to the 2019 Report by the International Center for Water Hazard and Risk Management (ICHARM, UNESCO Public Works Research Institute), Climate Change in the DRB means:

Table 9.5 Top 10 flood-prone *barangays* in terms of potentially affected population which is computed as the product of % area prone to flooding × total land area × 2015 population density. (Lifted from "Sustainability Science Demonstration Site on Enhancing Resilience to Disasters of Urban Water Systems of Mindanao in the Philippines" Final Report. 2016. UNESCO-Jakarta)

River system	Barangay	Potentially affected population
Davao River Basin	Ma-a	27,639
Davao River Basin	Tigatto	23,435
Talomo Watershed	Calinan	18,174
Davao River Basin	Bucana	17,322
Davao River Basin	19-B	8,939
Davao River Basin	8-A	7,463
Davao River Basin	5-A	7,381
Davao River Basin	Mandug	5,930
Lipadas Watershed	Lizada	5,539
Lipadas Watershed	Crossing Bayabas	3,133

Table 9.6 Top 10 most vulnerable *barangays* in terms of water district infrastructure. Vulnerability was scored using certain indices like exposure to 100-yr return flood. For details see "Sustainability Science Demonstration Site on Enhancing Resilience to Disasters of Urban Water Systems of Mindanao in the Philippines" Final Report. 2016. UNESCO-Jakarta

River system	Flood-prone barangay	Vulnerability ranking	Description
Davao River Basin	Mandug	1	Very high vulnerability
Lipadas Watershed	Lubogan	2	
Davao River Basin	Tigatto	3	
Talomo Watershed	Calinan	4	
Lasang Watershed	Pañalum	5	High vulnerability
Lipadas Watershed	Crossing Bayabas	6	
Davao River Basin	1-A	7	
Davao River Basin	Ma-a	8	
Davao River Basin	2-A	9	
Davao River Basin	8-A	10	

1. The July–September average monthly rainfall of about 150 mm could swell between 180 and 280 mm (Lifting from the Davao River Basin Management and Development Plan, the DRB 2014 Water Balance Study estimated the Mean Annual Rainfall at 2,652 mm or a monthly mean of 221 mm). For appreciation, the flash flood in Matina Pangi in June 2011 brought some 200 mm of rainfall that fell in about five days, 60 mm of which was in the last three hours before the flash flood.

2. The Intergovernmental Panel on Climate Change (IPCC) along with a good number of supporting literature (Stocker et al. 2013, Hirabayashi et al. 2013, Flaviana et al. 2009), acknowledge that our warming planet can cause more water-related extreme events such as frequent big floods, shifts in rainfall where some places will be wetter while others will be drier, and increased waterborne diseases. All these will impact rivers and their communities the hardest.
3. In 1/50 events of extreme rainfall, that is, higher than the monthly average of 150 mm, one will be about 200 mm. Again, that 200 mm flood does not mean a 50-year flood; it simply means a higher probability than a 1/100 that such a hazard can occur in any single year.

The most destructive flood in the DRB in terms of affected families and damaged properties was a super typhoon, *Vinta,* in 2017. *Vinta* did not directly hit the city but passed by its northern tip, causing the headwater to swell, and flash flood downstream. The three other typhoons, like *Vinta,* also did not hit the City directly. The floods in the DRB inundate common areas like 8-A, Ma-a, Mandug, and Tigatto riverbanks. Notwithstanding the repeated traumatic experience, the four *Barangays* remain densely populated. Climate change will continue to alter the frequency and intensity of floods and droughts in the DRB. At the current rate of land use conversions and business/settlement developments along its river, natural ground spaces that absorb floodwater shrink just as fast.

Warning-level torrential rains have become a common experience in settlements living by the riverbanks in the floodplain and the mouth of the DRB and other rivers, such as the Talomo and Lasang Rivers. Moreover, 80% of floods in the DCDRRMC data (Table 9.1) happened at night when people were less alert. Flooding has also been recorded in 'new' places where some resident senior citizens experienced none. This 'first flood' is an early warning in Climate Change emergency.

Although incomplete, Table 9.1 shows the range of direct cost of flood damage to properties (only) ranging from PhP 1 Million to PhP 160 Million/flood. Costs associated with climate-induced events may be classified as follows:

1. Up-front cost—examples: evacuation efforts, disaster relief. Aid to *Vinta* victims was about PhP 83 Million (Rappler News, Dec 27, 2017).
2. Economic losses—examples: property, productivity. In July 2018, the summer after *Vinta*, DPWH Davao Region reported it would complete some PhP 2.9 Billion worth of drainage and flood control projects in the City. This economic cost of floods is the traditional focus in disaster assessment. For instance, the DCDRRMC data (Table 9.1) include a column on physical damage such as roads, houses, and vehicles.
3. Social, cultural, and personal costs—examples: displacement and mental health. This type of cost cannot be estimated easily and immediately after a flood event, but for the Asian Development Bank (ADB) the value of this type of cost exceeds its monetary equivalent.

For comparison on cost estimate, an ADB case study (2022) on nature-based solutions (NbS) for two rivers in Mindanao, Buayan–Malungon, and Tagum-Libuganon, used Cost-benefit Analysis (CBA) and Multi-criteria Analysis (MCA) to estimate the expected annual flood damage in those basins. The estimates were PhP 1.1 Billion and PhP 1.947 Billion, respectively. Nature-based solutions include green spaces such as tree parks which not only help regulate runoff but can be enjoyed for recreation, the latter is an essential in urban areas. A new approach in risk management is to include the negative impacts of floods on mental health as a leading economic business case. A recent study in England (Public Health England, 2020) shows that victims who are at home during a flood may be prone to suffer from anxiety, depression, and post-traumatic stress disorder. At the household level, this mental health issue would mean cost for the health service and lost days at work. For a daily wage earner in the Philippines, the cost of treatment and medication can be several times higher than his wage. For the employers, sick personnel lessen individual and corporate productivity.

The third type of cost which includes psychosocial interventions does not get much attention in government reports. NGO (Non-Governmental Organization) reports, like those by Save the Children (2021) and ChildFund Alliance (2021), give it more pages. Beefing up the information, beyond the who, what, where, when, of an extreme event, is another area that needs improvement in government data gathering and monitoring.

9.5 What is Being Done?

Flooding, the ENSO, and global warming are natural events. But Climate Change in this century is anthropogenic. In the past 50 years, the economic boom in Davao City has led to the expansion of commerce and settlements, consequently converting once-green patches into bustling communities. Two good examples are the residential developments in Ma-a Ridge and *Brgy.* Tigatto. When one rides up the river from the mouth, one notices heavy settlement along the banks up to *Brgy* Mandug, about 30 m ASL. From thereon, the banks are lined mostly by banana plantations way up to *Brgy* Bantol. The rest of the uplands until the headwaters in Bukidnon are a mosaic of forests, open patches of logged areas, and plantations.

After WWII, the Davao River's meander toward the mouth has considerably changed (Fig. 9.3). The original segment is now occupied by settlements and various infrastructures (Jubilo et al. 2019). The 2019 flood hazard map shows that some of the flood-prone *barangays,* like SIR Matina and Bucana are where the old segment used to run. Furthermore, the extent of inundation has spread even wider. Three possible nature-based solutions in this case are restoring the meander, zonation and elevating houses. Engineering solutions may include the construction of gabions. However, SIR Matina and Bucana are crowded not only with settlement but also with business infrastructure. The decision-makers must include other sectors that have stakes in the Barangays. They need to understand the impacts, costs, and trade-offs of each solution.

To date, some technological interventions exist to address flooding in the City:

Fig. 9.3 Change in the meander of the Davao River. Red—original path (pre 1944); Blue—current meander in Google Map (Lifted from Jubilo et al. 2019)

1. Gabions and Levees

Government-built gabions (Fig. 9.4) and levees stretch in selected flood-prone segments of the river down to its mouth. These structures were built to hold and maintain a fixed water course and prevent erosion of the riverbanks. As stated in an earlier, in one *barangay*, the river did not overflow on the side where the gabion structure held strong. However, it flooded the other side, where it gave way.

2. Deployment of Early Warning Systems (DEWS)

In close coordination with the Department of Science and Technology, the CDRRMO had installed the DEWS (Deployment of Early Warning System) in selected flood-prone settlements (UN-OCHA, 2021). DEWS is a hydromet device with water level monitoring systems and automated rain gauges. The CDRRMO monitors the device and assists the *barangay* in operating the unit. The accurate uploading of the field recordings into an information management system and the translation of that information to feed the decision-makers needs capacity enhancement on the part of the technical monitors and data handlers.

3. Disaster Radio

With the aim to provide real-time updates and warnings to the people about emergencies, the city government of Davao has put up a radio station with the identity

Fig. 9.4 Construction of a gabion wall along a flood-prone lowerbank of Davao River. Photo by RUGamboa

"Davao City Disaster Radio—87.5" (Fig. 9.5). It is an emergency-centric community station that serves as information dissemination medium on disaster and emergency preparedness (Palicte 2020).

4. Flood Plain Zoning

The City government has revised its CLUP (Comprehensive Land Use Plan 2018–2022) incorporating the reclassification of the flood-prone areas derived by the DREAM project and areas not suitable for settlements using geo-hazard maps and the Geographic Information System (GIS). DREAM used high resolution 1-m Light Detection and Ranging (LIDAR)-Derived flood modeling techniques in locating flood-prone areas. The top ten flood-prone *barangays* in the DRB received

Fig. 9.5 Davao City Disaster Radio 87.5 Facebook Page. (Date: February 2020)

their respective flood maps through the HELP-Davao Project in 2016. Feedback on the implementation and monitoring of this nature-based solution has to reach the decision-makers as part of enriching the information system that will feed succeeding decisions.

9.5.1 Indigenous Resilience

With 80% of the DRB as their home, the four Indigenous Peoples (IPs) or *lumads* are the most significant land owners of the Basin, and have their own appreciation of climate change. For those *lumads*, everything is natural; even big floods are a part of the natural cycle and so are extreme events. But when logging (Fig. 9.6) came to their land, they started to experience extreme floods. As their forests got thinner and, in some areas, bald, they came to realize that "nature" (their environment) has changed (e.g., the landscape and its uses). Moreover, natural cycles have altered. The *lumads* have no concept of climate change—for them nature has spirits that control the cycles which take their own courses. Instead, they believe that man's activities, such as logging and plantations, are responsible for changing the environment. PD 705 (Revised Forestry Code) is a vague concept for them since they see economic activities as the main culprit for the loss of their forest, not the changing climate.

Fig. 9.6 The big tree is manually felled using a long saw (below). Photos taken by RUGamboa with permission from Dabao Museo, a public facility

There is among the Obu-Manuvu a concept of interconnectedness. The *puwas* (forests) and rivers, the natural and spiritual systems all form the lifeblood of the tribe. The *puwas* need protection because they connect to all other life support systems. This connectedness is what they call, *Pusaka. Pusaka* is a way of life that preserves balance and harmony with nature. *Pusaka* means many things about living in harmony with nature and keeping environmental balance and protection. It can be physical, referring to sacred places, hunting grounds, fishing areas, spiritual worship places, and burial grounds of ancestors that are respected and protected. *Pusaka* is also about biodiversity and their natural or spiritual roles such as sources of natural foods and medicine. Some animals serve as early warning of calamities. For example, the bird *kalaw* speaks of time, where it would appear at 5:00 AM to wake people up. Another example is the *alimokon* (pigeon). It brings good and bad omens depending on where it comes from—if it comes from behind a person, he has forgotten something, or someone has come to his house; if in front, it is an alarm that something good or bad may happen; if from the right, something good is about to happen; and if from the left, nothing good will result from his transaction. *Pusaka* is also an environmental philosophy that includes the conservation of no-touch zones (e.g., burial grounds) and the protection and sustainability of agricultural areas. The tribes reflect these philosophies and values in their internal governance, including protecting, conserving, and managing their forests and other natural resources.

For the Obu-Manuvu, *Pusaka* is also a cultural approach to being an effective steward under PD 705. It is the framework for the tribe's political organization, the Indigenous Political Structure (IPS). As the highest governing body, the IPS consists of the tribal council of elders and leaders. This IPS created the Indigenous Political Office (IPO) as its Technical and Economic Arm. In contrast, the IPO MUADTCELI is responsible for regulating the entry of other IP tribes and migrants and the formation of "quick tribe groups" who stake illegitimate claims in the ancestral domain.

The Obu-Manuvu has seen many government projects come and go. The most recent is the National Greening Program under the Climate Change Commission. The tribe claims that the tree plantlets do not go to the IPO of each CADT holder but rather to *dayuhan* (migrant) tribes that appear overnight for convenience and compliance. The mushrooming of migrants and IP settlers is causing disturbance in their Protected Area, thus, violating their *Pusaka*.

Amid a micro-climate change in their domain, how does the Obu-Manuvu show resilience? The Ancestral Domain Development and Sustainable Protection Plan (ADSDPP), as adopted by the national and local governments, is a planning tool that expresses the vision of the tribe. The IPS is a platform for the voice and competence of the tribe toward empowerment and self-governance. Their most significant challenge to date is the *nakisawsaw* (tribes from outside the region), who compete with them in their partnership with the government.

To cite an example: in the implementation of the NGP, the tribes found out that the DA (Department of Agriculture) and the DENR (Department of Environment and Natural Resources) organized new POs (Peoples Organizations) and had them accredited as "tribal groups", bypassing the IPO of the CADT holders. To date, the two agencies have not yet resolved the authenticity of the POs. The IPO reiterates its organic jurisdiction in caring for the nurseries and maintaining and protecting the growing trees and planted areas. They ask the two agencies to respect their indigenous political structure by recognizing them as legitimate stewards. They also expect to get specific skill training deemed necessary by the DA-DENR for the success of the NGP.

The elders interviewed for this study enumerated four grievances about their co-stewardship with the government:

1. The 2016 ADSDPP is violated many times, mainly on the government side.
2. RA 8371 (Indigenous Peoples' Rights Act 1997 or IPRA Law) has been there for 22 years, yet the *Barangay* LGU does not recognize this unique law nor uphold IP rights/jurisdiction.
3. The Free Prior Informed Consent (FPIC) in many development projects is not a transparent process. In the case of mining, there were instances when investors who met disagreements between the tribes would shortcut the process by creating a fictitious tribal leader who would issue a quick FPIC.
4. Some forest areas were converted to cacao plantations instead of being protected.

9.5.2 Success Stories

1. The declaration of *Pusaka* has led to confiscating several trucks of illegally logged trees, and cases were filed against a *ka-tribo* (tribe-mate) who violated the *Pusaka.*
2. A *Pusaka* council took to task the documentation of unwritten practices and produced a primer in collaboration with the Philippine Eagle Foundation.
3. The University of the Philippines Mindano-Obu-Manuvu *Bantay Bukid* partnership in the UP Mindanao Land Grant is getting support from the city government. The *bantays* filed cases against squatters doing *kaingin* (slash and burn) in the area. As part of its public service, the UP Mindanao built a training center in the Land Grant and has spearheaded livelihood trainings and the maintenance of a naturally grown nursery.

9.5.3 Civil Groups and Initiatives

There are several "green" advocate groups in the City. Three are highlighted here.

1. The ***Purok* Disaster Action Team (PDAT)** is an organizational strategy formed by SUGAR, an intervention project of an NGO, *Mindanao Land Foundation* (MinLand). The program aimed to enhance the self-help capacity of selected poorest of the poor *barangays* along the DRB. It found out that in times of disaster, say flash floods, the *barangay* officials sometimes get overwhelmed by responsibilities and can respond only so much with limited resources. More often, clusters of households or *puroks*, end up fending for their selves. With

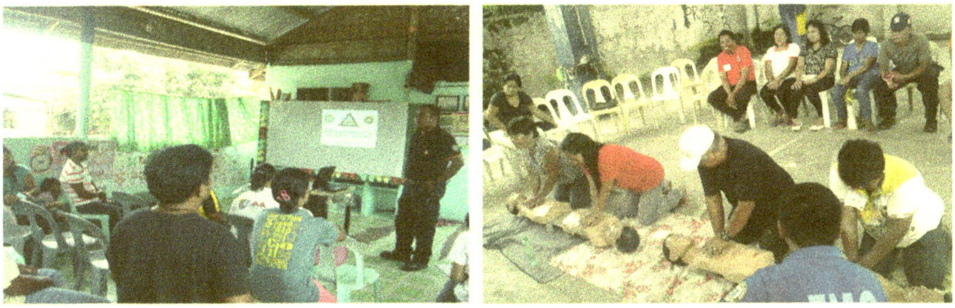

Fig. 9.7 Purok Disaster Action Team (PDAT) training on disaster survival. (Photos from SUGAR Project, MinLand. With permission)

the facilitation of MinLand, member *puroks* along the riverbank formed their respective PDATs who act as "leaders" during disasters. Each PDAT received first aid and survival kits aside from rescue training. It drew up its own *purok* evacuation route map. It underwent simulations and drills (Fig. 9.7). During a flood warning, an agreed upon protocol is: the PDAT in the most upstream is responsible for relaying updates, through text messages, on the condition of the river to all PDATs downstream. The downstream PDATs will cascade the information to all the *purok* households. This self-help capacity was tested spontaneously during *Vinta*. Those *puroks* with working and functioning PDATS evacuated in time, saving their lives and some essential belongings. One PDAT was non-functioning, and that *purok* suffered the severest damage vis-a-vis the other member *puroks*. Recognizing the strategic local contribution of such a team, all the member PDATs formed an association and registered themselves as a workers' association under the Department of Labor and Employment (DOLE).

2. The ***Bantayo Aweg*** (water guardians) and ***Bantay Bukid*** (forest guardians) volunteers (consisting of youth and *barangay* leaders) were initiated by IDIS (Interfacing Development Interventions for Sustainability), an NGO advocating for the protection and management of life-sustaining watersheds, from ridge-to-reef, in South-Central Mindanao. The *Bantayo Aweg* community researchers have been monitoring water quality in the last 15 years, and their findings have led to calls for organic farming and plantation in the buffer zones. The *Bantay Bukid*, on the other hand, has about 200 forest rangers. In 2017, the city government and the DENR deputized this group as guardians of the Panigan-Tamugan watersheds (soon a source of drinking water) from destructive and illegal activities. To build capacity, the volunteers received trainings, gadgets, and materials for patrolling and monitoring. As an incentive, the rangers receive monthly honorarium. IDIS itself has played front roles in advocating for a sustainable environment, such as passing the Watershed Code of Davao City and the Davao City Organic Agriculture Ordinance. It hosts a Watershed Summit annually as a form of regular assessment of the health of the critical watersheds in the city.
3. **The Hydrology for Environment, Life and Policy Davao Network (HELP-Davao Network)**. HELP is the policy component of the Intergovernmental

Hydrological Program (IHP) of UNESCO and HELP-Davao is the only network in the Philippines. The local network was established in 2004 and is unique from the rest of the global networks in the sense that it exists without any direct government subsidy—it is a SEC-registered (Securities and Exchange Commission) non-government organization which is composed of technical or lay individuals, groups, and institutions, all voluntarily engaging in *improving the links between hydrology (the hard core science) and the needs of society (community science)*. Among its engagements are doing scientific investigations or publishing literature for specific purposes. HELP-Davao pioneered the publication of a customized IWRM (Integrated Water Resource Management) Planning Manual for micro watersheds and has conducted planning writeshops for a number of flood-prone *barangays*. One of its volunteers has been facilitating the transfer of an environment-friendly technology for filtering waste water by using helophytes or water-loving plants. In 2019, HELP-Davao together with UNESCO-Jakarta produced a scorecard for rating the contributions of SETI (Science Engineering and Technology Innovation) projects in the attainment of the water-related Sustainable Development Goals (SDGs). That scorecard was adopted and expanded by UNESCO to include all the 17 SDGs. In 2023, UNESCO-Jakarta partnered with HELP-Davao again for a second SDG Scorecard, SECCI (Social Sciences, Human Sciences, Education, Culture, Communication, and Information). HELP-Davao is an example of collaboration under limited resources. It engages competing sectors and strive to positively influence the behavior of stakeholders by creating evidence-based frameworks for good water governance (Hearne et al. 2008).

9.6 Policy Interventions

As of 2020, the city government has its hands complete with a number of big plans, like the Local Climate Change Action Plan (LCCAP), the Metro Davao Urban Development Master Plan 2040, and the updated Davao City Comprehensive Land Use Plan 2013–2022. All these plans, including the 2014 DRB Management and Development Plan, contain recommendations about the flooding events in the city. The DRB Management and Development Plan alone contains a long list of action points. Each of those plans claims to have (a) science-backed recommendations and (b) community-based implementation, and there is no need to repeat them here. Instead, the authors will expand and provide specific examples for the two approaches, as follows.

9.6.1 Science to Support Decision-Making and On-The-Ground Actions

There is no substitute for sound evidence, but the soundness depends to a large extent on the quality of data and information—how they are chosen, gathered, updated, analyzed, integrated, and popularized so they can serve more stakeholders through time, not just for a particular period.

The ADB is at the forefront of deploying nature-based solutions (NbS) for their flood risk management projects. The solutions are science-based and are arrived at using multicriteria analysis (MCA) and cost-benefit analysis (CBA). NbS combines gray (hard) and green solutions with soft (nonstructural) measures such as land use planning, policymaking, early warning, and awareness raising. The City government and the DPWH have the fiscal power and a track record on hard solutions. Still, severe weather events in a climate change emergency may need more than default engineering solutions.

It is about time that the Climate Change Commission (national government) establishes **LCCDIES** for Climate Change Emergency for the 18 major river systems in the country. This information sharing system will enable the country to think as one in an emergency but act locally according to place-based knowledge and collective experience. LCCDIES hubs linked to NICCDIES can also be a support tool for the "whole-country, whole-community" approach the national government has been buzzing about during the various 2023 disasters.

With LCCDIES, all concerned regions and LGUs have knowledge of a standard minimum set of information needed to populate the data system including qualitative metrics. This means, regardless of the river system, there is only one monitoring template that needs to be filled out, whether by the *purok* leader, the DRRMC, or the Climate Change Commission. The template may be divided into boxes designated for each level of monitoring: a box for the *purok* leader (using the PDAT model), another box for the *barangay* captain as the first layer of data consolidation, and up the channels until the entire info/data about the flood is now in the NICCDIES. Competent technical teams manage the system at the regional and national levels. If the integrity of the LCCDIES is established and maintained, stakeholders, including researchers and policymakers, can retrieve information anytime for coordinated and harmonized policies.

In a quick and short survey the authors conducted for this chapter, there surfaced three common concerns:

1. There is a need to review existing LGU policies and programs and harmonize them with the DRB Management and Development Plan and the Comprehensive Land Use Plan. But who should be doing the inventory and harmonizing? This is where the LCCDIES will come in handy. It features a dedicated page for each of the 8 city's watersheds, providing updated information on the progress timeline in the management of each river system.
2. Service agencies, like the DENR and the DRRMC, must do integrated planning and synchronized implementation in coordination with the local government.

While these can be argued as a given in public service, there is still fragmentation in the delivery of services (Sales, et al. 2016), even in agencies supposedly working together in the same Presidential Cabinet Cluster (e.g., DILG, DENR, DPWH). This is where the LCCDIES comes in handy again. Actual events, trends, and projections will ensure integrated, leveled-up efforts by each agency, which generally operates with limited resources. Working on the same set of evidence, the different agencies can develop well-targeted interventions that avoid duplication and weather political terms by elected officials.

3. Provide retooling and capability enhancement for the technical personnel, especially the Monitoring and Evaluation team in the concerned agencies and offices. Part of the retooling is on conducting simple investigations, determining what to look for during monitoring, evaluating information through space and time, and basic knowledge on certain quantitative tools like the Multi-Criteria Analysis. The technical staff will also be involved in directly populating the LCCDIES and, thus, the need to enhance their IT skills. Because of the investment to enhance their competence, the technical team needs to occupy plantilla positions.

When we, the authors, reviewed some data obtained from government agencies, we found incomplete, sometimes inconsistent statistics and information. There were raw indices and matrices still in Word or Excel files. The need is to upgrade the gathering and handling of all related information into a more complete and reliable database ready for public sharing and with links to associated networks within the country and overseas. If some money is invested into the purchase of climate/environmental monitoring stations and early warning systems, then there must be qualified personnel dedicated to religiously transcribing and uploading the data in the LCCDIES.

Another area for improvement is reading and analyzing the data set, then making connections with data across space (e.g., provinces and countries) and time (e.g., 100-year trends). We found that the descriptive part of the tables or graphs contains no connections with previous similar events and no attempts to describe what is likely to happen in the near future (based on what is known at the moment). This 'incomplete story' can be a a handicap for policy makers. The local scenarios, along with data from the national agencies (e.g., PAG-ASA, NDRRMC) and global institutions (e.g., UN-Intergovernmental Panel on Climate Change) are useful for building robust large-scale model systems needed in understanding climate change systems. In NbS, the adoption of the best-fit model, needs a local but integrated context.

With a growing body of literature on scientific investigations about the DRB, both produced by researchers and environmental groups, the City Government is engaging the expertise of JICA to study the river flooding trends of Davao River and two more rivers, including the street and coastal floodings. Likewise, the academe can be called in to assist the technical team in making some sense of the database, provide complementary training, or give further recommendations. Again, the data and evidence from the ground that are fed to the policymakers need dedicated and competent staff responsible for maintaining a robust information management system for public use.

The type of science suggested here is not limited to formal science, which is the domain of technical officers and specifically trained "experts." It includes community science—"scientific research and monitoring driven and co-implemented by local communities and characterized by place-based knowledge, social learning, collective action, and empowerment" (Anthony et al. 2020). Community science is also considered in NbS. Davao City has more than 30 institutions of higher education and more number of environmental groups that contribute to the high public awareness of environmental issues. The recognition of the PDATs by the DOLE and the *Bantay Bukid* by the city government are examples of enabling community volunteers to engage in local actions. It is empowering them to take control of some things for themselves and be less dependent on external help.

9.6.2 The Four Indigenous Groups as Co-Managers

The CADT holders and the NCIP have authority under existing laws to share governance of the DRB. The passion of the four *lumad* tribes to protect their ancestral lands is enshrined in the IPRA (RA 8371) and needs no highlighting. After all, their domain extends 80% of the river basin. Unfortunately, the strategic contribution of the *lumads* to environmental protection and their indigenous wisdom about the interconnectedness of nature remains undervalued. There is little truthful documentation of their wisdom, which may be one reason why they are not appreciated as much in mainstream governance.

In community science, recognizing indigenous knowledge empowers and enhances local stewardship. Most often, that body of knowledge has a lot of common sense and can be sources of green solutions and technology. Taking the NGP as an example, the Obu-Manuvu is most familiar with the forest's topography, the tree species that grow in specific patches, the flowering seasons, and the territories of wildlife, among others. But if their indigenous way or *Pusaka* is skirted for some bureaucratic/dogmatic reasons, then the greening program becomes another government initiative that lasts until the budget is exhausted. The program becomes another top-down action introduced to the tribes who have been suffering from project fatigue.

In 2011, the NGP aimed to reforest 1.5 million hectares of degraded forestland around the country in six years. While the DENR has reported having reseeded two million hectares using some 1.7 billion seedlings, Sen. Ralf Recto asked the agency for maps showing the "carpet of green" (14 November 2019 Senate Hearing). Moreover, the 2019 COA findings stated that *reforestation remains an urgent concern, but fast-tracking its process without adequate preparation and support by and among stakeholders led to wastage of resources.* Flash floods in built-up areas are partly due to the lack of vegetation in the uplands that can help absorb precipitation. It follows that mainstreaming community science with the biggest land owners in the DRB may provide the "best fit" collaboration and sustainable solutions to this climate change emergency.

When decision-makers can hardly make a contextual appreciation of the flood-related events and interventions end up as palliative and sometimes cannot stand the test of another event, a combination of S&T and community science may serve a valuable role. A good example is the gabion structure meant to contain flood along the Diversion-Bacaca segment. The DPWH proceeded with the project without consulting with the supposed recipient barangays. One barangay was already reaping the protection of their community-initiated mangrove belt along their riverbank when the DPWH fell the mangroves without even sending any notice to the concerned barangay. When *Vinta* hit, the gabion on the side of this barangay gave way failing to protect the community from severe flooding.

9.7 Conclusion

The Davao River Basin as the biggest river system in Davao City plays several critical roles in the city's life, from geo-ecological to socio-political. But 20 floods in the past decade, or an average of two floods per year, is a climate change emergency. Of the 106 DRB *barangays* in the city's political jurisdiction, nine are most prone to flooding of which 8-A, Ma-a, Mandug, and Tigatto suffer the most in frequency, affected population, and water interruption.

There are several science-based literature on the DRB that policymakers can use as sources of evidence to improve climate change adaptations. These scientific investigations need strengthening and dedicated funding, and their results should become a prominent fixture in the planning and development exercises of the local government. The City can tap the academe for the conduct of those investigations as well as laymanizing the results for the policymakers and other stakeholders. But beyond hard science, nature-based solutions may efficiently complete the approach toward mitigating flood problems in the Basin.

Geographically and historically, the four Indigenous Tribes (Obu-Manuvu, Matigsalog, Ata, Bagobo-Klata) govern 80% of the DRB. Their domains are mainly in the uplands where most flash floods start, while the flood-prone areas in the plains and the mouth are occupied mostly by non-*lumads*(christians). In this sense, the *lumads* must be organic co-planners in addressing the climate change emergency. However, their local knowledge and wisdom, which can be sources of nature-based solutions, are marginalized over Western science and technology at the decision table. Considering their geographic distribution in the uplands, they are supposedly the ancestral keepers of the forests. But, as in the case of the National Greening Program, they are treated as local implementers of a program planned by people "at the top."

Aside from the indigenous tribes, the City also has a good number of civil groups that are passionate in their advocacy to keep Davao flood and landslide safe. Their efforts, along with those of the tribes can be tapped for collaborative undertakings, which consider community science as an integral part of using science as a tool for decision-making.

A healthy and clean Davao River is the lifeblood of 123 *barangays* spread across Bukidnon, North Cotabato, Davao del Norte, and Davao City. Flood mitigation programs for this river basin may have better success when "Western" science is coupled with community science, where local knowledge is acknowledged, community efforts recognized, and social learning with local and external experts solicited. With extreme weather events getting more frequent, hard infrastructure solutions need to integrate other approaches such as nature-based solutions that build green infrastructure and incorporate nonstructural measures. Onward to a "healthy Davao River Basin sustainably managed by empowered and responsible stakeholders" (HELP-Davao's vision).

References

Anthony C, Loucks L, Berkes F, Armitage D (2020) Community science: a typology and its implications for governance of social-ecological systems. https://doi.org/10.1016/j.envsci.2020.01.019

ChildFund alliance. https://childfundalliance.org/about-us/. Accessed 2021

Climate change commission programs. Climate Change Commission of the Philippines webpage (2024). https://climate.gov.ph/our-programs/national-integrated-climatechange-database-and-information-exchange-system. Accessed 2024

Davao City Comprehensive Development Plan 2012–2021. Lungsod ng Dabaw

Davao City Disaster Radio, 2020. DCDR 87.5 image. DCDR facebook page. www.facebook.com/davaocitydisasterradio. Last Accessed 12 July 2020

Department of Science and Technology (2017) Deployment of early warning devices in disaster prone areas (DEWS) project. Department of Science and Technology. www.dost.gov.ph. Last Accessed 12 July 2020

Department of Science and Technology (2012) DOST's DREAM project takes off. Department of Science and Technology. www.dost.gov.ph. Last Accessed 12 July 2020

DRB Management and Development Plan (2014) Department of Environment and Natural Resources, Region XI

Flaviana H, de Guzman R, Ortega D, Hayman P, Alexander B (2009) El niño southern oscillation in the Philippines: impacts, forecasts, and risk management. Philipp J Dev 36(66):9–34

Hearne D, Gamboa R, Monsanto V (2008) From ideas to action: a review of implementing HELP Principles in river basins with limited resources and capacity. Water SA (S Afr) 34(4). http://www.wrc.org.za

Hirabayashi Y, Mahendran R, Koirala S, Konoshima L, Yamazaki D, Watanabe S, Kim H, Kanae S (2013) Global flood risk under climate change. Nat Clim Chang 3:816–821

Jubilo KGA, Meryl RLA, Elexis Mae AT, Zayda DPA, Ari I-E, Maricor NS (2019) Lost waterways: clues from digitized historical maps of Manila and other Philippines cities

Mental health and psychosocial support (MHPSS) for children, adolescents and their families. Save the Children Child Rights Resource Cente. https://resourcecentre.savethechildren.net/document/mental-health-and-psychosocial-support-mhpss-children-adolescents-and-their-families/. Accessed 2021

Nature-based solutions for flood risk management: revitalizing Philippine rivers to boost climate resilience and enhance environmental stability. Asian Development Bank. https://www.adb.org/publications/revitalizing-philippine-rivers-climate-resilience. Accessed 2022

Palicte C (2020) Davao city gov't opens 'Disaster radio'. Philippine News Agency. www.pna.gov.ph. Last Accessed 12 July 2020

Public Health England (PHE) (2020) A method for Monetising the mental health costs of flooding. Published 12 June 2020 https://assets.publishing.service.gov.uk/government/uploads/system/uploads/attachment_data/file/890295/A_method_for_monetising_the_mental_health_costs_of_flooding_-_summary.pdf. Last Accessed 26 July 2020

Sales AC, Ruth UG, Ludivina MP, Lourdes RS, Joseph EA, Hydie RM, Ryan PC, Della Grace GB, Camilo DV, Gerardo BR, Ricardo PC, Cristy GG, Kenneth DB, Mayan Jane JI, Maria Lourdes OM, Mabel Diana MC, Reynaldo MP, Arnel MR (2016) Sustainability science demonstration site on enhancing resilience to disasters of urban water systems of Mindanao in the Philippines. Final report, 2016. UNESCO-Jakarta

Stocker TF, Qin D, Plattner G-K, Tignor M, Allen SK, Boschung J, Nauels A, Xia Y, Bex V, Midgley PM (eds) (2013) IPCC climate change 2013: the physical science basis, WG1. In: The fifth assessment report of the intergovernmental panel on climate change, p 1535. Cambridge University Press, Cambridge, UK

United Nations Office of the Coordination of Humanitarian Affairs (2017) DEWS to aid Davao City's hazard-prone areas. OCHA-Reliefweb. www.reliefweb.int. Last Accessed 12 July 2020

Water quality management (DENR-EMB XI). http://r11.emb.gov.ph/water-quality-management/

Ruth U. Gamboa Ph.D., is a pioneer in the ridge-to-reef approach in research and extension at the University of the Philippines Mindanao. Her collaborative works include a report on the state of the Talomo-Lipadas Watersheds and vulnerabilities of the urban water system in Davao City. She had served as President of the Board of Trustees of the Philippine-Canada Economic and Environmental Management-Davao City (PCEEM-Davao). She is a pioneer member of HELP-Davao (Hydrology for Environment, Life and Policy-UNESCO) and a member of the Technical Working Group of the Davao River Basin Management Alliance.—E-mail: rugamboa@up.edu.ph.

Ludivina M. Porticos is the Director of the Science Resource Center, University of the Immaculate Conception in Davao City. She is an environmental analytical chemist and professional food technologist. Her research undertakings include water (ground/surface/waste) quality and safety planning, and food safety. She is an active member and researcher of HELP (Hydrology for Environment, Life and Policy)—Davao Network under UNESCO. She is a pioneer member of the Davao River WQMA (Water Quality Management Area) and serves in the Executive Committee and Governing Board.—E-mail lporticos@uic.edu.ph.

Joseph E. Acosta Ph.D., is Associate Professor of Applied Mathematics in University of the Philippines Mindanao. His research works include operations research, energy and environment, agro-hydrology, flood modeling and generation of fine hazard maps using LIDAR Technology to more than 35 major watersheds. He has conducted extension services in training on flood disaster preparedness and risk assessment to about 40 LGUs and 5 HEIs in the Davao regions.—E-mail: jeacosta@up.edu.ph.

Chapter 10
The Post-Yolanda Relocation and Climate Emergency: The Case of Tacloban North

Ladylyn Lim Mangada

Abstract Following the devastating impact of Super Typhoon Haiyan in 2013, which decimated informal settlements along Tacloban City's coastline, the government decided to relocate thousands of families to the previously undeveloped food basket of the city, Tacloban North. This chapter presents the results of a case study undertaken three years after Super Typhoon Haiyan. Primary data were collected via 20 key informant interviews and three focus group discussions (FGDs). Relocatees maintained that the dwelling units are poorly constructed/overcrowded and lacked water and sanitation facilities. Most of the internally displaced persons (IDPs) faced severe socioeconomic challenges after relocation and these have affected their lives and compromised their ability to prepare for and respond to climate emergencies. Vulnerable populations are especially susceptible to a lack of access to resources in adapting during and after disaster events, making post-disaster mitigation a climate emergency. Poor communication between the local government and the affected communities further exacerbates the latter's situation, becoming a concerning byproduct of disasters brought about by severe climate change. This chapter asserts that encouraging stronger communication lines and incorporating the voices of affected communities in forming disaster-resilient communities can help mitigate this aspect of climate emergency.

Keywords Yolanda · Post-disaster relocation · Tacloban city · Resilience · Climate emergency

Key Points

- Communities that have relocated to Tacloban North in the aftermath of Super Typhoon Haiyan (local name: Yolanda) struggle with poverty, insufficient or lack of livelihood opportunities, and insubstantial housing.

L. L. Mangada (✉)
University of the Philippines Tacloban, Tacloban, Philippines
e-mail: llmangada1@up.edu.ph

K. B. Berse et al. (eds.), *Climate Emergency in the Philippines*, Disaster Risk Reduction, https://doi.org/10.1007/978-981-99-7804-5_10

- The climate emergency aggravates their vulnerable condition, as they remain stuck in the rehabilitation or reconstruction phase.
- Policymakers must review mass relocation strategies for disaster risk reduction and management as relocation sites lack the social facilities and economic opportunities needed to help people recover.
- Community risk assessments are important to ensure that those most affected by disasters and/or relocation are included in decision-making.

Summary for Policymakers

Resettled communities affected by Super Typhoon Haiyan in terms of poverty, livelihood, and insubstantial housing, are still recovering and cannot build themselves back up. The climate emergency aggravates their vulnerable condition, as they remain stuck in the rehabilitation or reconstruction phase.

There are many factors that undermine relocatees' ability to prepare and adapt to climate emergency and poor governance capacities is an important and often-cited example.

- Political leaders lack understanding of the losses and damage brought about by climate emergency and its impact on the human security of Haiyan survivors. There is very limited space available for citizens' voice.
- Preparing and adapting to a climate emergency means giving enough consideration to what really matters to the resettled communities.

To address this, policymakers at the national level must review mass relocation as a strategy to protect residents in coastal communities, contractor-driven housing, and interagency coordination.

There are many factors that undermine relocatees' ability to prepare and adapt to climate emergency and poor governance capacities is an important and often-cited example. There is very limited space available for citizens' voice.

- Mass relocation sites lack social facilities and economic opportunities further deteriorating their quality of life in a climate emergency. Areas for relocation do not allow agriculture-related activities, thus hindering residents' resilience at the onset of climate emergency.
- For contractor-driven houses to be durable (being able to withstand category 4 typhoon, extreme weather temperatures and intensity 7 earthquake), it needs a technical body, coming from the beneficiaries, local engineers/architect, and development advocates, that will supervise construction on the ground.
- There must be a limit to the number of housing units per contractor and subcontracting should be banned. A functional coordinative structure for the staff of the different government agencies (national and local) involved in housing and livelihood provision must be established.
- The projected damage of climate change impacts on agriculture is huge. Estimates of damage costs from modeling studies have reached up to about 145 billion pesos per year until 2050 (Rosegrant et al. 2016).

While adaptation planning at the national level is imperative, place-based vulnerability and risks due to changing precipitation patterns, increasing temperature, and increasing and/or intensifying extreme climate events, coupled with continuous environmental resources degradation need to be examined at the local level. Within the frame of landscape approach, vulnerability and risks analysis of local areas is imperative as basis for a risk-based strategic adaptation plan to ensure food security for the country.

Mass relocation sites lack social facilities and economic opportunities further deteriorating their quality of life in a climate emergency. Areas for relocation do not allow agriculture-related activities, thus hindering residents' resilience at the onset of climate emergency.

10.1 Introduction

The Philippines is one of the top countries in the world at risk of climate emergency. Most of the country's coastal population is constantly exposed to frequent and intensifying climate-related disaster, and lives in poverty, undermining the populace's ability to achieve resilience. Following Super Typhoon (ST) Haiyan in November 2013, which destroyed 50,000 homes in Tacloban City (NEDA 2018, TRRP 2014), the government organized an off-site and contractor-driven reconstruction program to reduce the vulnerability of people living on the coastlines to future disaster risk. Overall, this was the strategy adopted aimed at enhancing the resilience of ST Haiyan survivors against future disasters, which are considered to be the new normal, or the "climate emergency."

Climate emergency is a poverty multiplier, meaning that its impacts will further knock down vulnerable populations living in low-lying areas near the coastline, and affect those displaced and resettled persons. Climate emergency will simply exacerbate the uncertainty and insecurity of the people who have fewer resources to prevent, cope, and adapt to continuous/succession of adverse events.

People living on the poverty line are often displaced by disaster (Yee 2018). Relocation or resettlement becomes a recourse because some communities are found in geographical locations more prone to disasters than others. These communities are transferred from high-risk areas, such as coastal communities, to spaces that have less risk of exposure to hazards. Although this should be the last option for disaster risk reduction (Jha 2010), relocation becomes the default for developing countries with a number of vulnerable populations (Pelagi et al. 2019).

Further research needs to be made on the effects of the climate crisis on these vulnerable groups. Yong (2017) explores how post-disaster resettlement is financed, and the research reveals that financial support for relocation is limited, apart from the fact that displaced households are usually not able to choose where they will be relocated. Physical environment, the relationship of the new sites and the previous locations, as well as the resettlement community's ability to adapt, are important for the survival of resettled populations (Barrios 2014; Coburn et al. 1984; Oliver

Smith 1992). Due to an unempowered and unparticipative displaced population, among other factors, such as poor housing design and layout, unviable sites, post-disaster resettlement is more often than not set up for failure (Oliver smith 1991). Muggah (2011) notes that livelihood vulnerability is a controlling factor in disaster resettlement. Being able to find and sustain livelihoods is important for communities that are moved from their area of origin into new sites. However, more research needs to be made on livelihood vulnerability in post-disaster resettlement projects.

In addressing the concerns posed by the threat of climate change, scholars have forwarded the concept of resilience, recognizing that the vulnerability of communities can be reduced (Ayyoob 2016).

Resilience is often tied to policy and research. The ability to reorganize in the wake of disaster has been a topic of interest for scholars and policymakers whose long-term concern is to find solutions to such issues. Due to the notion of bridging prevention (proactivity) and response (reactivity) to hazards, resilience is tied to security and emergency planning (Welsh 2014). Resilience is also associated with build back better and sustainable socioeconomic development framework at its core.

This paper presents the result of a case study three (3) years after the overwhelming devastation of Super Typhoon Yolanda in Tacloban City. Of particular importance is the issue of relocation. Many believed that moving ST Haiyan survivors away from the coast through relocation would improve the well-being of the affected survivors and communities. The socioeconomic changes that occurred among the resettled as a consequence of offsite and contractor-driven housing approach is presented. Fieldwork was conducted for 6 months from August 2016 to February 2017 in Tacloban City, Philippines. Data were derived from key informant interviews with NGOs and relocates, and two focus group discussions with community leaders and government officials (i.e., city/barangay officials).

The main purpose of this paper is to gain a better understanding of how post-disaster relocation, given the scale of Super Typhoon Yolanda, affected seaside residents and communities. The socioeconomic opportunities and challenges of the resettled are presented and examined in the light of climate emergency.

10.2 Study Sites: Tacloban City and Tacloban North Resettlement

Tacloban City is located in the northern portion of Leyte Island (CDP 2017). It is built around a series of bays. Prior to ST Haiyan, Tacloban was known as a rising city and is one of the 33 highly urbanized cities (HUCs). It serves as the trading hub of the Eastern Visayas region (City Government of Tacloban 2014). The Philippine National Highway cuts through Leyte, and Tacloban City, and a large amount of goods are transported in the province. However, it is important to note that most of these developments are concentrated on the southern coast of Tacloban, with the north coast being far less developed (Oopdyke 2017). The city has a land area of

201.72 sq. km., and is divided into a total of 138 communities, called barangays. (National Statistics Office [NSO] 2010) In 2013, Tacloban was also named one of the most dangerous vulnerable cities (Muir-Wood 2013).

The Tacloban North Resettlement is located in the northern part of Tacloban City. It is 40 km from the city center. Several decades ago, the city purchased the large agricultural land, which was previously envisioned as an Eastern Visayas Agri-Industrial Growth Center. As a result of post-ST Haiyan reconstruction, farmers inhabiting the land were displaced, the land was rezoned as a relocation site, which displaced farmers who had previously inhabited these areas (Tukhanen et al. 2018). The Tacloban North Resettlement houses at least 21 housing communities. The National Housing Authority (NHA) plans to construct close to 14,000 units in the resettlement area. It takes about an hour ride to reach the area via public utility vehicle, but the Tacloban North Resettlement is poorly served by public transport and basic utilities. Relocatees are challenged with poor basic services (e.g., water, health care, educational facilities) and the unavailability of livelihood.

10.3 The "Challenging Relocated Life" in Tacloban North Resettlement

Relocation of disaster survivors is promoted as a way to provide new land and livelihood opportunities to the affected households and communities by ST Haiyan. It often occurs through the re-housing of displaced communities on unused vacant land.

Relocation induced by mega-disasters such as Haiyan follows a state planning approach. Such an approach focuses on organizational and administrative aspects of government policies and programs in recuperating physical losses experienced by devastated communities (see Badri et al., 2006; Samuels, 2012). Its planning process is often top-down oriented, giving minimum opportunities for the affected community to voice their needs. With insufficient avenues for meaningful community consultation and participation, state planners often fail to "integrate local characteristics and pre-disaster context of resettled communities" since the initial planning stage (Carrasco et al. 2016). How did the relocation site known as the Tacloban North Resettlement fare? Here are the accounts of the resettled in the government-housing program cum relocation site.

There are 21 relocation sites in Tacloban North Resettlement. The National Housing Authority has 17, while non-government organizations (NGOs) have nine. In total, there are 16,327 housing units, which have been planned for construction. Six years after Haiyan, only 8,375 housing units have been occupied by resettled families. They used to live in fragile dwelling units with no security of tenure in the coastal districts of San Jose, Sagkahan, Magallanes, and Anibong. Most of them depended on fishing as their primary form of livelihood, and its allied enterprises. In the first quarter of 2017, the resettled and other stakeholders raised the issues of poorly constructed and small housing units, poor provision of basic services, increased livelihood vulnerability, politicization, new risks, muted voices/marginalization of Haiyan survivors, and the difficulty of returning to the pre-Yolanda location.

10.3.1 Housing Units Are Small And Poorly Constructed

As a step toward resilience, the dwelling units were supposedly designed to withstand an earthquake. Unfortunately, the units have cracks on the walls, unaligned roofing, broken sinks and pipes, and overflowing septic tanks during bad weather condition (Pelagi 2019).

Leaders of various homeowners associations observed that there are plenty of unoccupied fully constructed shelter units. Some of them wondered who the recipient of these vacant units will be, and the criteria that will be used in selecting the recipients.

> *"Tam housing unit may mga crack it bongbong. It septic tank puno na. Ginbabaha kami kun nauran." (My housing unit has cracks on the wall. The septic tank is already full. We are flooded when it rains).*
>
> *Gudti man tam balay. Diri man kami naigo kay 4 tak anak. It kusina ngan CR ada sakob, waray kwarto. Kada balay 22 sq.m it floor area. Semento man tam balay yana pero guti-ay. Harayo la kami hin bagyo ngan storm surge". (Our houses are small. Cannot accommodate us because we have 4 children. We have comfort room and kitchen inside but no bedrooms. It is concrete but small, far from typhoon and storm surge).*

10.3.2 Inaccessible Relocation Site And Poor Basic Services

While many families gained new permanent housing away from climate risks, such as storm surge, several challenges emerged. These include the lack of drainage, electricity, transport, school and health centers, and piped water. City government tankers deliver water to relocation sites daily, although some residents must purchase from privatized sources to meet their needs. (Tukhanen et al 2018).

Relocatees asserted that the distant location of the relocation site from the commercial center and the inadequate provision of social services/utilities have disrupted their daily lives, making the survivors' lives doubly difficult.

> *"Kinahanglan drainage kay ginbabaha kami."(We need good drainage because we are being flooded).*
>
> *"Waray man tubig, it iba la it may kuryente. Sige manla it saad nira. maiha pa magkatubig dinhi..waray pa pipes." (There is no water, but others have electricity. They kept on promising that they will provide water, but it will take time because no pipes have been installed yet).*
>
> *"Mayda man mga taker nga nadeliver tubig ha mag resettlement. Pero nakapalit la gihap kami hin mineral water." (Water is delivered using the water tanks to the resettlement site but is not enough to meet our needs, so we buy mineral water).*

"Harayo man ini ha siudad. Na lelate ak ha trabaho pati mga bata ha eskwelahan. Napila para makasakay, makuri gud ngan magastos it pamasahe." (This place is very far from the center. Most of the time I am late for work as well as my children's schooling. In order to get a ride we fall in line because it is difficult and expensive to get a public utility vehicle.)

"Harayo pag eskwela tak anak kay high school na. Waray pa eskwelahan para han mga nagbalhin. An iba na college waray la anay umiskela. (This site is far, especially now that my child is already in high school. There is no school yet for those who are resettled. Some college students stopped going school due to distance and cost.)

10.3.3 Increased Livelihood Vulnerability

The resettled households lamented for the substantial decrease in income after displacement.

"Waray man pakabuhi dinhi, puro pangisda kami. Adto tak asawa nag lelabor hit mga balay na ginhihimo pa. Pero kulang la gihap it iya kita para hit amon panginahanglan ngan matat-apos na ini na pabahay. Mas dako an amon kita han nakadto pa kami ha Magallanes."(There is no livelihood here, especially for us fishers. Now my husband works as a construction worker in other housing sites. He earns less for what we need each day. I worry because the construction of the housing units might be finished soon. Our income was better when we were still in Magallanes).

"Maupay pa an nakadto pa kami ha San Jose di kami nagugutom kay pwede am magtinda, labot la hin isda. Pwede maglaba-laba. Yana waray gud pa kuro kwarta." (Life was better in San Jose. We did not experience hunger/missing meals because I would sell food/goods other than fish. I can do laundry. Today, we experience having no money."

"Napa Tacloban pa kami para magka mayda income. Waray man dinhi pwede mahimo nga pakabuhi, diri pwede mag baboy.manukan/hayupan." (We have to go to the city in order to have income. There are no livelihood opportunities here. Raising hogsor chicken is not allowed.)

"Sari-sari store man la it pwede dinhi." (Only sari-sari store or vending is allowed as a source of income)

Sige la lista mga ngaran/panimalay para daw pakabuhi… waray man. Puros nala kami dinhi utang. (They kept on listing our names for livelihood assistance but nothing happens. We end up indebted.)

Nak bana adto ha Manila linakat kay han una nagbabaligya man kami isda, yana waray na." (My husband left for Manila. We used to sell fish, but when we transferred here, we stopped.)

Makuri man tam pagkaon yana dinhi ha permanent shelter. Waray naman nahatag na INGOs. Mayda nanlista para sari-sari store or livelihood training pero wary man pan capital. Ako nakareceive hin pan sari-sari store. Kun waray trabaho tak aswa yana, ada kuhaa an amon pan adlaw-adlaw." (To be able to have food in our table is a challenge now. INGOs no longer give food packs. Somebody lists down names for livelihood training but no capital is provided. I was given goods for sari-sari store but when my husband is out of work, we use the goods for our needs).

According to the block leaders in the different housing sites, the success rate of the different livelihood assistance is below 20 percent. This is because households would use the capital or goods for their daily sustenance. The capability to cope with future risks is related to what livelihood assets resettled households have and how they cope with hazards. Sustainable livelihood recovery, therefore, is a top priority to sustain resettled families and communities.

10.3.3.1 New risks

Relocation and resettlement are approaches for hazard risk reduction. It should be cautiously weighed against possible hazards or risks. For the resettled families/households, security of tenure remains elusive.

> *Tam balay harani la ha dumpsite. Pag gin abre it purtahn ngan bintana, mabaho it nasulod ngan damo it kalangawan. Magkakasakit kami." (Our new home is close to the garbage dumpsite. A very foul smell comes in when we open the door and windows. The flies swarm the village and makes us sick).*
>
> *Waray man kami papel na ginkakaptan nga pruweba na kami it tag-iya hini na housing unit. (We do not have anything to prove that we own this house).*

10.3.3.2 Politicization

Key informants highlighted the influence of partisan politics in the distribution of aid, such as livelihood and assignment of housing units. Community leaders shared that beneficiaries of livelihood assistance were perceived to be close to persons of authority. Others revealed that in other resettlement sites, ineligible families get the "corner" units but with a "fee/grease money." Tukhanen (2018) pointed out that lobbying of local officials and timing of requests influenced the distribution of benefits and national-level disbursements, which may or may not have been based on need. This resulted in duplication of support for some, particularly those with close political or personal connections with local officials, while other intended beneficiaries received little to no support.

> *Nauuna pagtagi hin hinabangan an mayda connection. An iba waray and iba nag dodoble, or sobra pa." (Those who have political connections are prioritized in aid distribution. Some do not anything while others have more.)*

In 2017, President Duterte ordered the immediate relocation of 10,000 families. Some felt this rushed process ignored the original prioritization of beneficiaries, led to poorer housing quality, and overlooked important social preparation, which may have facilitated community acceptance of relocation (Tukhanen et al 2018).

10.3.3.3 "I shall return" to pre-Yolanda settlements

During the FGDs, the survivor-participants unanimously voiced out on the practice of some housing beneficiaries to return to their previous location in the No Build Zone or No Dwelling Zone. These beneficiaries spend most of the time in their previous settlement and would go back to their dwelling units in Tacloban North during bad weather. The main reason for going back to the pre-Yolanda residence/community is due to the absence of income-generating opportunities in the resettlement site. Other reasons forwarded include inaccessibility to basic services and to the commercial district. Some of them "defied" the housing policy by constructing temporary shelters on the coast despite the hazards/risks it posed to them.

The focus group discussions with the local officials and community leaders strongly echoed the issues raised by the relocates, such as inaccessibility to social services, socioeconomic resources, and employment opportunities, and a number of relocatees abandoning their new homes and returning to their original settlement. In addition, they articulated the onset of petty crimes and early pregnancies among teenagers.

Based on these findings, it can be inferred that the location of the resettlement critically determines the retention of the survivors in the new site.

10.3.3.4 Muted voices and the Marginalization of Haiyan Survivors

The survivor-participants who moved to the resettlement sites strongly reiterated the absence of mechanisms for them to clearly communicate their interests, concerns, and suggestions to post-Haiyan recovery and resilience. They are not represented in decision-making bodies. The resettled hardly know of the distinct function of national and local government agencies in the post-disaster relocation. Hence, they hardly influence decision-making.

> *"Waray man kami pakianhi ngan sumati kun hain kami igbabalhin." (Nobody consulted us where we want to go or which relocation site we were assigned).*
>
> *"Hira man la nagboboot han design ngan kadako nam unit. Pagbalhin nam waray man may nagpakiana kun ok or diri ini na mga balay." (They decide the size and design of the house. When we transferred no one bothered to ask whether the house was good or not).*

They excitedly recalled that government agencies and donors simply implemented their own housing and livelihood programs. Housing units were constructed without consultation and coordination from the end-users. Some survivors were forced to evacuate following orders of a politician, but their needs and complaints were not attended to. Livelihood activities were centrally planned and did not match the skills of the relocatees who were mostly fishers or whose income depended on the sea.

> *"Siring ko tak kinhanglan pan parlor kay asya nak pakabuhi dati ngan asya tak talent, gin tagan ak mga kaldero lugod." (I told them I need materials for parlor because that used to be my source of income and talent but they gave me cooking wares).*

Some community leaders claimed to have gathered strength to visit government offices to present their issues. However, they are only warmly welcomed, without any concrete action taking place.

For the past 5 years, the voices of those who resettled remain in the margins of the decision-making process. This is evidenced in the many programs and projects that were not tailored to pre- and post-disaster vulnerabilities, as well as the real needs of the survivors. The different leaders of the homeowners' associations repeatedly said: "Government does not listen to our pleas. Political authorities do not feel our plight. They were just contented that we were removed from the coast and transferred to a safe place. Our houses are concrete but we have empty stomachs."

10.4 Relocation, Post-Super Typhoon Haiyan Resilience, and Climate Emergency

The numerous problems confronting ST Haiyan relocatees is attributed to poor resettlement planning and management. In fact, as early as 2010 (Fernando et al) warned that human resettlement may end up with failure if it is not well planned and handled. As the results of the study show, ST Haiyan relocatees have not recovered from the socioeconomic losses in the aftermath of the devastating super typhoon. Relocation, as a post-disaster intervention, entails better understanding of pre-existing vulnerabilities in the planning and implementation of permanent settlement. If these causes are not considered in post-disaster strategies, there is danger that the same vulnerable state will be rebuilt, or new vulnerabilities will be created.

While ST Haiyan survivors were relocated away from the sea, they are now vulnerable to flooding and landslide due to poor drainage, substandard housing, and hunger caused by poverty. It is unfortunate that post-disaster relocation has not addressed the underlying inequalities or vulnerabilities that may have caused losses (Jain et al. 2017; Puriya and Biswas 2018). Post-disaster relocation needs to emphasize resilient housing and sustainable livelihood/resources that should be given to displaced people. Cernea (1997) pointed out that one of the main requirements for a successful relocation is the availability of adequate livelihood opportunities. It should be an opportunity for those affected to enhance their living standards (Cernea, 1988; Oliver Smith and Sherbinin, 2014). The issues on inaccessible relocation, inadequate and unsafe housing, lack of access to basic services and sustainable livelihood and opportunities, and insecure land tenure compromise the relocatees' capacity to move forward and respond to climate emergency. How should they cope and survive post-disaster conditions?

With irregular and insufficient incomes and the onset of climate emergency severely threatening their sources of livelihood, such as fishing, that could be lost completely if exposed to repeated disastrous events within short intervals, leaving insufficient time for recovery. Chambers and Conway (1999) and Jones et al. (2010)

argue that local people's livelihoods and their access to and control of resources can be affected by events largely beyond their control, such as climatic extremes.

The risks of extreme climate exposure on human health, particularly for the resettled poor, are well known. With government's underfunding on health, how will ST Haiyan relocatees survive with pandemic vector-borne or infectious diseases?

The onset of climate emergency severely challenges the relocatees in Tacloban North Resettlement through the following:

Tropical cyclones: This means that increasingly frequent and intense tropical and extra-tropical cyclones will cause severe wind damage, destroying low-quality housing units. Cyclones with storm surges compounded with a rise in sea level will cause severe problems for low-lying coastal communities (e.g., frequent evacuation and house repairs).

More funds for emergency response: Relocatees working in ports and other coastal infrastructure are especially at risk. Negative impacts will be felt, especially for subsistence fishers and seasonal farmers whose sources of income will be destroyed, which most of the resettled depend on. Hence, more relocatees will suffer from hunger and malnutrition.

A rise in sea level: This is the most fundamental challenge that Tacloban City faces from the climate emergency. The threat will increase due to the return of the Haiyan survivors to their previous coastal villages. At risk of erosion are on the entire sections of coastal Tacloban and their infrastructure, parks, and beaches. This will deprive relocatees of their livelihood, and increase food prices, which will in turn lead to hunger and poor health conditions.

Water quality and shortage: Frequent flooding in the relocation sites damages flood wells, pit latrines, and septic tanks. Sewage treatment systems and solid waste disposal areas can also be affected, contaminating water supplies. As overall rainfall decreases, potable water will increasingly fail to reach Tacloban North Resettlement. Severe droughts will compromise the replenishment of the water tables, the normal sources of water supply.

Flooding and landslides: Increases in the scale, intensity, and frequency of rainfall in post-Haiyan environments severely strain or overwhelm the storm drainage systems in Tacloban North and the city proper. Landslides, which have been causing loss of lives due to settlement within the area, have been experienced in the post-Haiyan area.

Truly, the climate emergency is causing more severe hazards and catastrophes. Coupled with persistent poverty and governance failures, this means disaster risk is likely to be aggravated. Thus, effective state-led disaster risk reduction measures are a necessary part of a broader effort to reduce climate risk. Although relocatees try to cope and adapt to post-Haiyan climate events, this does not guarantee coping with successive extreme events that will be similar to Yolanda (the local name for Haiyan). Over time, such coping responses might prove inadequate due to the magnitude of the climate event's impact. This means it will take relocates a longer time to recover and bounce forward from climate-related disasters.

Relocation may have provided relocatees safety from typhoon and storm surges but their resilience to severe climate-related events hinges on the livelihood and other resources available in the new site. Aside from fishing and community vending, relocatees lack capacities and skills which are limiting factor for effective local adaptation to hazardous events. Government needs to aggressively address these issues that compromise the resilience of the resettled.

The impact of the climate emergency on the resettled ST Haiyan survivors depends on their socioeconomic circumstances. The survivors' ability to mitigate threats rests on a better understanding of climate change risks, (its underlying causes) by the political leadership at all levels. The adequate attention to post-disaster sustainable livelihood and the political leaders' degree of receptiveness to scientific evidence and argumentation will make a significant difference in overcoming and facing climate emergency. Implementing and managing relocation within the wider context of long-term development can markedly reduce vulnerabilities and mitigate the impact of future disasters.

10.5 Conclusion

Despite the urgency and centrality of the country's vulnerability in a climate emergency, it largely remains unaccounted for in decision-making, Coastal areas remain to be ravaged by storms and resettled communities become prone to increased vulnerability brought about by poverty, unemployment, distributional inequalities, population growth, and lack of disaster reduction and management response at the institutional level (Smith 1996; Anderson 1995; ECLAC/IDB2002).

In view of several development and recovery plans, Tacloban City has formulated that appears to be disconnected with one another (Atienza 2019) and seriously lacks risk assessment. It is therefore essential to look into revisiting and reformulating these plans and frameworks. In 2011, Jones and Preston 2011, 305) found that a risk-based approach "stands as the most appropriate overarching framework for assessing climate change adaptation" (30 A comprehensive all-hazard approach to risk assessment and management is useful for considering how climate change mitigation policies might affect the impact of disasters on people. This approach is increasingly seen as crucial for the assessment and management of climate impacts at the global scale (IPCC 2012; PROVIA 2013). At the country level, the all-hazard approach risk assessment provides government and organizations a basis for prioritization of disaster risk reduction activities. It is an essential prerequisite to a full and comprehensive array of disaster risk reduction plans and policies that contribute to the government's objective of reducing people's vulnerability and enhancing their resilience against disaster and climate change.

To address the persistent challenges of the Haiyan survivors in the Tacloban North Resettlement, effective legislation or policy measures should be developed urgently. Political leadership at all levels is called upon to consider the following:

1. A participatory and inclusive process must be promoted, valued, and institutionalized in the relocation sites to involve relocatees of different risk groups (e.g.,

PWD, fishers, carpenters. sari-sari storeowners, LGBT, youth, etc.) in the planning and decision-making process. This is also particularly important because climate change impacts "will fall most heavily on the poorest and most vulnerable communities" (Sachs 2008). Community participation should be used to exchange accurate and realistic information, including on available resources. This also reduces disaster risk by enabling beneficiaries to understand the vulnerabilities, which led to the disaster, and how to avoid recreating these or other vulnerabilities.

2. Local and national governments must strengthen the intergovernmental and collaborative capacities of their personnel. Government offices should not work in silo. Overcoming risks and impacts of climate emergency necessitates working with different offices. This requires collaboration and communication across sectors and disciplines. Strengthening collaborative skills facilitates partnership between relevant actors while problematic coordination hampers the delivery of adequate services on time.
3. Contractor-driven housing units must be seriously evaluated. The unsafe housing, which was constructed with low-quality materials, compromised the well-being of the beneficiaries. Visual or community-led inspection of housing conditions and other infrastructures in the relocation sites must be conducted.
4. The consciousness of officials at all levels of government on climate emergency must be raised. Successful resilient planning is "highly dependent on the level of commitment and leadership of national and local authorities." However, the underlying causes of social vulnerability and drivers of risk are not well-understood and/or addressed by policymakers and practitioners (Zou and Thomalla 2010; Veland et al. 2013; Wang et al. 2012a,b). Disaster risk reduction must also be recognized as a national and local priority with a strong institutional basis for implementation.
5. Societal awareness and preparedness for risk reduction can be done by use of local knowledge, innovation, and education to build a culture of safety and resilience at all levels. For example, the public may be informed of the risks and possible consequences of climate change, as well as the key roles of government offices involved.
6. For better adaptation, climate services must be made available to farmers and fishers. Key government agencies must be visible to these sectors through education and learning from their views and practices on coping to risks and disasters. Climatic events must be adequately explained to them to ensure that they understand this as the effect of rising and changing extreme temperatures. The integration of local knowledge, science, and technology in the climate services can reduce or prevent the impact from climate disasters.
7. The need to review mass relocation as a strategy to protect vulnerable residents in low-lying coastal communities. Local government units need to strategically use their physical and land resources to address climate emergency challenges.

References

Adger WN, Hughes T, Folke C, Carpenter S, Rockström J (2005) Social-ecological resilience to coastal disasters. Science 309(5737):1036–1039. https://doi.org/10.1126/science.1112122

Aldrich DP, Meyer MA (2015) Social capital and community resilience. Am Behav Sci 59(2):254–269. https://doi.org/10.1177/0002764214550299

Anderson M (1995) Vulnerability to disaster and sustainable development: a general framework for assessing vulnerability. In: M. Munasinghe, C Clarke (eds.), Disaster prevention for sustainable development: economic and policy issues, pp. 41–59. World Bank, Washington, D.C.

Ali BS, Asgary A, Eftekhari AR, Levy J (2006) Post-disaster resettlement, development and change: a case study of the 1990 Manjil Earthquake in Iran: post-disaster resettlement, development and change. Disasters 30 (4): 451–68. https://doi.org/10.1111/j.0361-3666.2006.00332.x

Bang HN, Few R (2012) Social risks and challenges in post-disaster resettlement: the Case of Lake Nyos, Cameroon. J. Risk Res. 15 (9): 1141–57. https://doi.org/10.1080/13669877.2012.705315

Bankoff G (2003) Cultures of disaster: society and natural hazards in the Philippines. London; New York: Routledge

Barenstein J (2013) Post-Tsunami Relocation Outcomes in Sri Lanka: Communities' Perspectives in Apara and Hambantota. In: Barenstein J, Leemann E (eds) Post-Disaster Reconstruction and Change: Communities' Perspectives. CRC Press, Boca Raton, FL, pp 215–239

Barrios RE (2014) 'Here, I'm Not at Ease': anthropological perspectives on community resilience. Disasters 38(2):329–350. https://doi.org/10.1111/disa.12044

Bosher L (2014) Built-in resilience through disaster risk reduction: operational issues. Building Research & Information 42(2):240–254. https://doi.org/10.1080/09613218.2014.858203

Bowman LJ, KB Henquinet (2015) Disaster risk reduction and resettlement efforts at San Vicente (Chichontepec) Volcano, El Salvador: Toward Understanding Social and Geophysical Vulnerability. J Appl Volcanol 4(1):14. https:// doi.org/https://doi.org/10.1186/s13617-015-0031-0.

Carrasco S, Ochiai C, Okazaki K (2016) A study on housing modifications in resettlement sites in Cagayan de Oro, Philippines. J Asian Architect Build Eng 15(1):25–32. https://doi.org/10.3130/jaabe.15.25

Cernea MM (1988) Involuntary resettlement in development projects: policy guidelines in world bank-financed projects. World Bank Technical Papers. The World Bank. https://doi.org/10.1596/0-8213-1036-4

Cernea M (1997) The risks and reconstruction model for resettling displaced populations. World Dev 25(10):1569–1587. https://doi.org/10.1016/S0305-750X(97)00054-5

Chandler D (2014) Resilience: the governance of complexity. critical issues in global politics. Abingdon, Oxon; New York, NY: Routledge

Coburn AW, Leslie JDL, Tabban A (1984) Reconstruction and resettlement 11 years later: a case study of Bingol Province, Eastern Turkey. In: Schuppisser S, Studer JA (Eds.), Earthquake relief in less industrialized areas. A.A. Balkema, Rotterdam, pp 49–58

Cote M, Nightingale AJ (2012) Resilience thinking meets social theory: situating social change in socio-ecological systems (SES) research. Prog Hum Geogr 36(4):475–489. https://doi.org/10.1177/0309132511425708

Cutter SL, Barnes L, Berry M, Burton C, Evans E, Tate E, Webb J (2008) A place-based model for understanding community resilience to natural disasters. Global Environ Change 18(4):598–606. https://doi.org/10.1016/j.gloenvcha.2008.07.013

Cutter SL, Barnes L, Berry M, Burton C, Evans E, Tate E, Webb J (2008) A place-based model for understanding community resilience to natural disasters. Global Environ Change 18(4):598–606. https://doi.org/10.1016/j.gloenvcha.2008.07.013

Eadie P (2019). Typhoon yolanda and post-disaster resilience: problems and challenges. Asia Pacific Viewpoint 60(1):94–107.

Ebi KJ, Mastrandrea GK, Mach SK, Plattner MA, Tigno M, Midgley PM, 1–19. IPCC, UK and New York

ECLAC/IDB (2000) A matter of development: how to reduce vulnerability in the face of natural disasters. In: Confronting Natural Disasters: A Matter of Development. New Orleans

Faas AJ, Jones E, Whiteford L, Tobin G, Murphy A (2014) Gendered access to formal and informal resources in postdisaster development in the Ecuadorian Andes. Mountain Res Develop 34(3):223–34. https://doi.org/10.1659/MRD-JOURNAL-D-13-00100.

Fernando, Nishara, Koko Warner, and Jörn Birkmann. 2010. "Migration and Natural Hazards: Is Relocation a Secondary Disaster or an Opportunity for Vulnerability Reduction?" In Environment, Forced Migration and Social Vulnerability, edited by Tamer Afifi and Jill Jäger, 145–56. Berlin, Heidelberg: Springer Berlin Heidelberg. https://doi.org/10.1007/978-3-642-12416-7_11.

Folke C (2006) Resilience: The Emergence of a Perspective for Social– Ecological Systems Analyses. Glob Environ Chang 16(3):253–267. https://doi.org/10.1016/j.gloenvcha.2006.04.002

Folke C (2006) Resilience: The Emergence of a Perspective for Social– Ecological Systems Analyses. Glob Environ Chang 16(3):253–267. https://doi.org/10.1016/j.gloenvcha.2006.04.002

Gaillard J-C (2008) Differentiated Adjustment to the 1991 Mt Pinatubo Resettlement Program among Lowland Ethnic Groups of the Philippines | PreventionWeb.Net. The Australian Journal of Emergency Management 23(2):31–39

Ganapati NE, Ganapati S (2008) Enabling participatory planning after disasters: a case study of the world bank's housing reconstruction in Turkey. J Am Plan Assoc 75(1):41–59. https://doi.org/10.1080/01944360802546254

Holling CS (1973) Resilience and stability of ecological systems. Annu Rev Ecol Syst 4(1):1–23

IPCC (2012) Summary for policymakers: a special report of working groups I and II. In: V. C.B., T.F. Barros, D. Stocker, D., D.J. Qin, K.L. Dokken, M.D. (eds.), Managing the risks of extreme events and disasters to advance climate change adaptation

IPCC (2014) Climate change 2014: synthesis report: contribution of working groups I, II and III. In: R.K. Pachauri, L.A. Meyer (eds.), Fifth assessment report of the intergovernmental panel on climate change, 151. IPCC, Geneva, Switzerland

Jha AK, Duyne JE, Barenstein (2010) Safer homes, stronger communities: a handbook for reconstructing after natural disasters. World Bank, Washington, DC

Jones RN, Preston BL (2011) Adaptation and risk management. Wiley Interdiscip Rev Climate Change 2(2):296–308. https://doi.org/10.1002/wcc.97

Klein RJT, Nicholls RJ, Thomalla F (2003) Resilience to natural hazards: how useful is this concept? Environ Hazards 5(1):35–45. https://doi.org/10.1016/j.hazards.2004.02.001

Lewis J (1999) Development in disaster-prone places: studies of vulnerability. Intermediate Technology Publications, London

Muggah R (2011) A unified approach to conceptualizing resettlement. In: Koser K, Matin S (eds) The migration-displacement nexus: pattern, process, and policies. Berghahn Books, New York, NY, pp 29–60

Muir-Wood R (2013) The dangerous city of Tacloban. Risk management Solutions

Muncke P, et al. (2016) World risk report. United Nations University – Institute for Environment and Human Security

National Statistics Office (NSO) (2010) Census of Population and Housing, v1.1. Philippines, Manila

NEDA reconstruction assistance for Yolanda National Economic and Development Authority (2013)

O'Keefe P, Westgate K, Wisner B (1976) Taking the naturalness out of natural disasters. Nature 260(5552):566–567. https://doi.org/10.1038/260566a0

Oliver-Smith A (1991) Successes and failures in post-disaster resettlement. Disasters 15(1):12–23. https://doi.org/10.1111/j.1467-7717.1991.tb00423.x

Oliver-Smith A (1992) Problems in post-disaster resettlement. In: Aysan Y, Davis I (eds) Disasters and the small dwellings: perspectives for the UN IDNDR. James and James Science Publishers Ltd., London, pp 58–66

Oliver-Smith A, De Sherbinin A (2014) Resettlement in the twenty-first century. Forced Migration Review 45(February):23–25

Palagi S, Javernick-Will A (2019) Institutional constraints influencing relocation decision making and implementation. Int J Disaster Risk Reduction 33(February):310–320. https://doi.org/10.1016/j.ijdrr.2018.10.016
PROVIA (2013) PROVIA guidance on assessing vulnerability, impacts and adaptation to climate change. Consultation document. Nairobi, Kenya: United Nations Environment Programme. www.unep.org/provia
Puriya A, Biswas A (2018) A modified needs assessment method in post disaster recovery processes. In: 8th International Conference on Building Resilience, ICBR, Lisbon
Samuels A (2012) Moving from great love: gendered mobilities in a post-tsunami relocation neighborhood in Aceh, Indonesia: moving to a post-tsunami relocation neighborhood in Aceh, Indonesia. Int J Urban Reg Res 36(4):742–756. https://doi.org/10.1111/j.1468-2427.2012.01125.x
Sharifi A (2016) A critical review of selected tools for assessing community resilience. Ecol Ind 69(October):629–647. https://doi.org/10.1016/j.ecolind.2016.05.023
Singh C, Jain G, Malladi T (2017) Rethinking post- disaster relocation in Urban India. IIED Briefing. Urban Policy and Planning
Smith K (2013) Environmental hazards: assessing risk and reducing disaster, Sixth edition. London; New York: Routledge Taylor & Francis Group
Tacloban City (2014) The Tacloban Recovery and Rehabilitation Plan
Tuhkanen H, Boyland M, Han G, Patel A, Johnson K, Rosemarin A, Mangada LL (2018) A typology framework for trade-offs in development and disaster risk reduction: a case study of typhoon Haiyan recovery in Tacloban, Philippines. Sustainability 10(6):1924. https://doi.org/10.3390/su10061924
Usamah M, Haynes K (2012) An Examination of the resettlement program at Mayon Volcano: what can we learn for sustainable volcanic risk reduction? Bull Volcanol 74(4):839–859. https://doi.org/10.1007/s00445-011-0567-8
Veland S, Howitt R, Dominey-Howes D, Thomalla F, Houston D (2013) Procedural vulnerability: understanding environmental change in a remote indigenous community. Global Environ Change 23(1):314–26. https://doi.org/10.1016/j.gloenvcha.2012.10.009
Wang M-Z, Amati M, Thomalla F (2012b) Understanding the vulnerability of migrants in Shanghai to typhoons. Nat Hazards 60(3):1189–1210. https://doi.org/10.1007/s11069-011-9902-9
Wang M-Z, Amati M, Thomalla F (2012) Understanding the vulnerability of migrants in Shanghai to typhoons. Nat Hazards 60(3):1189–1210. https://doi.org/10.1007/s11069-011-9902-9
Welsh M (2014) Resilience and responsibility: governing uncertainty in a complex world: resilience and responsibility. Geogr J 180(1):15–26. https://doi.org/10.1111/geoj.12012
Whiteford L, Tobin G (2004) Saving lives, destroying livelihoods: emergency evacuation and resettlement policies. In: Castro A, Singer M (eds) Unhealthy health policies: a critical anthropological examination. AltaMira Press, New York, pp 189–202
Wilkinson O (2015) Faith and Resilience after Disaster: The Case of Typhoon Haiyan. Misean Cara. Misean Cara. 2015. http://www.miseancara.ie/wp-content/uploads/2016/02/Faith-Resilience-After-Disaster.pdf
Wisner B, Blaikie P, Cannon T, Davis I (2004) At risk: natural hazards, people's vulnerability and disasters, 2nd edn. Routledge, London
Yee DK, Kim P (2018) Constructing reconstruction, territorializing risk: imposing 'no-build zones' in post-disaster reconstruction in Tacloban City, Philippines. Critical Asian Studies 50(1):103–121. https://doi.org/10.1080/14672715.2017.1407663
Yong C, Tan Y, Luo Y (2017) Post-disaster resettlement and livelihood vulnerability in rural China. Disaster Prev Manage Int J 26(1):65–78. https://doi.org/10.1108/DPM-07-2016-0130
Zou L, Thomalla F (2010) Social vulnerability to coastal hazards in Southeast Asia: a synthesis of research insights." In: Chu TH, Szuster B, Kam S-P, Ismail A, Noble A (eds.), Tropical deltas and coastal zones: food production, communities and environment at the land-water interface. Comprehensive assessment of water management in agriculture series, v. 9. Wallingford, Oxfordshire, UK; Cambridge, MA: CABI

Zulauf A (2012) The impact of landlessness on rural livelihoods after post-Mitch resettlement in Nicaragua. In: Barenstein J, Leemann E (eds.), Post-disaster reconstruction and change, 259–80. CRC Press. https://doi.org/10.1201/b13027

Ladylyn Lim Mangada is an Associate Professor in Political Science from UP Visayas Tacloban and an Affiliated Researcher of the Stockholm Environment Institute.

Chapter 11
Small Island Community Resilience: Building the Capacities of Isolated Communities to Address Climate Risks and Disasters

Jorge S. Ebay, Josephine Tajo-Firmase, Darlene Joy D. Calsado, Lovella Mae M. Magluyan, and Marianito M. Ramirez Jr.

Abstract Small island communities located in the open sea are exposed to different kinds of hydro-meteorological and climatological hazards such as typhoons, monsoon winds, micro-climate anomalies, and dry spells. Isolation from basic services, insecurity over land tenure, economic deprivation, and cultural inferiority exacerbate this exposure. This chapter discusses the public service initiative of the University of the Philippines Visayas (UPV) carried out in partnership with the local communities of Gigantes Islands to mainstream disaster risk reduction and climate change adaptation into local plans and practices by way of community development. Three objectives linked the programmatic and multi-year effort: (1) to expand community and local government knowledge about small island community vulnerabilities and risks; (2) to establish functional links between the islands and the mainland government services; and (3) to enhance local capacities to gain recognition and voice and mainstream small island community concerns into local planning and policy. Local governments and small island communities around the Philippine archipelago may adopt the "Small Island Community Resilience Framework" developed by the UPV team and the good practices and recommendations discussed in this chapter.

Keywords Resilience · Climate change · Disaster risk reduction · Small island communities · Community organizing

Key Points

- Given their size and location, the isolated, fishing-dependent barangays in Gigantes Islands face enormous limitations in addressing disaster and climate

J. S. Ebay (✉) · J. Tajo-Firmase · D. J. D. Calsado · L. M. M. Magluyan · M. M. Ramirez Jr.
University of the Philippines Visayas, Miagao, Philippines
e-mail: jsebay1@up.edu.ph

K. B. Berse et al. (eds.), *Climate Emergency in the Philippines*, Disaster Risk Reduction, https://doi.org/10.1007/978-981-99-7804-5_11

risks. They are exposed to different kinds of hydro-meteorological and climatological hazards.

- Policymakers must address the layers of marginalization or the social, economic, and political characteristics of small island communities that render them vulnerable to climate change impacts and disasters.
- Local governments may adopt the Small Island Community Resilience Framework, which encompasses (a) risk assessment, (b) capacity building, (c) addressing isolation, (d) resource management, (e) community education, and (f) mainstreaming disaster risk reduction and climate change adaptation into local plans and policies.

Summary for Policymakers
On the Climate Emergency Faced by Small Island Communities

The small island communities of Asluman, Gabi, Granada, and Lantangan, in Carles, Iloilo, are highly exposed to climate risks and disasters such as typhoons, monsoon winds, dry spells, and microclimate anomalies that disrupt livelihoods and threaten food sources. Their vulnerability result from a combination of geographic, environmental, economic, social, and political conditions, producing layers of marginalization that overtake their capacities and resources to manage, thus, the need for external support and guidance.

On Climate Change Impacts

A warming trend for the Philippines, as projected by climate models, is expected to alter climate patterns that could inflict more trouble for small islands. In Gigantes' context, this projection could mean more extreme events, stronger typhoons, longer *halakay* , harsher *pugada*, or more acute freshwater supply shortages. Hydrological and meteorological disasters have already inflicted a significant amount of damage on the island before. Without adequate safeguards, erratic temperature and rainfall patterns triggered by the changing climate could worsen future disaster impacts and magnify their erosive effects on people and the ecosystem. Rising sea level and surface temperature are climate change variables that have yet to be monitored in Gigantes.

To build the resilience of small island communities to climate change impacts and disasters, the government needs to address the layers of marginalization or the social conditions that exacerbate their vulnerabilities.

Biophysical and socio-economic factors compound the exposure and vulnerability of small island communities to climate change impacts.

- Biophysical characteristics including insularity, small land area, and physical seclusion, makes small island ecosystems naturally exposed and vulnerable to climate change.

- Unfavorable social, economic, and political conditions of small islands exacerbate exposure and vulnerability due to sectoral exclusion, poverty, and poor governance.

In Gigantes, climate change projections could mean more extreme weather events, stronger typhoons, longer *halakay*, harsher *pugada*, or more acute freshwater supply shortages

To address this, policymakers at the national level must:

- Expand efforts to map and identify small island communities throughout the Philippine archipelago;
- Sustain the science-based information databasing initiatives on small islands to inform planning and policymaking;
- Integrate archipelagic and maritime perspectives into national planning frameworks and guidelines;
- Integrate CCA and DDR into policies and programs that aim to protect and conserve the country's natural resources for sustainable development; and,
- Establish a capital investment program dedicated to isolated and impoverished small island communities to support their adaptive capacities.

Local government must customize risk assessment tools sensitive for small islands to map maritime risks and hazards and safe navigational routes using updated bathymetric maps

To address this, policymakers at the local level must:

- Support the establishment of inter-barangay alliances among communities located in small islands;
- Link small island concerns with the broader subnational development issues and agendas through community education and advocacy;
- Customize risk assessment tools sensitive for small islands to map maritime risks and hazards and safe navigational routes using updated bathymetric maps;
- Build capacities to manage hazards and risks, secure people, livelihood, and resources, and institute sound mechanisms to facilitate effective response when disaster happens;
- Network with and tap intermediary links and facilitating agents such as non-government organizations, the academe, community-based groups, and alliances, among others, to assist in facilitating appropriate development programs for small islands; and,
- Continuously support and update science-based understanding of small islands' biophysical, social, political, and economic conditions.

The Small Island Community Resilience Framework

Based on UPV's experience, the application of the Small Island Community Resilience Framework was pivotal in enhancing insular capacities to address disasters and climate risks. The framework begins with a comprehensive assessment to uncover the layers of marginalization that include sectoral exclusion, poverty, tenurial insecurity, isolation from basic services, and cultural inferiority. The information database generated from the assessments guided the development of relevant programs, projects, activities, and policies to resolve complex marginalization issues. The use of archipelagic, maritime, or ecosystem-based perspectives is vital in developing the necessary sensitivity to respond to small island communities' special needs programmatically.

11.1 Introduction

11.1.1 Disaster and Climate Risks in Small Islands

Tropical Storm Warning Signal Number 3 was raised over their area that evening, so Helen closed her sari-sari store earlier than usual. She took her Noche Buena meal and went to bed after cleaning the dishes. Strong winds and heavy rains came just as she was about to fall asleep. Then the electricity went out. Worried about her newborn piglets, she got out of bed, reached for her flashlight and raincoat, and walked toward her small piggery. She was relieved when she saw them safe. She dodged flying debris as she made her way to her sister's sturdier house nearby to let the storm pass. Several hours later, she received news that the huge, decades-old silk cotton tree crushed her house when it fell into her bedroom, right into her bed. She was thankful that she was not there when it happened.

Rene shuddered as he imagined the pandemonium outside. Over the din of howling winds and pelting rains, he heard an ominous creaking sound, like a massive door being ripped off its hinges. He stood still to listen, but his children's piercing screams grabbed his attention. He rushed towards them and panicked when he saw that strong winds had blown off their bedroom's roof. He quickly ushered them out. As they sought shelter under the kitchen sink, their old cabinet containing cooking utensils crashed to the floor. The kids screamed some more. He tried his best to console them as they shivered in fear and cold. They huddled under the sink until the conditions stabilized. The typhoon finally let up just as the first ray of sunshine broke the next day. None of them was hurt, but their house was a total wreck.

For the second time in the last 6 years, the four barangays of Gigantes Islands—Asluman, Gabi, Granada, and Lantangan—experienced Christmas under extraordinary circumstances.[1] Typhoon Ursula (Phanfone) made its fourth landfall in the area at 2:30 a.m. on 25 December 2019, but blustering winds and torrential rains that upended Christmas Eve celebrations began hours before midnight. The Category 2 typhoon battered the islands for more than 6 hours. It brought pain and despair, not Yuletide cheers, to residents. Sitio Langub in Barangay Granada, where Helen and

[1] In November 2013, Typhoon Yolanda (Haiyan), the strongest typhoon to make landfall in the Philippines, hit Gigantes with almost PhP200 million worth of damages. The islands were within the typhoon's 50-km radius. Yolanda severely devastated natural and built assets and disrupted lives, livelihoods, and community flow. The humanitarian emergency created by the Category 5 storm triggered widespread and extended distress that forced people to forgo Christmas celebrations that year.

Rene[2] reside, suffered tremendous destruction. Ursula wrecked houses across the four barangays, felled powerlines, uprooted trees, and destroyed community infrastructures. The estimated cost of damage reached almost PHP 90 million (Carles MDRRMO 2020).

The inordinate exposure and vulnerability of small islands to climate and disaster risks have been widely investigated and documented. Susceptibility is usually due to a combination of biophysical (e.g., limited land area and natural resources, hazard-prone location) and socio-economic constraints (e.g., poverty and high population densities) (Wong et al. 2005; GFDRR 2018). Research suggests that increasing sea surface temperatures, rising mean sea levels (UNISDR and UNDP 2012; Church et al. 2013), and wave-driven flooding (Storlazzi et al. 2018) have severe implications for the natural capital and assets of island systems. They present dire risks to biodiversity and people's lives and livelihoods (Pelling and Uitto 2001; Jamero et al. 2017; Lenderking et al. 2020). While existing climate change models are not granular enough to objectively predict long-term scenarios, climate-related events and variability (Franta et al. 2016; Kumar et al. 2018) have disrupted the flow of life and services in small island communities in the short-term (Jamero et al. 2019), making safety and resilience programs urgent. Resilience is understood as the "ability of a system, community or society exposed to hazards to resist, absorb, accommodate and recover from the effects of a hazard in a timely and efficient manner" (UNGA 2016). Investing in resilience can help secure people and assets, both natural and built, in small islands.

As an archipelago with more than 7,000 islands, the Philippines can draw insights from the experiences of Small Island Developing States (SIDS) in resilience building. Many small island communities in the country have limited capacities to deal with disasters and climate risks, including everyday risks. In Gigantes, monsoon winds (*amihan* and *habagat*) bring localized weather disturbances like *halakay* (extended sea storms) and *pugada* (squalls) that cause sea accidents, halt fishing activities, and threaten wide-scale hunger with greater frequency. Since the islands are off-grid from the weather bureau's radars, they do not receive adequate warning information about these micro disturbances. They rely mainly on observation and experience to keep themselves safe.

For three consecutive years, from 2017 to 2019, dry spells affected Gigantes Islands from January to May. Rain's prolonged absence caused widespread freshwater supply shortages. To secure water for drinking and other domestic needs, residents walked distances, went to neighboring barangays, or traveled to nearby islands and the mainland. These dry spells coincided with peak tourism months and tourists' demands for freshwater aggravated shortages.

[2] Not their real names.

11.1.2 UPV and the Small Island Community Praxis

Future climate scenarios in the Philippines present more travails for small island communities. Projections predict rising temperatures in the next decades (CCC 2011)[3] that could alter the characteristics of climate risks and disasters (Cruz et al. 2017; Jamero et al. 2017) and further complicate small island living. Placed within the context of Gigantes Islands, these scenarios could bring more extreme events, stronger typhoons, longer *halakay*, harsher *pugada*, or more acute freshwater supply shortages. It is paramount for communities and households on the island to adapt to these challenges to reduce threats.

Small islands share many underlying characteristics. Physically, they have land areas of less than 10,000 square kilometers (Hess 1990; Falkland and Custodio 1991) that translate to resource constraints, heightened by the dwindling natural resources (Lewis 2009; Balzan et al. 2018), chronic poverty, resource dependence, poor market links, and limited sustainable livelihoods options (CCS et al. 2011; Kelman 2015; Duvat et al. 2018). Many small island communities endure the "Isola effect," or the physical isolation that exposes them to different sea- and climate-related disturbances vis-a-vis shortfalls in safe spaces, products, and services (Wong et al. 2005). Social and political conditions, including high population densities, detachment from mainland-based services (Brizmohun 2019), inadequate representation, and low disaster risk reduction and climate change adaptation capacities (Kelman 2015), make the process of achieving small island community resilience difficult.

Research interests on fishing communities in the Philippines tend to focus on resource management and policy development and not on vulnerability and exposure to disasters and climate change. The demand to improve technology (Barut et al. 2004), enhance productivity and economic competitiveness (Carolino and de la Cruz 2013), promote sustainable use of resources (Pomeroy and Carlos 1997), strengthen governance (Ferrer 2016; Ferrer et al. 2018b) and develop cohesive and strategic fisheries and marine policy (Garcia 2005) buoyed these interests. It is crucial to examine marine resource governance and determine its adaptability to the changing climate to guide ecosystem protection and conservation programs toward resilient, inclusive, and sustainable growth (UP CIDS 2016) for small island communities.

The Community Development (CD) faculty of the University of the Philippines Visayas (UPV) began public service work in the Gigantes Islands in 2012 focused on building small island community capacities for disaster risk reduction. UPV's vision to be at the forefront of fisheries and aquatic sciences education, research, and development inspired the initiative. The faculty's desire to fulfill academic obligations and provide students with a platform for practice added further encouragement. Demand for praxis is constant in the field of CD to enrich theory, content, and analysis, and to set standards and expectations for younger scholars to hone their skills as prospective development practitioners.

---

[3] Using a mid-range emissions scenario, the Philippine Atmospheric, Geophysical and Astronomical Services Administration (PAGASA) projects temperature a rise by 0.9–1.1°C in 2020 and 1.8–2.2 °C in 2050 (CCC 2011).

Participatory action research (PAR) served as the primary approach for the multi-year programmatic work. It placed the community at the core of the research process (Denzin and Lincoln 2018), where their inputs and analysis were recognized and valued (Kelman et al. 2011). Community organizing (CO), a social development tool (Batistiana and Murphy 1996), functioned as an overarching strategy for program execution. It highlighted people's experiences and aspirations (Manalili 1990) for sustained collective action (Christens and Speer 2015) and furnished communities with collective voice and power (Schutz and Sandy 2011). Local leaders and community stakeholders stood at the forefront of implementation, steering the process of small island community resilience. Overall, the engagements contended that biophysical and socio-economic factors must be considered in resilience programming for public safety and sustainable small island development.

11.2 Undertaking the Small Island Community Resilience

The local government unit of Carles, a second-class municipality located 146 km northeast of Iloilo City, the province capital, exercises administrative jurisdiction over Gigantes Islands. Its land area, measured at 10,405 hectares, is small compared to its expansive 367,625-hectare municipal waters (Carles MPDO 2015). The municipality's coastline stretches up to 159 km. Carles asserts a territorial claim to almost two-thirds of the Visayan Sea (ICODE and FPE 2015). The town's economy thrives on fisheries and tourism (Fig. 11.1).

The four barangays in Gigantes Islands, Asluman, Gabi, Granada, and Lantangan, are among the 18 island barangays of Carles.[4] Their 2019 combined population reached 16,273 in 3,395 households (see Table 11.1). High birth rates and in-migration bump up the population size. According to reports, migrants compose 20% of the islands' total population (ICODE and FPE 2015). With a small land area of approximately 1,000 hectares, the island's population density is one of the highest within the municipality.

Since 2013, Gigantes Islands emerged as a major tourism draw in Carles due to its natural attractions. White sand beaches, clear blue waters, unique limestone formations, and an anchialine pool (locally known as Tangke) have attracted more than a quarter of a million tourists between 2013–2017 (Ferrer et al. 2018a) and generated millions of revenues. There are endemic species of frog (*Platymantis insulatos*) and lizard (*Gecko gigantes*) (Brown and Alcala 2000) found in crevices of limestone walls and caves. Endemism, a result of the island's bio-geographic isolation (Wong et al. 2005), suggests the urgency of natural resource protection and

[4] Carles has a total of 33 barangays. Fifteen of these barangays, including the Poblacion (town center), are on the mainland (i.e., the bigger island of Panay) and the rest are situated across different island groups including Gigantes, Binuluangan, Calagnaan and Sicogon islands.

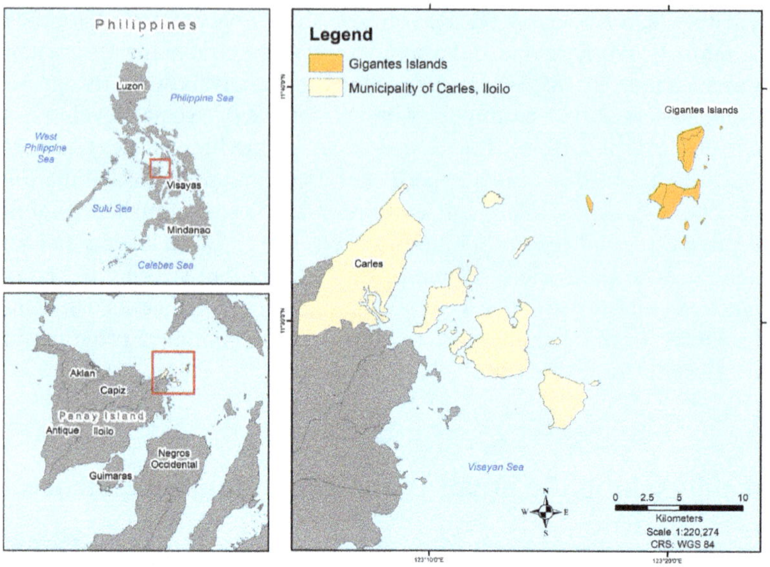

Fig. 11.1 Location of Gigantes Islands and the Municipality of Carles (Copyright: AD Moscoso, UP Visayas)

Table 11.1 Gigantes population data per barangay by sex, household, and family, 2019

Barangay	Male	Female	Total	Households	Families
Asluman	2015	1826	3841	884	910
Gabi	1486	1288	2774	619	670
Granada	2373	2160	4533	790	833
Lantangan	2643	2482	5125	1102	1144
Total	**8517**	**7756**	**16273**	**3395**	**3557**

Source: Barangay records, Asluman, Gabi, Granada & Lantangan, 2019

conservation and adds to its tourism allure. The municipal government and island-based tourism stakeholders positively look at tourism as an income and revenue earner (UPVFI 2019).

Very little information about Gigantes' vulnerabilities, risks, and socio-economic conditions were available before UPV's entry. Qualitative and quantitative assessments established the islands' information database, identifying the simultaneously occuring layers of marginalization enumerated below. The database provided critical insights into the small island ecosystem's exposure and vulnerability critical in roadmapping small island community resilience.

Sectoral exclusion. The assessments noted the non-inclusion of the basic sectors and other marginalized groups in local decision-making bodies. Venues for the participation of small-scale fisherfolks, children and youth, women, the elderly, and persons with disabilities were almost absent (UPVFI 2016). They were not represented in local councils, nor were their concerns sufficiently addressed. The education sector was most visible only during special occasions or celebrations.

Cultural inferiority. People from the bigger island have perceived the people from Gigantes with indifference and sometimes with animosity and denigration. This perception has changed through the years, but in the past, island dwellers have been described as less refined, poorer, and more inferior (e.g., *baho palata* or smell like fish) by their counterparts on the mainland. And since the bigger island has been traditionally understood as the seat of economic, political, and social power, small island communities are seldom perceived as opportunity hubs, creating cultural tension and reinforcing the feeling of differentiation and isolation.

Economic deprivation. The four barangays have been classified as geographically isolated and disadvantaged areas (GIDA). In 2016, the average poverty incidence rate stood at 80%, with the age dependency ratio at 70% (UPVFI 2017). Almost 70% of households were dependent on fishing, shell diving, and other fisheries-related trades. Other identified income sources were mostly informal, like motorcycle driving, charcoal production, upland farming, retail store operation, food vending, cosmetology, general labor, and domestic work. Poverty coexisted with dwindling resources due in large part to extractive and destructive activities.

Tenurial insecurity. With karst-dominated topography and limited land for cultivation, people in Gigantes depended on the surrounding Visayan Sea for sustenance and livelihood. However, competing users and market interests stifled access and control and skewed economic benefits against them. Feudal relations put command over the means of production to the hands of the few, perpetuating inequality. Encroachment of commercial fishers using destructive gear hastened environmental degradation and the decline of fish stocks. Many small-scale fishers engaged in unsafe livelihoods, e.g., dynamite fishing or compressor diving, to survive. With deficient wherewithal to secure well-being, many families endure precarious daily existence. Inadequate safe spaces for habitation force the construction of almost 70% of houses in vulnerable zones. Informal settlers, composed mainly of migrants who live in makeshift homes on the shore, regularly contend with tidal flooding and storm surge (UPVFI 2017).

Isolation from basic services. Distance prevents many small islanders from getting mainland-based services due to cost and physical risk, especially during monsoon seasons. Prohibitive overheads make the provision of satellite government services limited. The fundamental rights of vulnerable and marginalized groups are mostly underserved. It is difficult to avail of health, higher education, and welfare services, including emergency relief, in the islands. Many cases of violence against women and children (VAWC) were unreported and unresolved (UPVFI 2016). Nutrition and health programs are needed to address the prevalence of stunting (at 27.9%), wasting (4.85%), and underweight (14.6%) cases among under-five children (Municipality of Carles 2020). Out-of-school children and youth comprise 46% of the school-age population, while more than 50% of secondary school graduates do not proceed to tertiary schools (UPVFI 2017).

High exposure to disasters and climate risks. The four major typhoons that devastated Gigantes Islands in the last 11 years—Typhoon Frank (2008), Typhoon Yolanda (2013), Typhoon Ruby (2014), and Typhoon Ursula (2019)—demonstrated vulnerability to natural hazards. These typhoons destroyed assets, crippled livelihoods, reversed development gains, and weakened risk reduction initiatives. Natural hazards interact with social, economic, and political conditions, heighten risk exposure, and compound community development challenges. Fishing households on the island also endure debilitating everyday risks triggered by the microclimate phenomena of *halakay* and *pugada*. These anomalies happen more frequently than typhoons and have higher tendencies to disrupt fishing activities, complicate survival woes, present physical dangers, and threaten access to food resources. The most recent episode of *halakay* lasted for more than two weeks in August 2019. Associated with Typhoon Hanna (Lekima), the sea storm halted fishing activities and disconnected people from the mainland. Faced with depleting food stocks, local leaders declared the State of Calamity, which had no immediate use due to the improbable procurement of emergency goods from the mainland. While the LGU Carles was ready to help, its small wooden sea vessels could not get through the big waves. The Regional Disaster Risk Reduction and Management Council (RDRRMC) of Region VI eventually dispatched the Philippine Coast Guard (PCG) to deliver food aid to the islands to mitigate hunger and relieve the anxiety of residents (Yap 2019).[5]

Projected temperature increases could trigger more dry spell episodes and threaten freshwater resources. A 2010 natural resource assessment already revealed the critical state of the islands' freshwater resources, dropping from 0.5 to 1 m (ICODE and FPE 2015) during March, April, and May. Aquifers recharge slowly during dry months, resulting in supply shortages and longer queuing lines in public wells or springs, made graver by tourists' spiking demands.

There are no existing attempts to monitor the sea-level rise and sea surface temperature. However, the Bureau of Fisheries and Aquatic Resources (BFAR) conducts regular red tide monitoring. In 2017, almost 1,500 fishing families were affected by the red tide outbreak (Municipality of Carles 2017).

11.2.1 Resolving the Layers of Marginalization

UPV partnered with the communities, government agencies, and other civil society organizations to promote small island community resilience through a program of action that assembled scientific data, technology, and policy to enhance local

[5] This episode of halakay made the national news but it did not get as much media mileage as the Iloilo-Guimaras boat tragedy that claimed 31 lives. See https://newsinfo.inquirer.net/1151183/final-death-toll-of-3-boat-mishaps-at-iloilo-strait-31#ixzz6OS77OXcW.

capacity. The introduction of RA 10121, the National Disaster Risk Reduction and Management (DRRM) policy in 2012, laid the foundations of this initiative. The program[6] constituted, equipped, and trained four Barangay Disaster Risk Reduction and Management Committees (BDRRMCs) (UPVFI 2014) in disaster preparedness and response. It proved to be very timely. When Typhoon Yolanda struck in 2013, they achieved zero casualties. Local leaders credited this accomplishment to the program.

Typhoon Yolanda proved to be a moment pivotal for UPV's succeeding engagements for three reasons. First, it gave UPV an opportunity to facilitate timely post-disaster response and rehabilitation programs that were markedly different from the islander's previous experiences.[7] Second, it validated the relevance of risk reduction and adaptation work. Local leaders realized that Gigantes' physical and socio-economic conditions are compounding sources of risk that need serious attention. Third, it prompted sectoral mobilization to address marginalization and vulnerability issues. Encapsulated within the Small Island Community Resilience Framework (see Fig. 11.2) and executed through PAR and CO, the program enabled the four barangays to bounce forward post-disaster.

PAR set the small island community resilience into operation by breaking down its components into activities and linking them with desired outcomes (see Table 11.2). CO mobilized participation that rebuilt communities, confronted the layers of marginalization, and empowered local leadership for sustained collective action.[8]

1. Risk Assessments

Activities under this component established the community information database, resolved fundamental data gaps, and contributed to the body of knowledge about Gigantes Islands. Risk assessments used qualitative and quantitative instruments and computer software. Local leaders, sectoral representatives, and students joined the participatory capacity and vulnerability analysis (PCVA), school safety assessment (SSA), and child-centered participatory risk assessment (CCPRA) workshops

---

[6] Approved in 2010, RA 10121, the new policy on DRRM in the Philippines, was unheard of in the islands. UPV was credited as the first organization responsible for promoting risk reduction and climate change adaptation programs in Gigantes.

[7] When Typhoon Frank hit Gigantes in 2008, local officials recounted that they hardly received emergency assistance. Relief goods that reached them more than one month after the typhoon were selectively distributed since not enough was provided. Post- Yolanda, UPV sought Christian Aid's (CA) support to facilitate the rapid response and early recovery program for the four barangays that conform with international standards. The program complemented the humanitarian support provided by other international humanitarian organizations that responded to the emergency.

[8] Three sequential and complementary programs set the small island community resilience agenda into operation: (1) RISE (Bangon) Gigantes Project: Rehabilitation for Island Sustainability and Empowerment supported by Christian Aid (CA); (2) Child-Centered Disaster Risk Reduction Program for Western Visayas supported by UNICEF; and, (3) Building Resilient and Sustainable Small Island Communities supported by the Foundation for the Philippine Environment (FPE). Work in Gigantes Islands also provided inputs to the CA-backed advocacy project entitled Scaling Up Resilience in Governance (SURGE): Inclusive Community-Based Disaster Risk Reduction (ICBDRR) that elevated the small island community resilience concerns to the level of national policy.

Fig. 11.2 Framework for small island community resilience

to fill out disaster timelines, seasonal calendars, hazard maps, stakeholder analyses, institutional diagrams, asset pentagons, and the school safety checklists. Qualitative assessment results complemented quantitative data gathered through an island-wide survey and tabulated through the Barangay Management Information System (BMIS) software. GIS digitized and improved the precision of hand-drawn hazard maps. The assessments identified typhoons, *halakay*, *pugada*, and dry spells as top hazards. Protection of coastal and marine environments and terrestrial and freshwater resources were cited as urgent concerns.

The database underscored the need to broaden the conception of vulnerability and exposure not only as outcomes of natural and physical factors but as manifestations of small island communities' unique social, economic, political, and cultural characteristics. The layers of marginalization and risk profiles of the four communities were drawn from this database. Data guided the community resilience undertaking and provided evidence for the formulation of DRRM plans, early warning systems, Barangay Development Plans, and resource protection and conservation action plans.

Table 11.2 Components, activities, and outcomes of small island community resilience

Component	Activities	Outcomes
Risk assessment	• Participatory capacity and vulnerability analysis (PCVA) • Disaster timeline • Seasonal calendar • Asset pentagon • Institutional analysis • Stakeholder analysis • Hazard mapping • School safety assessment (SSA) • Hot spots assessment (HSA) • Community-wide survey • Barangay Management Information System (BMIS) • Digitization of hazard maps	Body of knowledge about small islands recognizes biophysical and socio-economic riskfactors
		Regular updating of community ecological profiles to guide local planning and policymaking
Addressing isolation	• Installation of duplex communication system • Installation of the Automated Weather System (AWS) • Organization and training of the School Hydrometeorological Information Network (SHINe) • Training on restricted land mobile operation • Formulation and testing of community-based early warning system • Partnership and linkage building through the Island Sustainable Development Alliance (ISDA)	Functionally established the interface and connection of the four barangays with each other and with the mainland LGU functionally established
		Democratized and improved access to warning information on hydro-meteorological has been democratized and improved
		Convergence of the four barangays to unify small island voice to gain decision-making leverage
Capacity building	• Organization and training of Barangay Disaster Risk Reduction and Management Committees (BDRRMCs) • Training on Water Search and Rescue (WASAR) • Training on Evacuation Management (ECM) • Evacuation drills and simulation exercises • Training on Child-Centered Disaster Risk Reduction (CCDRR)	Functional local administrative structures for disaster risk reduction and management (DRRM) and climate change adaptation CCA
		Local DRRM bodies are organized, trained, and equipped
	• Organization of school-based DRRM Councils • Organization of sectoral groups including tourism stakeholder groups • Provision of livelihood support and training for women and men	Local decision-making bodies recognize the participation of sectoral and marginalized groups

(continued)

Table 11.2 (continued)

Component	Activities	Outcomes
Resource management	• Mangrove rehabilitation and reforestation • Formulation of Solid Waste Management Policy • Formulation of the Gigantes Island Unified Marine Protected Area • Drafting of action plans to protect the natural heritage (e.g., caves and endemic species) and built tourism assets (e.g., century-old lighthouse) of the island	Enforced protection mechanisms for the life-supporting natural resources in the islands
		Regulated access and use of tourism assets and resources
Community education	• Documentation of endogenous knowledge on DRRM/CCA • Multi-stakeholder forums on DRR/CCA • Environmental awareness campaigns • School-based activities • Earth Day Celebration • One Gigantes Day • Integration of environmental protection advocacies in barangay fiesta celebrations	Protection and conservation of small island ecosystem received broad support as communities developed small island pride
		Recognition and dissemination of local knowledge on risk reduction and community adaptation
Mainstreaming DRR/CCA	• Grassroots Participatory Budgeting • Formulation of Barangay DRRM Plans • Accreditation of sectoral groups for DRR, CCA, and local development • Integration of gender, children, and youth concerns in the Barangay Annual Investment Programs (AIPs) • Integration of DRR/CCA in the Barangay Annual Investment Programs (AIPs)	Legitimized spaces for sectoral participation in local decision-making bodies
		Integrated disaster risk reduction and climate change adaptation concerns into local development plans and programs

2. Addressing Isolation

This component tackled isolation issues and enhanced access to risk information through technology (duplex radio communication system and the automated weather station) and software (training, procedures, and protocols) solutions. The duplex communication system eased intra- and inter-barangay communication. Before, the four barangays were isolated from the mainland and each other. As detached entities, local officials had no direct means to discuss common concerns regularly. The technology synchronized the four barangays' decisions and actions and channeled key messages for disaster preparedness, public safety, environmental protection, and tourism to the public. It facilitated communication in real-time and connected the island-based constituents with mainland authorities, assuaging perceptions of neglect and exclusion.

The AWS installation resolved off-grid problems and gave the four barangays access to vital micro-climate data. Supervised by teachers, students interpret the data and disseminate these to the four barangays using the duplex communication system. The AWS localized the production and democratized access to weather forecasts and warning information, guiding people to safety as they interact daily with the sea. The School Hydrometeorological Information Network (SHINe), a student-led organization trained by the Philippine Atmospheric Geophysical and Astronomical Services Administration (PAGASA), operates the AWS.

3. Capacity Building

This component developed island-based skills and knowledge that transitioned the four barangays from humanitarian response and rehabilitation to local development. It also highlighted the strategic value of environmental protection and community-based tourism. There were two interrelated streams of capacity building activities. The first stream improved community mechanisms for disaster risk reduction and climate change adaptation. The second stream established platforms for the management and protection of the small island ecosystem. Both strengthened barangay governance structures and processes .

The reconstitution of the Barangay Development Councils (BDC) and the Barangay Disaster Risk Reduction and Management Committees (BDRRMC) was a fundamental reform. Sectoral groups (small-scale fisherfolks, children and youth, women, the elderly and PWDs, tour guides, motorcycle drivers, and resort owners) were organized, trained, accredited, and involved in barangay affairs. They were given seats in the BDCs, allowing them to participate in barangay planning and decision-making. They were likewise involved in the BDRRMCs. The education sector, represented by students and teachers, was also integrated into the reconstituted structures. A training series carried out with the government and civil society partners, upskilled local officials and organized groups, imbuing community governance with accountability and transparency.

To strengthen women's productive capacities, food technologists from UPV gave training on alternative food development using indigenous raw materials. With modest start-up capital, the trainees further ventured into home-based income-generating activities like fish drying, and sari-sari store and food stall operation. Initial gains were sustained through the community-managed savings and credit associations (COMSCA). The positive outcomes of COMSCA encouraged the formation of other groups that included men, particularly motorcycle drivers and tour guides.

Guidance from the provincial government and civil society partners led to the creation of the Island Sustainable Development Alliance (ISDA), a meso-level governance platform that provided a venue to discuss island-wide issues, encourage mutual aid, and secure negotiating leverage with mainland-based authorities. Jointly funded by the four barangays, ISDA promoted convergence, complementation, and working together. It illustrated the benefits of partnership and collaboration on disaster risk reduction, climate change adaptation, marine, and terrestrial resource protection, tourism regulation, solid waste management, health and sanitation, infrastructure improvement, and promotion of small island pride. It facilitated inter-barangay dialogs and strengthened unity with "one voice and one action" to argue for recognition and inclusion. The LGU of Carles recognized ISDA as a model for inter-barangay alliances useful in overcoming insularity and addressing other isolation-related issues.

Overall, capacity building activities enabled the four barangays to confront sectoral exclusion, economic deprivation, and high exposure to disaster and climate risks. The platform also bridged isolation through the help of convergence partners that systematized the delivery of humanitarian and development assistance. Providing opportunities for participation led to the realization that local development gains more meaning and traction if it is relevant, inclusive, and secures people, livelihood, and resources.

4. Resource Management

Efforts to manage and protect the small island ecosystem have two primary objectives: (1) to address tenurial insecurity issues and threats to food resources; and (2) to ensure that the supporting, regulating, provisioning, and cultural services of the ecosystem are maintained to support small island resilience and sustainable development. To accomplish these, the four barangay undertook the following activities: (1) rehabilitation of the mangrove ecosystem and protection of marine resources; (2) greening of upland areas and biodiversity conservation; and (3) protection of heritage assets through tourism regulation. The collective decision to pursue these concerns did not happen overnight or autonomously. It was a product of a protracted process involving extensive dialogs, expert guidance, and island-wide education campaigns.

ISDA tapped convergence partners to pursue the agendas mentioned. To improve existing programs for fisheries resource management, the Zoological Society of London (ZSL), the Northern Iloilo Alliance for Coastal Development (NIACDEV), and experts from UP Visayas provided scientific and technical expertise for mangrove and beach forest rehabilitation and the establishment of the Gigantes Islands Unified Marine Protected Area (UMPA). Reforestation, biodiversity conservation (especially of endemic species), and protection of cave and cave resources received inputs from the Department of Environment and Natural Resources (DENR) and the Western Visayas Caving Association (WVCA). Experts from the Iloilo Science and Technology University (ISATU) assisted in crafting local solid waste management (SWM) policies to promote sanitation, mitigate contamination of freshwater resources, and support community-based tourism. The Foundation for the Philippine Environment (FPE) commissioned specialists' services to assess the island's tourism use-values and jumpstart tourism development planning.

Resource management is a continuing program that requires consistent guidance from the government and other partners. However, the initial gains generated with UPV's facilitation and ISDA's leadership showed that small island communities, with steady support, can take steps to confront their big problems. Modest gains allowed them to address economic deprivation, tenurial insecurity, isolation from basic services, and high exposure to disaster and climate risks.

5. Community Education

Public education and dialog with partners intensified environmental protection campaigns, cultivated small island pride, and patched cultural inferiority. Schools performed crucial roles here as they impart knowledge and influence people's mind-sets and attitudes. DRR and CCA concerns and other issues tackled by ISDA were routinely integrated into school-based activities, especially during the National Disaster Resilience, Nutrition, and Arts months. Along with barangay officials, schools also undertook special celebrations (Earth Day and One Gigantes Day) and cultural festivals (*Tikab* in Asluman, *Wasay-Wasay* in Gabi, and *Palawod* in Granada) that were blended with safety, resilience, and environmental sustainability themes. These activities enhanced public appreciation of ecosystem protection, encouraging them to hold the island way of life in high regard and showcase it to visitors and tourists.

Therefore, community education campaigns became a vehicle for tourism promotion, subsequently pulling Gigantes Islands from obscurity. Eventually, the island became the face of municipal tourism and earned recognition as the plum attraction of Northern Iloilo. A boon to local development, tourism did not only expedite post-disaster recovery and rehabilitation. It also briskly transformed the island.[9] And while they lauded the physical development, residents also maintained their firm commitment to community-based tourism that prioritizes people and environmental welfare over the mass market and commercialism.

By shedding light on the vulnerability and exposure of small island communities that risk reduction and climate change discourse in the Philippines seldom acknowledge, the engagement in Gigantes was recognized as good practice by DRR practitioners and advocates within and outside Iloilo Province. DRR champions from the island shared their experiences in national advocacy forums, argued for the inclusion of small island communities in contemporary discourses, and presented evidence that supported calls to review the national DRRM policy. They also validated the enabling benefit of evidence-based, in-situ risk reduction, and adaptation measures, particularly in generating strong stakeholder buy-in and solidifying community commitment to safety and resilience.

6. Mainstreaming DRR/CCA

The mainstreaming component manifested in (1) the formal participation of sectoral groups in barangay affairs, and (2) the integration of DRR and CCA concerns into

---

[9] As locally financed and managed resort establishments and food shops mushroomed, the Department of Public Works and Highways (DPWH) constructed a circumferential road network in Gigantes Norte, and the National Power Corporation (NAPOCOR) installed higher capacity generators to energize the four barangays fully. Moreover, the Iloilo provincial government, through the Carles LGU, set aside funds to establish a Birthing Clinic in Gabi, and the Department of Science and Technology (DOST) conducted livelihood training and provided a scallop shell crusher technology to convert the ubiquitous middens into usable materials. The Bureau of Fisheries and Aquatic Resources (BFAR) organized the small scale fisherfolks to support marine resource conservation efforts, and the Department of Tourism (DOT) conducted a series of training to hone the skills of tour guides for accreditation.

policy and local development plans and programs. The organization and recognition of ISDA and the accreditation of various sectoral groups proved that participatory development mechanisms work and produce results. Guidance from the Department of Budget and Management (DBM) on the administration of the annual Internal Revenue Allotment (IRA) shares clarified the intricacies of budgeting that enabled the barangay officials to systematize and prioritize allocation through the Annual Investment Plan (AIP). The disposition of barangay funds, including the 20 percent (20%) development fund, 10 percent (10%) SK fund, five percent (5%) children's fund, five percent (5%) DRRM fund, five percent (5%) gender and development fund, and tagging of climate change adaptation-related activities, became more participatory and transparent as a result. Programs, action plans, and policies on fisheries resource protection, biodiversity conservation, and community-based tourism were crafted to guide future decisions and actions.

The mainstreaming process envisaged an inclusive, resilient, and sustainable growth for the island. Granted that it was designed as a capstone of the multi-year programmatic engagements, its underpinnings were embedded in various activities across the five other components discussed. Risk assessments guided program identification and prioritization. Organizing and accreditation democratized decision-making processes and expanded the responsibility to address the simultaneously occurring layers of marginalization. Education and advocacy campaigns recruited broad public support as convergence, through ISDA, established the merits of a united front to gain recognition and voice. Partnership building enabled local leaders to secure external support to address technical and budgetary shortfalls while maintaining community-based processes and autonomy. Essentially, mainstreaming demonstrated the commitment of local leaders and stakeholders to sustain the various programs and initiatives implemented.

11.3 Conclusion

Conventional models and approaches cannot fully capture the complexity of exposure and vulnerability to disasters and climate risks of many small island communities. Many DRR and CCA programs in the Philippines tend to focus on big disaster events and use tools oriented to assess land-based hazards (e.g., flooding and landslides). But as the case of Gigantes Islands illustrates, new perspectives and lenses may be required to comprehend the dynamic conditions of small island communities and the taller order they face to secure lives, livelihoods, and assets. Wong et al. (2005) assert that "small size and isolation in relation to the mainland … [justify the] special and differential treatment" of small island communities.

Many small island communities in the Philippines remain unaccounted in statistics, policies, and plans. Most of the time, they are lumped with communities in big islands despite their distinct biophysical (seclusion, composite ecosystem) and socio-economic attributes (poverty, high population density). Accordingly, their unique problems and needs have not been recognized and remain unaddressed. The multi-year, programmatic engagement of UPV took on this concern using the Small Island Community Resilience Framework.

The framework explicitly addressed the risks and vulnerabilities of households and communities located in detached and isolated small islands. Essentially organic, it was influenced by partners' inputs and shaped by the evolving small island context. It assembled a multi-pronged program of action, based on scientific data, with six inter-linked components to address the layers of marginalization and the dynamic threats to people and assets (both natural and built) to ensure safety and build resilience.

The Small Island Community Resilience Framework identified three fundamental issues affecting small islands' safety and resilience: lack of scientific data, deficient links with mainland service providers, and low insular capacities to tackle natural and anthropogenic risks. UPV used this as a guide to envisage three gradated objectives: (1) to expand community and local government knowledge about small island community vulnerabilities and risks; (2) to establish functional links between the islands and the mainland government services; and (3) to enhance local capacities to gain recognition and voice and mainstream small island community concerns into local planning and policy.

PAR and CO reified the framework into operational components that provided a roadmap for action and implementation. Through PAR, critical information databases were generated and utilized as bases for program and activity design. Science-based understanding of hazards and risks also aided capacity building, community education, action planning, and policymaking. The purposive use of the small island lens uncovered unique issues and conditions, reinforcing calls for their special and differential treatment. CO was instrumental in strengthening community-based structures that reformed barangay governance processes and formalized sectoral support. Through convergence, multiple stakeholders were engaged, thereby securing multi-level cooperation. It empowered the small islanders, giving them the confidence to engage with policy and decision-makers.

UPV's work in Gigantes Islands has been lauded for its success in generating robust support from the community and agency partners. The following processes played out well:

Information databasing. The use of participatory methodology and qualitative and quantitative tools generated critical data that served as bases in crafting mandated local plans, safety protocols, action plans for natural resource protection, advocacy campaigns, and policies. This database enriched the documentation of the small island experience and contributed to the growing body of knowledge about small island communities in Western Visayas.

Capacity development. Organizing and training strengthened barangay governance structures through sectoral participation and upskilling. It collected disparate interests and converted them into a unified voices to gain better leverage in negotiations and better handle in dialogs with policy makers and government authorities at the municipal, subnational, and national levels. The establishment of ISDA provided a horizontal platform for inter-barangay cooperation that put Gigantes on the map of regional agencies, the provincial government, and other development partners.

Bridging isolation. Technology solutions built a communication loop that virtually bridged insularity and isolation. It facilitated intra- and inter-barangay interface and expedited information dissemination. Likewise, it allowed students to produce and reproduce weather data, thus, resolving the inaccessibility of critical warning information created by the off-grid location.

Natural resource management. Action plans for protecting the natural resources sustained the supporting, regulating, provisioning, and cultural services of the small island ecosystem. The use of marine, terrestrial, and tourism resources must be regulated so that they can become reliable post disaster fallbacks. Through natural resource management, traditional food resources are secured, which could assure supply sufficiency during critical times, e.g., during monsoon seasons.

Inclusion. The premium given to sectoral participation, upskilling, local governance reforms, and resource management provided a template for the inclusion of small island community concerns into policies and plans. It also places them within the orbit of government services. Sensitivity to the uniqueness of small island ecosystems can influence prevailing planning and development regimes and correct the preponderant practices that obscure or exclude them.

To replicate the initiatives in Gigantes Islands and broaden the constituency of support for small island community resilience, the following recommendations are presented:

- Application of the archipelagic, maritime, or ecosystems perspectives to comprehend the special and differential dynamics of small islands;
- Development and utilization of risk assessment tools to map maritime risks and hazards or safe navigational routes using updated bathymetric maps;
- Integration of small island DRR and CCA concerns into the national marine policy and frameworks for fisheries resource management, biodiversity conservation, and tourism development;
- Review the criteria for computing the IRA to consider a budgetary allocation to manage municipal waters, especially for fishing-dependent LGUs;
- Manage changing social behavior that values commercialism and consumerism at the expense of sustainable development plans and maintenance of small islands' ecological integrity;
- Network with and tap intermediary links and facilitating agents such as non-government organizations, the academe, and community-based groups and alliances, to share and coordinate responsibility; and,
- Continuously support and update science-based understanding of small islands' biophysical, social, political, and economic conditions.

References

Balzan M, Potschin-Young M, Haines-Young R (2018) Island ecosystem services: insights from a literature review on case-study island ecosystem services and future prospects. Int J Biodivers Sci Ecosyst Serv Manag 14(1):71–90. https://doi.org/10.1080/21513732.2018.1439103

Barut NC, Santos MD, Garces LR (2004) Overview of Philippine marine fisheries. In: In turbulent seas: the status of Philippine marine fisheries by DA-BFAR department of agriculture-bureau of fisheries and aquatic resources. Coastal Resource Management Project of the Department of Environment and Natural Resources, Cebu

Batistiana MB, Murphy D (1996) Rural community organizing in the Philippines. Community Organization Training and Research Advocacy Institute, Quezon City

Brizmohun R (2019) Impact of climate change on food security of small islands: the case of Mauritius. Nat Resour Forum 1–10. https://doi.org/10.1111/1477-8947.12172

Brown R, Alcala A (2000) Geckos, cave frogs and small land-bridge islands in the Visayan Sea, pp 19–22. Haring Ibon.

Carles MPDO (2015) Carles comprehensive land use plan, 2015–2025. Municipality of Carles, Iloilo

Carles MDRRMO (2020) Sitrep #3: municipal-wide declaration of state of calamity due and in response to the devastating effects of Typhoon Ursula (International Name Phanfone), per Sangguniang Bayan Resolution No. 063–2019 dated 26 December 2019. Municipality of Carles, Iloilo

Carolino A, de la Cruz L (2013) Rethinking livelihoods in coastal communities. In: Siason I (ed) Coastal resource management: perspectives from the social sciences. UP Visayas, Department of Agriculture–Bureau of Agricultural Research, Miagao, Iloilo

CCC (2011) National climate change action plan, 2011–2028. Philippines Climate Change Commission, Manila

CCS, MaCEC, SAC-Northern Quezon (2011) A voyage to disaster resilience in small islands: a guide for local leaders. Christian Aid Philippines Country Office, Quezon City

Christens BD, Speer PW (2015) Community organizing: practice, research and policy implications. Soc Issues Policy Rev 9(1):193–222

Church JA, Clark PU, Cazenave A, Gregory JM, Jevrejeva S, Levermann A, Merrifield MA, Milne GA, Nerem RS, Nunn PD, Payne AJ, Pfeffer WT, Stammer D, Unnikrishnan AS (2013) Sea level change. In: Stocker TF, Qin D, Plattner G-K, Tignor M, Allen SK, Boschung J, Nauels A, Xia Y, Bex V, Midgley PM (eds) Climate change 2013: The physical science basis. Contribution of Working Group I to the Fifth Assessment Report of the Intergovernmental Panel on Climate Change. Cambridge University Press, Cambridge, United Kingdom and New York, NY, USA

Cruz RV, Aliño P, Cabrera O, David CP, David L, Lansigan F, Lasco R, Licuanan W, Lorenzo FM, Mamauag S, Peñaflor E, Perez R, Pulhin J, Rollon R, Samson M, Siringan F, Tibig L, Uy N, Villanoy C (2017) 2017 Philippine climate change assessment: impacts, vulnerabilities and adaptation. The Oscar M. Lopez Center for Climate Change Adaptation and Disaster Risk Management Foundation, Inc. and Climate Change Commission, Pasig City

Denzin N, Lincoln Y (eds) (2018) The SAGE handbook on qualitative research, 5th edn. Sage Publications Inc., Los Angeles

Duvat V, Magnan A, Wise R, Hay J, Fazey I, Hinkel J, Stojanovic T, Yamano H, Ballu V (2017) Trajectories of exposure and vulnerability of small islands to climate change. Wiley Interdiscip Rev Clim Chang. https://doi.org/10.1002/wcc.478

Falkland A, Custodio E (1991) Hydrology and water resources of small islands: a practical guide. UNESCO, Paris

Ferrer AJ, Flores MA, Galaura EG, Hopanda J (2018a) Analyzing the recreational use value of Gigantes Islands, Carles, Iloilo. University of the Philippines Visayas Foundation Inc., Miagao, Iloilo

Ferrer AJ, Garces L, Perez M, Pomeroy R (2018b) Inter-local bay alliances in Northern Mindanao, Philippines: experiences and lessons learned in establishment and organisational sustainability. Mar Policy 81–88
Ferrer AJ (2016) Institutions and policies in fisheries management. In: Olewiler N, Francisco H, Ferrer AJ (eds) Marine and coastal ecosystem valuation, institutions, and policy in Southeast Asia. Springer Science + Business Media Singapore, Singapore
Franta B, Roa-Quiaoit HA, Lo D, Narisma G (2016) Climate disasters in the Philippines: a case study of immediate causes and root drivers from Cagayan de Oro, Mindanao and Tropical Storm Sendong/Washi. Belfer Center for Science and International Affairs, Harvard University, Cambridge, Massachusetts
Garcia ML (2005) Progress in the implementation of the Philippines national marine policy: issues and options. United Nations Nippon Foundation Fellows, United Nations. http://www.un.org/depts/los/nippon/unnff_programme_home/fellows_pages/fellows_papers/garcia_0506_philippines.pdf
GFDRR (2018) Disrupt. Communicate. Influence. In: Proceedings from the 2018 understanding risk forum. International Bank for Reconstruction and Development, The World Bank, Washington, DC
Hess A (1990) Overview: sustainable development and environmental management of small islands. In: D'Ayala W, Hein P (eds) Sustainable development and environmental management of small islands. UNESCO, Paris
ICODE and FPE (2015) An island way of life: resource assessment and plan development for Gigantes Islands, Carles, Iloilo. Iloilo CODE NGOs, Iloilo
Jamero ML, Onuki M, Esteban M, Billones-Sensano XK, Tan N, Nellas A, Takagi TH, Thao ND, Valenzuela VP (2017) Small island communities in the Philippines prefer local measures to relocation in response to sea level rise. Nat Clim Change 1-6. https://doi.org/10.1038/NCLIMATE3344
Jamero ML, Onuki M, Esteban M, Chadwick C, Tan N, Valenzuela VP, Chriton R, Avelino JE (2019) In-situ adaptation against climate change can enable relocation of impoverished islands. Mar Policy 108:1–8
Kelman I, Lewis J, Gaillard JC, Mercer J (2011) Participatory action research for dealing with disasters in small islands. Isl Stud J 59–86
Kelman I (2015) Disaster risk governance for Pacific Island communities. Asia-Pac J 1–20
Kumar L, Eliot I, Nunn P, Stul T, McLean R (2018) An indicative index of physical susceptibility of small islands to coastal erosion induced by climate change: an application to the Pacific islands. Geomat, Nat Hazards Risk 9(1):691–702. https://doi.org/10.1080/19475705.2018.1455749
Lenderking HL, Robinson S, Carlson G (2020) Climate change and food security in Caribbean small island developing states: challenges and strategies. Int J Sust Dev World. https://doi.org/10.1080/13504509.2020.1804477
Lewis J (2009) An island characteristic: derivative vulnerabilities to indigenous and exogenous hazards. Shima Int J Res Isl Cult 68–82
Manalili A (1990) Community organizing for people's empowerment. Kapatiran-Kaunlaran Foundation Inc., Manila
Municipality of Carles (2017) Resolution no. 68–2017. Resolution declaring the Municipality of Carles, Iloilo under the State of Calamity in response to an outbreak of red tide, Carles, Iloilo.
Municipality of Carles (2020) Operation Timbang plus 2020. Municipal Health Office, Carles, Iloilo
Pelling M, Uitto J (2001) Small island developing states: natural disaster vulnerability and global change. Environ Hazards 3(2):49–62. https://doi.org/10.3763/ehaz.2001.0306
Pomeroy R, Carlos M (1997) Community-based coastal resource management in the Philippines: a review and evaluation of programs and projects, 1984–1994. Mar Policy 445–464
Schutz A, Sandy MG (2011) Collective action for social change: an introduction to community organizing. Palgrave MacMillan, New York
Storlazzi CD, Gingerich SB, van Dongeren A, Cheriton OM, Swarzenski PW, Quataert E, Voss CI, Field DW, Annamalai H, Piniak GA, McCall R (2018) Most atolls will be uninhabitable

by the mid-21st century because of sea-level rise exacerbating wave-driven flooding. Sci Adv 4:eaap9741. https://doi.org/10.1126/sciadv.aap9741
UNGA (2016) Report of the open-ended intergovernmental expert working group on indicators and terminology relating to disaster risk reduction. United Nations General Assembly, Seventy-first Session, 1 December 2016
UNISDR and UNDP (2012) Disaster risk reduction and climate change adaptation in the Pacific: an institutional and policy analysis, p 76. UNISDR, UNDP, Suva, Fiji
UP CIDS (2016) National marine policy review and strategic direction: review and update of the 1994 national marine policy and the formulation of the national marine strategy. UP Center for Integrative and Development Studies, Quezon City
UPVFI (2014) Risk reduction and capability building program for the Gigantes Group of Islands: terminal report. Community Development Society of the Philippines, Quezon City
UPVFI (2017) RISE (Bangon) Gigantes project: rehabilitation for island sustainability and empowerment: terminal report. University of the Philippines Visayas Foundation Inc., Miagao, Iloilo
UPVFI (2019) Building resilient and sustainable small island communities 2: terminal report. University of the Philippines in the Visayas Foundation Inc., Miagao, Iloilo
UPVFI (2016) Child-centered disaster risk reduction (CCDRR) program for Western Visayas: terminal report. Center for West Visayan Studies, Iloilo City, UP Visayas
Wong PP, Morone E, Lana P, Fortes M (2005) Island systems. In: Hassan R, Scholes R, Ash N (eds) Ecosystems and human well-being: current state and trends, vol 1. Island Press, Washington DC
Yap T (2019) Hanna' leaves tourists stranded, cuts food supply in Northern Iloilo islands. Manila Bulletin. https://mb.com.ph/2019/08/10/hanna-leaves-tourists-stranded-cuts-food-supply-in-northern-iloilo-islands/

Jorge S. Ebay is an Associate Professor of Community Development at the University of the Philippines Visayas, currently pursuing a Ph.D. in Development Studies degree at UP Los Baños. As former Director of the Center for West Visayan Studies, he led the team that undertook public service programs for small islands that earned two UP Padayon Awards for Excellence in Public Service. He specializes on DRR-CCA action research and climate-induced migration in the Philippines.

Josephine Tajo-Firmase is an Assistant Professor of Community Development at the University of the Philippines Visayas. She manages a child-centered DRR project for small islands with support from UNICEF. Her extensive public service track record ranges from community-based DRR-CCA to building resilient local health systems. Prof. Firmase has published her research on DRR and women's studies in social development and social science journals.

Darlene Joy D. Calsado is a community development practitioner and a Research Associate at the Center for West Visayan Studies, University of the Philippines Visayas, currently pursuing a master's degree in Social Work. Her research and advocacy interests are in children and youth rights inclusion and integration in DRR-CCA programs. She applies her development and humanitarian experience in geographically isolated and disadvantaged areas (GIDA) in Western Visayas.

Lovella Mae M. Magluyan found her niche in doing research and working with fishing communities. She is currently the climate change adaptation-disaster risk reduction Field Assistant of Conservation International (CI) Philippines Foundation, Inc. She helps small island communities in Northern Iloilo in implementing green-grey climate adaptation solutions. She is currently pursuing a master's degree in Biology at the University of the Philippines Visayas with research interest in bivalve biodiversity.

Marianito M. Ramirez Jr. is a University Extension Associate of the Community Outreach Program - BIDANI, the public service office of the College of Arts and Sciences of the University of the Philippines Visayas. His research, development and humanitarian work on DRR and CCA were honed by his work with small island communities. He fused this experience with his passion for mountaineering to extend the same services to indigenous groups and hinterland communities within Panay Island.

Chapter 12
Coastal Areas as Safe Zones for Shelter? A Case Study on the Municipality of Daanbantayan, Cebu, as Basis for Coastal Community Vulnerability Assessment and Adaptation Strategies

Liza D. Corro

Abstract Flooding and storm surges are now experienced in coastal communities that were previously believed to be safe havens from such hazards and extreme climate events, manifesting signs that the Philippines is in a state of climate emergency. If nothing is done to address this, apart from the loss of lives and destruction of property from disasters, government resources will be dissipated on recovery and rehabilitation programs rather than the country's other priority areas for development. This chapter examines perceptions of former coastal residents in Daanbantayan, Cebu, who were hard hit by Super Typhoon Haiyan, on the danger of residing in coastal areas, their reasons for relocating, and the different modes of assistance they received to cope with the impact of the super typhoon. This chapter also covers problems encountered by the LGU in relation to the classification of coastal residents whether they are residing in an unsafe or danger zone vis-à-vis the guidelines on the implementation of the government's Emergency Shelter Assistance program. Recommendations are made, using the Building Sustainable and Resilient Coastal Communities Framework, by applying the governance and technological adaptive measures, which can be mainstreamed to address climate change problems.

Keywords Coastal areas · Vulnerability assessment · Adaptation strategies

Key Points

- The Municipality of Daanbantayan is located in the northernmost tip part of the Province of Cebu. From the period of 2008–2019, the three strongest typhoons on record to have hit Daanbantayan, with maximum sustained winds of 150 kph or

L. D. Corro (✉)
University of the Philippines Cebu, Cebu, Philippines
e-mail: ldcorro@up.edu.ph

K. B. Berse et al. (eds.), *Climate Emergency in the Philippines*, Disaster Risk Reduction, https://doi.org/10.1007/978-981-99-7804-5_12

more were Typhoon Fengshen (Frank), Typhoon Haiyan (Yolanda), and the most recent, Typhoon Phanfone (Ursula).
- The Case of Daanbantayan demonstrates the need for the coordinated passage and swift implementation of relevant laws (Comprehensive Land Use Plan, Water Code, Building Code)
- It is just as crucial to simplify these laws for ease of understanding and to educate, train, and build the capacities of LGU and personnel who will lead implementation.
- Community participation and involvement in coastal management plans, designing strategies, and decision-making is critical for climate change action and efforts to work on the ground where they will be applied.

Summary for Policymakers
If we are to prevent the destructive impacts of disasters and climate change on areas such as Daanbantayan, then our climate change action and adaptation strategies must cut across the individual, community, institutional, and systemic levels.

There is an imperative for a coordinated legislation by policymakers and their swift implementation by the different concerned agencies, that directly respond to the needs of affected communities.

Strengthen the Local Government Code to enable the preparation of Comprehensive Land Use Plans (CLUPs) among all local government units (LGUs). There are still many LGUs which have not complied with an approved CLUP, even among highly urbanized cities.

This just shows how constrained the LGUs are in coming up with this very important piece of document, wherein the Climate Change and DRRM adaptation strategies and measures that will address all these climate emergencies are supposed to be integrated.

- The government should recognize the needs of the LGU and must immediately put in place a process beyond what they currently have, because apparently this still does not help the LGUs in the preparation of their CLUPs.
- The national government must mandate, together with funding, the conduct of seminars, trainings, and activities at the provincial level, to make it manageable, where guided actual preparations of CLUPs can be conducted.

Specific provisions on water easements under the Water Code should be reviewed to take into account the kind of hazard, which the classification of land brings, and the nature and purpose of the structure constructed.

Simplify the relevant Water Code for ease of understanding and implementation. There is a need to revisit and simplify the Water Code, which provides the specific water easements, including the varying requirements such as distances depending on the land classification whether it is urban or rural, agricultural or forest.

This can facilitate easier understanding and implementation by those who intend to reside, or are residing along these areas.

- The rationale for the distance requirement should be correlated with the kind of hazard which the classification of land poses, as well as with the nature and purpose of the structure constructed.
- The distance of a house from the seashore or riverbank and other vulnerable areas does not matter as much as the high possibility of a disaster or calamity happening in certain identified areas. This variability should be considered in the setting of distance limitations in the revision of the Water Code, or other similar laws.

There should be community participation and involvement in coastal management planning and decision-making. Policymakers must consider what the community believes to be effective solutions, in order to come up with integrated coastal management plans, like in issues pertaining to unmitigated coastal usage.

The building code must take into consideration the materials and design of structures of buildings in areas which are identified as hazard and danger zones.

- Ordinances must be passed by local governments to ensure safety and resilient materials will be required to be used by residents, in the event that the LGU still cannot pass an ordinance which prohibits coastal residency.
- While waiting for passage of relevant laws and ordinances, there should be a measure to prevent or regulate further occupancy along coastal areas.

Create an inventory of all available technology on DRRM and educate and train LGUs to be able to utilize these technologies. The very purpose for which these technologies were created is for these LGUs to be able to utilize them and help in mitigating climate change emergencies and other calamities.

It is not enough that laws are passed. Legislation and enforcement must be accompanied by appropriate policy actions and programs responsive to the real and context-specific socio-economic and behavioral issues, taking into consideration what really works for the community where these mitigation and adaptation measures are going to be applied. There should be community participation and involvement in coastal management planning and decision-making. Policymakers must consider what the community believes to be effective solutions, in order to come up with integrated coastal management plans, like in issues pertaining to unmitigated coastal usage.

- The issue of uncontrolled coastal occupancy needs to be holistically addressed as it contributes to the vulnerability of coastal communities.
- Expediting the building of sustainable and resilient coastal communities is crucial to mitigating the impacts of climate change.

12.1 Introduction

12.1.1 Climate Change Problem

The belief that they are safe from the impacts of climate change and sea level rise (SLR) still abounds among many in coastal communities, despite the fact that they reside along coastal areas. Perhaps, this can be ascribed to the fact they have resided there for decades, and have never been victims of sea level rise, strong winds brought about by super typhoons, or extreme climate variable events (ECVEs). Being impervious to the impact of climate change and SLR, despite the frequency with which the Philippines is now visited by typhoons and its increasing strength, may be attributed to the fact that residential structures are still visibly proliferating along coastal areas and have never been strictly regulated or restricted by local governments. For coastal residents whose houses were devastated by super typhoons, still, they were able to survive and immediately regain their livelihood, under the government's recovery and rehabilitation programs, together with the consistent support from humanitarian groups and other non-government organizations (NGO).

According to a 2007 Intergovernmental Panel on Climate Change (IPCC) report, about 1.2 billion people across the world live in coastal areas. This was expected to increase to 1.8–5.2 billion by the 2080s with coastal migration and population growth. In the Philippines, about 2000 of the country's more than 7000 islands are occupied and more than half the population (over 64 million Filipinos), live in coastal areas (Licuanan et al. 2014). Among the country's LGUs, 60–70% are located along the country's coasts. As an archipelago, the Philippines does not have a lot of flat inlands where people and economic enterprises can be situated. Majority of the country's poor, particularly fisherfolks and subsistence farmers have chosen to settle down in coastal areas where they could access rich natural resources for their basic food provisioning and livelihoods (Batongbakal, 2002).

The Philippines is noted to have the fourth longest coastline in the world with a length of 36,289 km (Fig. 12.1). The country possesses coastal and nearshore resources of extraordinary extent and diversity. These coastal ecosystems serve as habitat to countless organisms and support an estimated 25% of Philippine fisheries production (Licuanan, 2019). The Philippines supports the world's highest biodiversity of marine shore fishes, with the Visayan Sea as one of the major fishing grounds with its wide and varied marine habitats, according to a Bureau of Fisheries and Aquatic Resources (BFAR) News.

Scientific experts widely agree that the heightened greenhouse gas emissions in the atmosphere result in elevated air and sea temperatures. Consequently, this results in ocean acidification due to the absorption of carbon dioxide. The warming also brings about unpredictable precipitation patterns and hydrological cycles. As a consequence, extreme weather events such as storms and storm surges become more frequent and severe. These adverse effects are seen to worsen in the years ahead, with sea level rise projected to double its rate of change from approximately seven inches (~178mm) (USAID 2009).

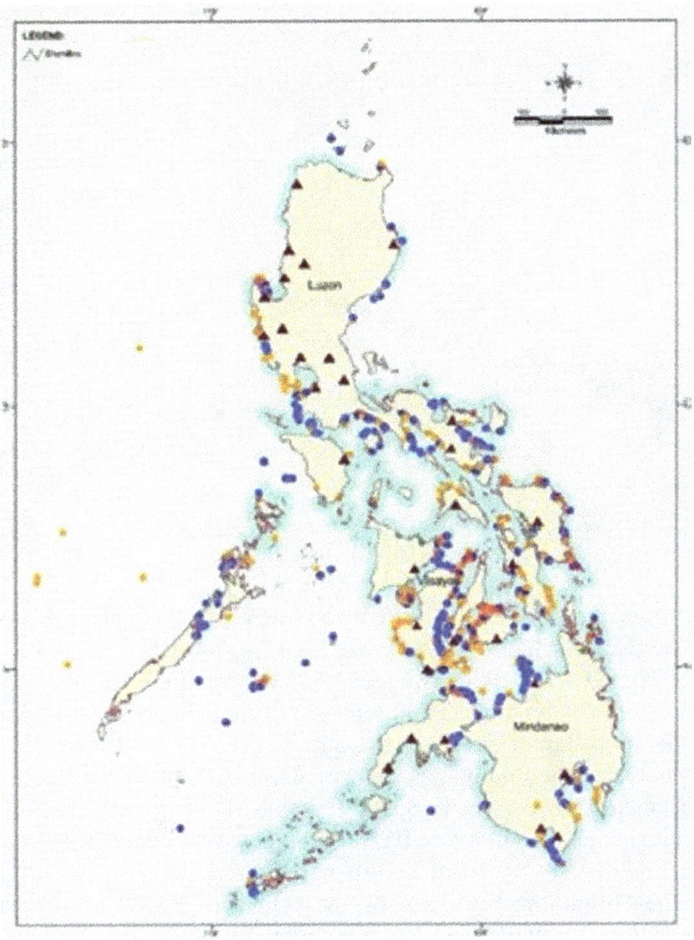

Fig. 12.1 Hot spot map—Philippine Environment Monitor 2005

In a video on "Understanding Sea Level Rise", the National Aeronautics and Space Administration (NASA) mentions that by the end of this century, sea level rise could be between 1 and 6 feet (0.3–1.8m). The alarming rate of consistent increase of the global sea level rise since 1993 is shown in Fig. 12.2.

Rising sea level also plays a role, among many others, in tidal effects, flooding, and saltwater contamination of freshwater streams and aquifers. Annually, about 120 million people globally are exposed to tropical cyclone hazards alone and storms killed 250,000 people between 1980 and 2000 (IPCC, 2007a). In the Philippines, sea-level rise is predicted to be more than 2–3 times more significant than it can occur globally (Licuanan, 2019).

Despite the risks and hazards that climate change brings, coastal areas continue to be congested and migration to this area by residents majority of whom belong

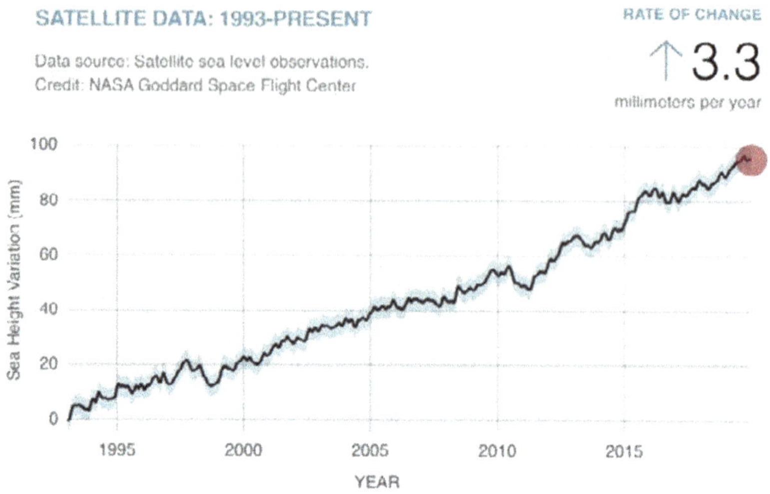

Fig. 12.2 Global sea level rise

to the poor strata continues to rise, without any regard to the danger that is brought about by rising sea level, strong winds and flooding brought about by storms and storm surges. The ECVE and the rising sea level in the country are now getting more frequent and more costly. In the Philippines, out of the 10 deadliest typhoons, 3 of these occurred in the last 10 years, with deaths ranging from 1,268 to 6,300. The 10 costliest typhoons in the Philippines happened within the last 15 years, 6 of which occurred in the last 10 years, with Super Typhoon Haiyan, the highest in terms of cost, causing damage amounting to 95.5 billion pesos (Wikipedia) (Figs. 12.3 and 12.4).

Despite the Philippines being among the countries expected to be hardest hit by climate change, the Philippine Congress allotted 63.6 billion Philippine pesos as budget for climate change in 2019, which is a mere 1.6% of the 3.757-trillion Philippine peso budget for the year 2019. (CNN Philippines, 2019).

The 2030 Sustainable Development Goals (SDGs) of the United Nations envision a safe, resilient, and sustainable community for all of humanity. SDG Goal 11 aims for making cities and human settlements safe, resilient, and inclusive with specific targets to enhance sustainable human settlement planning and management in all countries (SDG 11.3); and to significantly reduce the number of deaths and the number of people affected, and to substantially decrease the direct economic losses caused by disasters, with a focus on protecting the poor and the most vulnerable (SDG 11.5). SDG 2030 aims to substantially increase the number of cities and human settlements adopting and implementing integrated policies and plans towards inclusion, resource efficiency, mitigation, and adaptation to climate change, promoting resilience to disasters, in line with the Sendai Framework for Disaster Risk Reduction 2015–2030, towards a holistic disaster risk management at all levels (SDG 11.b).

Rank	Storm	Season	Fatalities
1	"Haiphong"	1881	20,000
2	Yolanda (Haiyan)	2013	6,300
3	Uring (Thelma)	1991	5,101-8,000
4	Pablo (Bopha)	2012	1,901
5	"Angela"	1867	1,800
6	"Winnie"	2004	1,593
7	"October 1897"	1897	1,500
8	Nitang (Ike)	1984	1,363
9	Reming (Durian)	2006	1,399
10	Frank (Fengshen)	2008	1,371

Fig. 12.3 Deadliest Philippines typhoons

Rank	Storm	Season	Damage	
			PHP	USD
1	Yolanda (Haiyan)	2013	₱95. 5 billion	$ 2.2 billion
2	Odette (Rai)	2021	₱51.8 billion	$ 1.02 billion
3	Pablo (Bopha)	2012	₱43.2 billion	$ 1.06 million
4	Glenda (Rammasun)	2014	₱ 38.6 billion	$ 771 million
5	Ompong (Mangkhut)	2018	₱33. 9 billion	$ 627 million
6	Pepeng (Parma)	2009	₱27. 3 billion	$ 581million
7	Ulysses (Vamco)	2020	₱20. 2 billion	$ 418 million
8	Rolly (Goni)	2020	₱20 billion	$ 369 million
9	Pedring (Nesat)	2011	₱15. 6 billion	$ 356 million
10	Lando (Koppu)	2015	₱14.4 billion	$ 313 million

Fig. 12.4 Costliest Philippines typhoons

This chapter studies the experiences of the local government unit (LGU) of Daanbantayan, a municipality in the Province of Cebu which was greatly devastated by Super Typhoon Haiyan in 2013. The chapter focuses on (1) the residential conditions of the typhoon victims who were provided with resettlement housing and their reasons for conceding to be relocated; (2) the key constraints in its rehabilitation and recovery programs as a coastal community vis-à-vis the sustainability of these programs in the context of climate change emergencies; (3) the implementation of the Emergency Shelter Assistance program and issues encountered on the identification of what is a safe and non-danger zone in the context of coastal residents, as well as the seemingly lacking government plans and policies which exacerbated these issues; (4) the government technology or knowledge that could have been utilized in order to assist efforts on the quick recovery and rehabilitation programs; and lastly, (5) the recommendations are made on key policies for effective integration of Climate Change Act (CCA) and Disaster Risk Reduction and Management (DRRM) initiatives with respect to laws on water easements, land use and building code for resilient housing in order to strengthen the adaptive capacities of coastal communities in coping with climate change impacts.

12.1.2 *Framework for Sustainable and Resilient Coastal Communities*

After Typhoon Haiyan caused great devastation in Daanbantayan, the call was to "build back better" with more resilient buildings and houses, and with this, residents were encouraged to relocate to safer and low-risk areas.

In analyzing Daanbantayan's climate change adaptation (CCA) and disaster risk reduction and management (DRRM) initiatives in its recovery phase of rebuilding its destroyed infrastructure, this chapter is anchored on the adaptation strategies framework adopted by the IPCC. IPCC defines adaptation as "the process of adjustment to actual or expected climate and its effects. In human systems, adaptation seeks to moderate or avoid harm or exploit beneficial opportunities."

For an effective and successful adaptation strategy, there needs to be a joint understanding and cooperation between and among inhabitants of the communities, local authorities, and other key stakeholders, like policymakers, implementing agencies, and even technical experts. There has to be a mutual understanding of the issues relating to climate change, coastal hazards and risks, and the needs of the different participants. Adaptation is a combined and integrated approach. (Briton et al. 2011) (Fig. 12.5).

Adaptation to climate change involves social, cultural, economic, and technical elements in order to build increased sustainable and resilient coastal communities and natural coastal ecosystems. With this collaborative effort, coastal communities and the environment can better cope with and adjust to the potential effects of climate change (Briton et al. 2011). Through proactive planning, policies, strategies, and/or

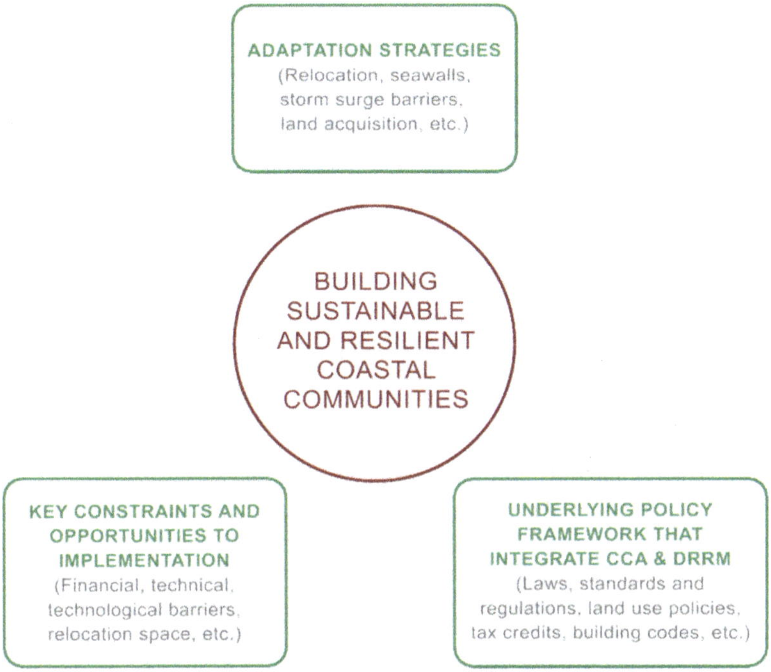

Fig. 12.5 IPCC adaptation strategies

actions are put in place in advance, to avoid or reduce the potential risks from coastal hazards. It is about involving communities and making strategic decisions for the future, similar to the concept of community-based adaptation (CBA) strategies.

At the global and regional levels, the IPCC has recommended some adaptation and mitigation options and responses, as well as the needed interrelationships for sustainable development. The adaptation measures for coastal zones identified under the IPCC 2017 report include coastal afforestation, such as for mangroves and integrated coastal zone management practices.

In the Philippines, the Department of the Interior and Local Government (DILG) has a three-component framework for CCA and DRRM for LGU compliance toward building more resilient and safer communities. The cycle framework includes (1) assessment; (2) capacity enhancement; and (3) enabling access to financing in order to achieve adaptive and disaster-resilient LGUs (Hornilla 2017). The CCA and DRRM are now required to be integrated into each LGU's Comprehensive Land Use Plan (CLUP).

12.2 Research Methodology

Perceptions among Daanbantayan residents on whether there is danger in residing in coastal areas, were elicited from specific sample respondents through structured interviews conducted in 2019, to supplement key informant interviews conducted since 2017, together with review of related literature and use of secondary data, documents, and reports.

Respondents of the survey were residents of relocation sites, particularly the three sites constructed when the Emergency Shelter Assistance (ESA) was ready for distribution. These are the International Red Cross Village, the French Village Habitat for Humanity, and the Unity Village of the Gawad Kalinga relocation centers. The housing beneficiaries in these relocation centers were known to have qualified either due to their low financial income status or as residents in danger or unsafe zones, particularly in coastal areas.

The situation under study was the aftermath of Typhoon Haiyan, how it impacted the community, and how the municipal administrators of Daanbantayan managed the residential communities in coastal areas. These residents were purposely chosen as respondents in this study based on the premise that their being beneficiaries of housing units in relocation centers represented their vulnerable status in 2014–2015, either as financially marginalized or as residents of "unsafe" or "danger zones" after Typhoon Haiyan hit Daanbantayan. Their identity as housing recipients had been processed and validated by the different NGO benefactors as well as by the LGU of Daanbantayan, who were then in charge of the housing distribution. Residents who currently reside in the coastal areas were not included as respondents since they may not have resided in the said area when Typhoon Haiyan had struck. Unlike those residing in the relocation centers, there might have been a difference in their income situation which qualified them as ESA beneficiary.

Questions were asked using the Cebuano language to elicit answers from respondents pertaining to their housing condition before their transfer to the relocation centers, how they regard their condition in terms of its vulnerability for those located in a coastal area, and their varying responses in times of extreme weather events.

Data collection and interviews were made from pertinent government agencies like the Department of Science and Technology (DOST), Department of Environment and Natural Resources—Mines and Geosciences Bureau (DENR-MGB), Cebu Province Planning and Development Office (CPPDO), and a City Planning Development Officer (CPDO) of a highly urbanized city, researchers who did modeling on hazard maps, to assess whether the LGUs of Cebu Province like Daanbantayan has the proper technical assistance or know-how in determining the safe and danger zones in their respective areas. Interviews were also conducted with former and present government officials of Daanbantayan. News and other online sources of secondary data were also used as references.

An ocular inspection was made on various coastal barangays of Daanbantayan, as to the extent of proliferation of human settlements along its seashores, and their proximity to the sea. Some of these photos taken are shown below (Fig. 12.6).

12.2.1 Study Site

The Municipality of Daanbantayan belongs to the Visayas island group, within the Central Visayas Region 7. It is located in the northernmost tip part of the Province of Cebu (Fig. 12.7). It is mainly a coastal municipality, with a total land area of 92.27 km^2 or 35.63 square miles, comprising of 20 barangays, (2 islands, 13 coastal, and 5 inlands.) The municipality's land area constitutes 1.87% of Cebu's total area. It is 128 kilometers from Cebu City and it takes around 3–4 hours of travel time in a private car but between 4 and 6 hours in a public bus transportation. Daanbantayan is bordered on the north by the Visayan Sea, to the west by Bantayan Island, on the east by the Camotes Sea, and on the south by the town of Medellin. It has several small islands/islets which are considered thalassic attractions (1) Chocolate; (2) Monad Shoal; (3) Malapascua; (4) Gato Cave & Islet; (5) Maria; and (6) Carnaza.

One of the conspicuous islets situated in the heart of the Visayas Sea is the Gato Cave & Islet, a small island characterized by its rocky terrain. The captivating ecosystem in this place serves as a habitat for seabirds and flying foxes. It is also home to soft coral canyons and rare nudibranchs.

A seamount with a depth of approximately 20 meters (~66ft) which can be found within the municipality's jurisdiction is called Monad Shoal. It has gained traction from recreational divers due to its thresher sharks, to include common thresher shark (alopias vulpinus) and pelagic thresher shark (alopias pelagicus). These specific types of sharks can go as deep as 350 meters, yet the topography of the shoal allows for their visibility in shallow waters.

Near the vicinity are two other island barangays, Carnaza and Malapascua (also known as Logon Island), which contribute to the captivating landscape of the place, both of which are frequented by tourists for their fine white sandy beaches and other coastal recreational activities.

12.2.2 Population and Economy

Daanbantayan is classified as a first-class municipality. Its population as determined by the 2015 Census was 84,430. This represented 2.87% of the total population of Cebu province or 1.14% of the overall population of the Central Visayas region. Based on these figures, the population density (as of 2015) is computed at 915 inhabitants per square kilometer or 2,370 inhabitants per square mile (Philatlas). It has an average household size of 5.15 based on a 2000 census. Its annual population

Fig. 12.6 Coastal residences in Daanbantayan, Cebu

Fig. 12.7 Map of Daanbantayan, Cebu, Philippines from the Municipal LGU

growth rate from 2010 to 2015 is 2.31%, and it is ranked 769 out of 1,634 local government units.

With its strong foreign and domestic tourism industry in its known island barangay of Malapascua, Daanbantayan envisions itself to be a world-class tourist destination by 2028. Its main source of livelihood in the whole of the municipality, however, is farming and fishing, owing to its mostly agricultural and coastal areas. Going by the Central Visayas average daily basic pay in 2014, these fisherfolks of Daanbantayan earn 199.32 Philippine pesos (more or less US$4.5) daily, while those in farming earn 184.53 Philippine pesos (more or less US$4.16) daily.

According to the Bureau of Local Government Finance, the annual regular revenue of Daanbantayan for the fiscal year of 2016 was 158,572,254.44 Philippine pesos (US$3,338,925.59 equivalent), both from locally generated income and from the Internal Revenue Allotment (IRA) share. As of 2017, the actual IRA share of Daanbantayan annually from the national government is 157,940,183 Philippine pesos (US$3,134,984.05 equivalent). Its poverty incidence based on the PSA survey as of 2012 is 28.5 percent with a lower limit of 23.4 percent and an upper limit of 33.3 percent.

12.2.3 Climate

The monthly number of precipitation days in Daanbantayan is lowest in the month of February and it is highest in the month of July. The average precipitation at its lowest is 56 mm, and its highest is 172 mm. The average of rainy days in February is 12.5 days which is the lowest, and in July is 28.4 days which is the highest. The monsoon creates steady strong winds from the months of December to April, and calm winds from June to October. These Meteoblue provided data are based on thirty years of hourly weather model simulations. The largest known tidal range at Daanbantayan is at 2.48m (8.14 feet), with no available date when this occurred. This study is constrained by a lack of data on the study site's mean sea level rise. But data on the SLR in the Philippines as a whole is projected to increase and exceed the global average SLR. The increase is expected to be approximately 20 cm by the end of the twenty-first century under the worst climate scenario (PAG-ASA, 2018).

The satellite-based observation and projection of the annual mean sea level (MSL) trend in Cebu, Philippines, is increasing and projected to rise at 1.74 mm per year (Mangoang et al. 2019).

12.2.4 Strongest Typhoons Which Hit Daanbantayan

From the period of 2008–2019, the three strongest typhoons on record to have hit Daanbantayan, with maximum sustained winds of 150kph or more were Typhoon Fengshen (Frank), Typhoon Haiyan (Yolanda), and the most recent, Typhoon Phanfone (Ursula). Typhoon Fengshen's date of impact in the Philippines was from June 20 to 23, 2008, causing in the whole Philippines 1,410 deaths and a total damage in the amount of 13.5 billion Philippine pesos. Frank's highest wind was 10-minute sustained at 165 km/h and highest wind speed at 204 km/h. It affected 4,784,634 persons in 6,377 barangays in 419 municipalities in 58 provinces all over the Philippines. It is said that 45,149 families were evacuated to 439 evacuation centers across the country during this Typhoon. In the country, about 105,374 houses were destroyed by Frank's flashfloods and the typhoon's strong winds (Mangaoang et al. 2019).

The sinking of the vessel MV Princess of the Star contributed to the high number of deaths. Daanbantayan also experienced death with the sinking of the vessel, with some of its residents working as its crew. Aside from damage to houses along coastal areas, several barangays in Daanbantayan were flooded with mud during this typhoon, which took weeks, even months to clean up and caused destruction of properties. There is however no available record on the total cost of damages caused to the town of Daanbantayan by typhoon Frank.

The most recent to have hit Daanbantayan is Typhoon Phanfone (Ursula), which hit the Philippines in December 2019. This was a relatively strong and deadly tropical cyclone which traversed through the Philippines on Christmas day. Its total damage caused in the Philippines was 67.2 million dollars (2019 USD) and with the highest winds at 10-minute sustained was 150 km/h (93 miles) per hour, with gusts of 195 kilometers (121 miles) per hour. In Daanbantayan, at least 9,000 people, or 60% of its population was heavily affected by the storm. Destroyed infrastructure and residential buildings, blocked roads, uprooted trees, destroyed crops, and debris prompted the declaration of state of calamity in order to hasten the delivery of services to its constituents. The LGU asked the Department of Social Welfare and Development (DSWD) to deliver 17,000 food packs to the town, of which the priority recipients were the residents of Malapascua and Carnaza islands. "We need food. Aside from food, our constituents also need housing materials. Although it is not like the effect of (super-typhoon) Haiyan, some houses were nonetheless ruined with their roofs ripped by strong winds," according to its Mayor in a news report by Cebu Daily News (2019). The latter also mentioned their difficulties in delivering services to the islets within the jurisdiction of the town because at least 90% of the town's motorized bancas have been destroyed.

In between Typhoons Fengshen and Phanfone, was Typhoon Haiyan, which greatly devastated the Municipality of Daanbantayan in 2013. This was not only the known deadliest and the costliest storm in Philippine history in the last century, but likewise the strongest tropical cyclone on record to ever make landfall.

The devastating impact of Super Typhoon Haiyan was seen across the Philippines, more particularly in the central and eastern Visayas regions last November 8, 2013. The typhoon's strong winds, heavy rainfall, destructive storm surges, widespread landslides, and flash floods resulted in over 6,000 fatalities and left more than 28,000 injured. (NDRRMC, 2013) More specifically in Daanbantayan, the typhoon left 50 individuals injured and 9 lives lost.

It is worth noting that Typhoon Haiyan exceeded the threshold for a Category 5 storm, with a wide span of approximately 600 kilometers and sustained winds reaching 315 kilometers per hour. With such high velocity, it traversed from Guian (Eastern Samar), Tolosa (Leyte), and lastly Daanbantayan (Cebu) before exiting over the South China Sea. Tacloban, in particular, experienced storm surges exceeding five meters in height. Meanwhile, floodwaters were recorded to extend over a half kilometer inland in the city, while low-lying areas to the south reached over two kilometers. (Licuanan et al., 2019.)

The disaster caused by Haiyan had no recorded precedent in the history of Daanbantayan or in the whole of the Philippines. The casualty in this town was low, however, 95% of its residential areas, government infrastructure, and agricultural resources, were destroyed, with damages totaling to almost 1.2 billion Philippine pesos as estimated by its town officials. This was reported in the "Post Disaster Needs Assessment for Super Typhoon Yolanda Hardest Hit Areas in Northern Cebu", as imparted by a former official then of Daanbantayan at the time Typhoon Haiyan hit its town, based on an interview made with the person.

Haiyan left several houses in Daanbantayan either totally destroyed or partially damaged. The Department of Social Welfare and Development (DSWD) came up with the Emergency Shelter Assistance (ESA) program wherein those qualified beneficiary residents whose houses were totally destroyed were given 30,000 Philippine pesos (more or less US$672 equivalent in 2013), while those whose houses were just partially damaged were given 10,000 Philippine pesos (more or less US$224.20 equivalent in 2013). The ESA was designed to speed up the recovery of communities devastated by Haiyan, by helping residents rebuild their houses. These communities were mostly in Western Visayas, Eastern Visayas, northern Cebu, and Coron in Palawan province.

The distribution of this ESA is coursed through the local government unit, with the mayor of the affected LGU taking the helm in the distribution of the ESA. However, controversies emerged surrounding the distribution of the ESA in the town amid claims by the allegedly intended beneficiaries classified as living in "unsafe" and "danger" zones, of their names having been unilaterally taken out from the list of beneficiaries by its mayor and the Municipal Social Welfare and Development (MSWD) representative.

As part of the various relief and recovery assistance which poured into the Philippines, Daanbantayan was a recipient of several and varying kinds of aid, like food packs, clothing, medicines, water bottles, water filtrations, money for work programs, working tools, temporary housing materials of tarpaulin and construction materials, including plywood and galvanized iron sheets for roofing. Humanitarian groups also came over to provide their own labor in clearing the municipality of debris left over by the typhoon. On top of the various aids provided, there were several free housing constructed by various international, national, and local private organizations for qualified residents of Daanbantayan. At the time when Daanbantayan started its ESA distribution, there were three housing resettlements clearly in place with identified beneficiaries. These were the International Red Cross Village, the Habitat French Village by the Habitat for Humanity, and the Unity Village of Gawad Kalinga (Figs. 12.8, 12.9 and 12.10).

There were still several other housing resettlements being planned, under the auspices of the National Housing Authority of the Philippines, but these were not yet then in place, nor were the beneficiaries identified when the ESA funds were being distributed.

The Department of Social Welfare and Development's (DSWD's) Memorandum Circular (MC) No. 24 provided that LGUs shall utilize their CLUP and multi-hazard maps in determining as to who will be provided with ESA, in order to ensure that the area of the resident claiming ESA is safe from any hazard. However, this memorandum circular became a big source of confusion and distraught among residents as well as town and city administrators of a coastal town like Daanbantayan. Despite the municipal administrators' desire to ensure compliance with the guideline, they admitted they did not have the proper technical know-how to determine what is a "danger" and "non-danger" zone for those who resided in coastal areas. Basing heavily on the then-recent experience of Tacloban City in the Eastern Visayas during the onslaught of Typhoon Haiyan, Daanbantayan's administrators treated all its coastal areas as prone to a storm surge, without regard to the water easement rule. The kind of storm surge, which the people in Tacloban City experienced at the height of the onslaught of Haiyan, was not experienced then by residents of Daanbantayan. Nevertheless, the latter thereafter considered residences in coastal areas in its town as unsafe and danger zones, for purposes of the ESA implementation. However, there

Fig. 12.8 The Unity Village of Gawad Kalinga

Fig. 12.9 The International Red Cross Village

Fig. 12.10 The Habitat French Village by the Habitat for Humanity

were residents who would not accept this classification. Daanbantayan, in the aftermath of Haiyan, wanted to build a resilient and safe town, knowing its vulnerability as a coastal town, and considered that the purpose of disqualifying from ESA those who resided in danger or unsafe zones, is to encourage them to transfer to safer places, not to mention that there were several relocation centers and more were still planned for construction.

Residents who survived Haiyan wanted to receive the ESA, considering that their houses were just as devastated by the super typhoon. Even if their residences were situated in a coastal area, for them this did not automatically mean they were residing in a danger zone. Residents argued that they had lived in these areas for decades and had never experienced any damage due to storm surge, and the only damage they experienced were caused by typhoons. These are reflected in the results of the survey conducted.

12.2.5 Survey Results

From a total of 307 respondents from all the three mentioned relocation centers, 176 stated that they previously resided in coastal areas before they transferred to the relocation centers. A total of 182 respondents agreed, with an increase of 6 persons coming from non-danger zone residents who supported the 176 coastal area residents, that coastal areas are danger or unsafe zones.

For those who previously resided in coastal areas among the 175 who responded, their distance from the sea (without distinction if it is high tide or low tide) varied between 1 meter to 100 meters. The highest number were living at a distance of 1–10 meters from the sea consisting of 72 residents, followed by 64 residents living at a distance of 21–40 meters from the sea, and 31 residents living at a distance of 11–20 meters.

Despite the danger of living in their previous houses, 163 respondents said they chose to reside there because they had no other place to live, while 81 of them said it is where they have their livelihood. Eighty six respondents said they are used to living in the area because it is where they grew up having inherited their homes from their parents, while 7 respondents said they were made to live there. Respondents were allowed to choose more than one answer.

Among those who had the perception that they are living in danger zones, the danger they perceived or had actually experienced consisted of natural hazards namely strong winds as the highest number with 126 respondents citing it, followed by flooding with 92 respondents and storm surge with 70 respondents.

These respondents secured themselves whenever there was a typhoon, flooding, or other calamities. A total number of 148 answered they would go and evacuate to the school identified as evacuation center by their barangay or village officials, followed by 14 who would go to their relatives who can accommodate them, and next are 12 respondents who would go to their friends with big houses which can accommodate them and lastly, 5 said they would just remain in their houses, despite the calamity. Even for those who did not believe their previous houses were in a danger zone, they still chose to relocate with 201 respondents agreeing to relocation because they were provided free houses, whereas 178 respondents cited the sturdy materials of the free housing as their reason for agreeing; 171 respondents cited safety from calamities as their reason for transferring to the new relocation site, 81 respondents wanted to have their very own house, 66 respondents lacked money to have their damaged houses fixed, 2 respondents wanted the independence gained from having a new house, while one respondent was asked by relatives to transfer. Respondents were allowed to give several answers to this question (Figs. 12.11, 12.12, 12.13, 12.14, 12.15 and 12.16).

Based on interviews with LGU officials at that time, in order to ensure compliance with ESA guidelines in the distribution of ESA Funds, they admitted to have adopted a "catch all" rule that if a victim's house is located in a coastal area, then it is temporarily declared as an "unsafe/danger" zone. They suspended the distribution of the ESA fund of 30,000 Philippine pesos or 10,000 Philippine pesos subject to a definitive ruling on whether the beneficiary is residing in a danger zone. The LGU requested

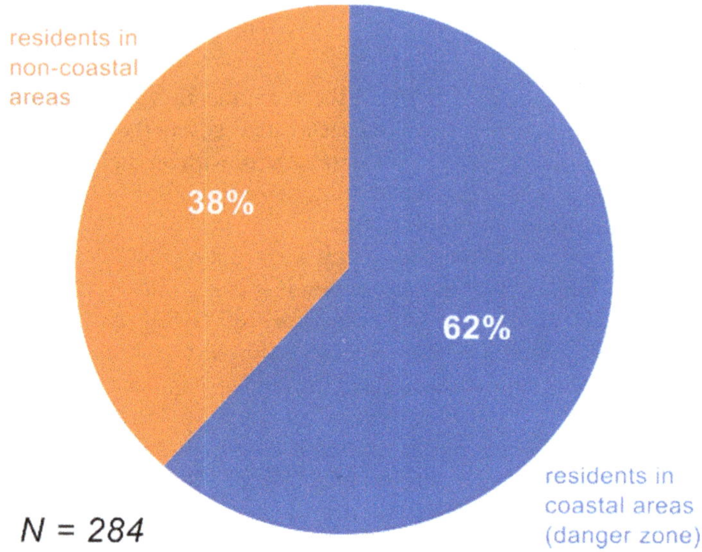

Fig. 12.11 Classification of resident's location

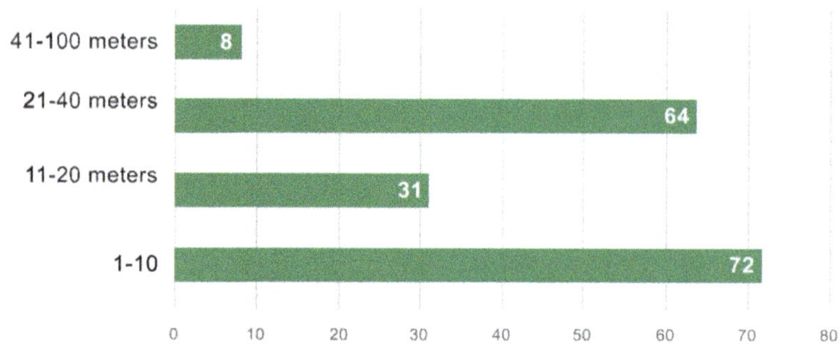

Fig. 12.12 Distance of residences from shoreline

for a technical assistance from the DENR to assist them in assessing whether the coastal area is a danger and unsafe zone, and up to what distance from the sea should a house be for it to be considered as located in a danger zone.

Those residents who still wanted to receive ESA, resisted their classification as residing in unsafe or danger zones and invoked the Hazard map from the DENR Mines and Geosciences Bureau (MGB) to claim that they do not reside in danger and unsafe zones. These maps showed that the only hazard in their areas is rain-induced landslide and rain-induced flooding. A letter from the Director of the DENR-MGB Region seven was used to support this.

Daanbantayan local authorities did not use any CLUP nor the DENR- MGB's hazard maps in determining what is "unsafe area" as stated under DSWD's MC No.

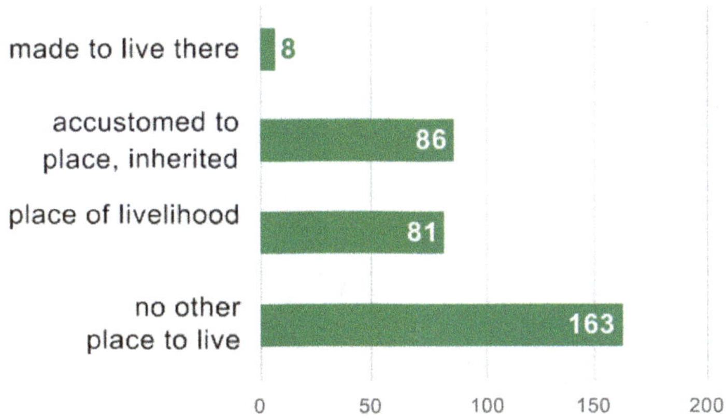

Fig. 12.13 Reasons for living in coastal areas

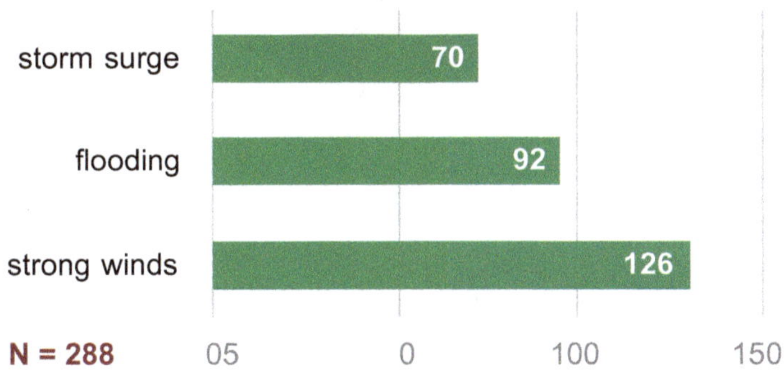

Fig. 12.14 Existing hazards in residents' locations

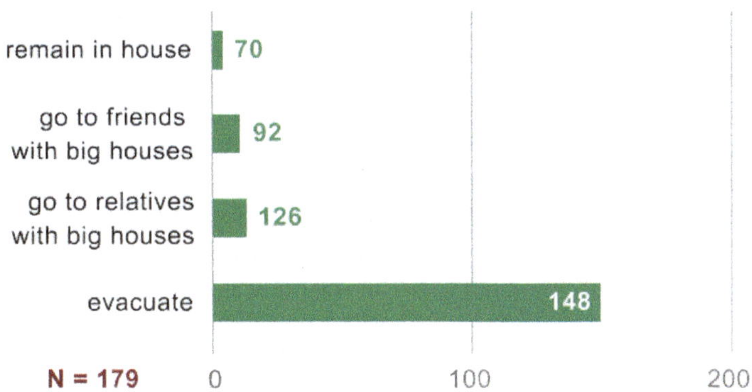

Fig. 12.15 Residents' responses to calamities

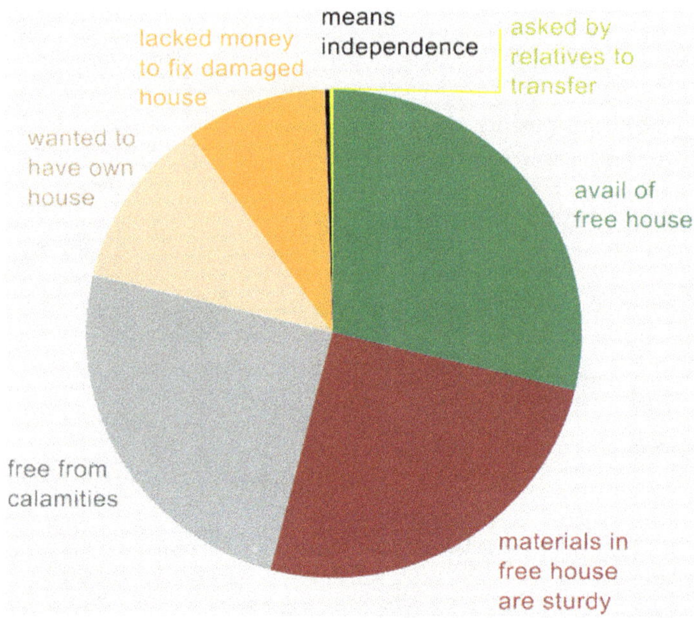

Fig. 12.16 Reasons for availing relocation centers

24. The LGU did not have any approved CLUP, and the only hazard map which it knew to be available only pertained to rain-induced landslide and rain-induced flooding. It did not address storm surge.

To ensure compliance with DSWD MC No. 24 guidelines, the LGU instead adopted the Office of the Presidential Assistant for Rehabilitation and Recovery's (OPARR) 40 Meter No-Build Zone Policy. The LGU adopted an easement of 40 meters from the seashore wherein no structures should be built, and all houses situated within the 40-meter "no build zone" policy along coastal areas, were considered as located within the unsafe or danger zone. The reason this 40-meter buffer zone was adopted by Daanbantayan was to avoid catastrophic damage caused by "storm surges". This was its learning, particularly from the case of Tacloban City.

The ESA was withheld from those households living near the shoreline, those living near river banks and swamps, pending the technical assistance it requested from the DENR in determining the exact unsafe area.

Daanbantayan acknowledged its difficulty in establishing the so-called "unsafe zone" or "no-build zones" due to the absence of technical personnel to undertake such tasks. The LGU nevertheless disbursed its ESA funds to other beneficiaries not residing along the seashore.

Daanbantayan officials then did not reclassify the status of those residing along shorelines as being in unsafe and danger zones, because aside from the fact that Daanbantayan had no CLUP, the MGB's geohazard map presented pertained only to a specific barangay. What was needed was an assessment for the entire Municipality of

Daanbantayan. Likewise, it addressed only rain-induced landslide and rain-induced flooding and did not address storm surge.

However, upon interview with the DENR-MGB for purposes of this study, the latter emphasized that their hazard maps do not include hydro-meteorological concerns, like storm surges, for the latter is already within the jurisdiction of the DOST to produce this kind of multi-hazard maps. An interview with the Regional Director of DOST, Region 7 confirmed the same.

The "Joint DENR-DILG-DND-DPWH-DOST Memorandum Circular No. 2014-01" dated November 05, 2014, addresses not only geohazards but also hydro-meteorological hazards, which was made the basis for the production of multi-hazard maps. The officials then of Daanbantayan were of the belief that there was no multi-hazard map which it could have had used as a reference in determining danger zones. And MGB did not redirect LGU officials to multi-hazard maps produced by DOST. A current reference to DOST's joint project with the University of the Philippines called Project Nationwide Operational Assessment of Hazards (Project NOAH), showed Daanbantayan as one of the towns prone to storm surges, based on the former's study.

As to the CLUP, it was confirmed by both the Municipal Planning and Development Officer of Daanbantayan and of the Province of Cebu that Daanbantayan does not have any CLUP. There was an attempt by Daanbantayan—prior to the assumption of its new municipal officers in 2013—to submit a CLUP, but this was not approved by the Sanguniang Panlalawigan (Provincial Board) of Cebu Province, pending an updated version which incorporates the CCA and DRRM strategies of the municipality. But this was never complied with by Daanbantayan LGU.

12.3 Discussion

Going by the total cost of damage brought about by typhoon Haiyan in 2013, at the national level (in the amount of 95.5 billion Philippine pesos as against the national budget in the amount of 63.6. billion Philippine pesos, allotted in 2019 for national climate change emergencies,) or at the local level, (the total cost of damage wrought by Haiyan in Daanbantayan at 1.2 billion Philippine pesos, as against its 2016 annual regular revenue of 158,572,254.44 Philippine pesos) it can be observed how glaringly insufficient or financially incapable and unprepared the Philippines is, to manage extreme climate variable events. Considering the country's economic, social, and geophysical conditions, there is an urgency to implement mitigating actions and adaptation strategies against all these vulnerabilities and risks, in order to immediately prepare for forthcoming climate change emergencies in the near future. This early, a proactive intervention is urgently needed to reduce or avoid the harm and destruction and other potential but very real risks which had been predicted to result from ECVEs and rising sea level.

Based on the survey results, regardless of the distance of their residence from the sea, Daanbantayan residents recognize they are residing in a danger zone due to the fact that they are living in a coastal area. Even some of those who do not reside within the coastal area, still agree that danger is inherent when residing in a coastal area. Going by these respondents' answers, this is a clear indicator that provisions on water easements should not just be about a specific number in terms of distance or setback from the shoreline as provided under the easement sections of the Water Code.

Sec. 51 provides the required setback as 3 meters in urban areas, 20 meters in agricultural areas, and 40 meters in forest areas. Adding to the confusion in the case of Daanbantayan, was the existence of a 40-meter buffer zone provided by OPARR, which does not reconcile with the Water Code's distance requirements' basis for classification.

Looking into the storm surge experience of Tacloban City, or storm surge scenarios as effects of sea level rise due to climate change, these setback distances cannot protect people, once storm surge and floodings occur. Several bases should be made in declaring coastal areas as a danger zone, not just due to their proximity to the sea. An important consideration should be made on the possibility of danger which might be inflected to the lives and properties of coastal residents based on the quality and purpose of structures constructed along the coastal area, and whether the coastal community is located in an open or sheltered area, in reference to the sea. It is observed among fishermen their practice to transfer their boats, if warned ahead that a typhoon is about to hit their areas, to locations they have ascertained their boats are protected and can withstand strong winds, on the basis of the location of the area and the contour of its shorelines.

Aside from the need to amend the Water Code's easement or setback requirements to make it more responsive to the real condition of the times, LGUs should be more proactive in monitoring and preventing proliferation of structures along areas where residents can be subject to more dangers from the sea, if for now, there is difficulty in providing a relocation center for them. Through this, increase of coastal residences will be minimized if it cannot yet be totally eradicated. The LGU should pass a local ordinance to regulate construction of coastal infrastructure, with consideration on the purpose and nature of the infrastructure, and how vulnerable it is in relation to the location of the coastal area. Local ordinances of this nature are intended to supplement LGU comprehensive land use plan, especially where it is still in the drafting process.

Despite all the perceived and actual dangers experienced, coastal residences continue to thrive. Even in those areas from where beneficiaries of Haiyan housing relocation centers used to reside, it can be observed that houses still exist. A mere ocular inspection of Daanbantayan's coastal areas will easily confirm that these are still fully occupied with residential infrastructures. Knowing that failure to give up a coastal residence will disqualify one from a housing benefit in relocation centers, no admission can be elicited from some of those interviewed, to confirm whether they still continued to maintain their previous residences in the coastal areas. When asked why their coastal residential structures continue to exist, varied reasons were given, some of which are that they are now occupied by their relatives whom they did not authorize, or it is where they kept their fishing gears or even an answer that it is just used for pig's shelter.

It is worth mentioning that the government, humanitarian organizations, and several non-government organizations extended livelihood programs among Haiyan victims in Daanbantayan, aside from the fact that in these relocation centers, the benefactors also provided livelihood centers. Based on the survey, the first two main reasons for residing in coastal areas are "they had no other place to live" and they had gotten accustomed to living there because they were born in those areas. But the third reason is it is where they have their livelihood. Noticeable in the two relocation centers, there are no clear livelihood activities going on in their livelihood centers. In fact, no clear answer can be gathered from Daanbantayan residents to informal queries on what, among livelihood programs extended by donors after Haiyan happened, continued to be sustained in the town. It is a very important observation from this study that even relocating these coastal residents to safer places, and despite the funding and facilities, without a clear and constant support for their livelihood endeavors, it will not guarantee a sustained livelihood for them, whether they are by the coastal areas or not.

Aside from the main reason (that the house is provided for free and that they want to have a house of their own), other reasons cited by residents to relocate include their being free from calamities, the sturdy materials of the house, and lack of money and resources to fix their damaged house. Being secure in their house and their preference for resilient housing is a common denominator among their answers for transferring to the relocation centers. Their common response whenever there is a storm is they rush to transfer to evacuation centers, which are usually public schools, or they stay with friends or relatives who have big houses who can accommodate them. Unique to two of the three relocation centers subject of the study, are the typhoon-resilient design of their house roofs. When interviewed, some of the respondents replied with a tone of pride in their voices, expressing that being far from coastal dangers, they are safe in their houses and are assured from strong winds or super typhoons, knowing they are residing in resiliently designed houses. Calamity funds are usually poured in to procure and provide temporary roofing materials, to typhoon victims whose houses are damaged. Instead of regular distribution of tarpaulins and galvanized iron sheets to replace damaged and blown away roofs each time a typhoon visits, it will be more economical for LGUs to require resilient roofing materials to those who insist living along coastal areas, in the meantime they cannot yet be made to transfer or cannot be provided with safer places away from these coastal areas, or if not, resilient designs must be required for coastal houses, similar to those of the relocation centers. These should be integrated in Building Codes and Office of Building Officials should require this. Furthermore, it is sound planning for an LGU to invest in the construction of an identified safe evacuation center, instead of displacing students when school buildings are converted into evacuation centers, every time there are typhoons and calamities.

One glaring deficiency which surfaced in relation to the Emergency Shelter Assistance fund distribution is the absence of an LGU's CLUP and their lack of technical know-how in establishing what is a danger and unsafe zone. It surfaced in the course of this study that there is a Project NOAH posted then as early as 2014 in the Local Government Academy (LGA) website, wherein Daanbantayan is stated therein as one of the municipalities which is prone to storm surge, and this could have been made as basis in classifying residential areas as within a danger and unsafe zone. Despite this data availability as early as 2014, Daanbantayan was not aware of it, and neither were they advised by any other government agencies at that time, when it was asking for technical assistance. This could have facilitated Daanbantayan's identification of who were living within danger and unsafe zones, which could have expedited the distribution of its ESA funds. There was clearly no coordination and support between the local government, and from the different national government agencies, which could have guided the LGU. It is in instances like this, that there should be proper turn-over and trainings to be made by government agencies, who together with academic institutions, are usually responsible in coming up with technology which can aid LGUs, so these can be fully utilized to aid their targeted beneficiaries in addressing climate change issues.

The CLUP is the required guiding document in relation to DSWD's Memorandum Circular No. 24, for the classification of beneficiaries to qualify if they are residing in safe and non-danger zones. The CLUP is required under the Local Government Code, and its contents are very crucial to the operations of an LGU. Under the DILG's required updated CLUP, it must incorporate CCA and DRRM strategies and initiatives in order to regulate and control physical development of the LGU (Sec. 20, R.A. 7610), it being the plan for management of local territories.

Under Memorandum Circular No. 2019-172, dated October 17, 2019, issued by the DILG, it now mandates compliance with the preparation/ updating of Comprehensive Development Plans (CDPs) and CLUPs by June 2020. Failure to comply will subject the LGUs to an appropriate administrative case filed against them pursuant to Sec. 20 of the Local Government Code. To support the development/updating of the CDP, LGUs are directed to DILG Memorandum Circular No. 2008-156 which provides a guide that puts into operation the concepts and processes enunciated in the Rationalized Planning System and other planning-related policies. Upon formal inquiries made, it turned out that it is not only the Municipality of Daanbantyan which does not have a CLUP, but it is actually the whole Province of Cebu. It is critical to point out the study made by Quitalig et al. in 2016, that capacity building is desperately needed by CLUP planners as LGUs are unable to craft this required document, which consists of three volumes. CLUP planners are not equipped to deal with the issues of CLUP formulation, not just on know- how but also regarding staffing, equipment, and database. The government should provide a packaged technical assistance on the whole cycle of CLUP preparation with corresponding funding to fully capacitate the cities and municipalities. The passage of a Memorandum Circular is not enough. It should be matched with technical, financial, personnel and database support, among others, to make a CLUP a reality for LGUs.

The importance of a CLUP for an LGU cannot be gainsaid, which is why it is provided under the Local Government Code, and the DILG is constantly monitoring its execution by LGUs. By the sheer number of LGUs not having complied with the preparation of their CLUPs, then releasing a memorandum compelling these LGUs to comply is not the solution. The DILG should have been alerted that there is a problem in preparing it, and it should have come up with better strategies on how to assist LGUs, complete with seminars and trainings in order to make their CLUPs a reality.

12.4 Conclusion

If we are to prevent the severe impacts of extreme weather events as exacerbated by climate change impacts and carry out climate action at the scale and speed necessary to meet global climate goals, then this requires everyone to proactively and effectively address socio-economic, behavioral, and technological dimensions. There is a need for effective financial, regulatory, and institutional frameworks, financial support, and significant capacity-building at the individual, community, institutional, and systemic level.

It is not enough that laws are passed. Legislation and enforcement must be coupled with addressing the real and context-specific socio-economic and behavioral issues, that take into full account what really works for the community where these mitigation and adaptation measures are going to be applied. One way this can be done is through the community's participation in strategizing what they believe as effective solutions for them, in order to come up with integrated coastal management plans, like in issues pertaining to unmitigated coastal usage.

The present mode of uncontrolled coastal occupancy, which is clearly hazard-prone, needs to be holistically addressed as integrated within the broader deprivations among vulnerable communities. There is a need for a coordinated passage of laws by policymakers and swift implementation of the same by the different concerned institutions and agencies, which takes into consideration the needs of these affected communities. When we expedite the building of sustainable and resilient coastal communities in this archipelagic country of the Philippines, this clearly leads us in the right pathway of mitigating the impacts of climate change emergencies.

References

Batongbakal J (2002) The crowded shoreline: a review of the Philippines' foreshore and shore land policy. Philipp Law J 76(3):321–368

Climate Change Act, Republic Act No. 9729 https://www.lawphil.net/statutes/repacts/ra2009/ra_9729_2009,html

Compliance of LGUs on the Preparation/Updating of CLUP and CDP https://www.dilg.gov.ph/PDF_File/issuances/memo_circulars/dilg-memocircular-20191017_2987a70ea4.pdf

Corro L (2016) Challenges of the Hyogo framework for action in the disaster relief response and early recovery in a municipality in Cebu, Philippines, in the aftermath of super Typhoon Yolanda (International Name: Haiyan). Public Policy J 14:167–198

Delos Reyes, Mario and Francisco, A., Building Sustainable and Disaster Resilient Informal Settlement Communities, Public Policy Journal Volume 13 (2014–2015), 173–200

Disaster Risk Reduction and Management (DRRM) Act, RA 10121, 2009

Emergency Shelter Assistance Project for Families with Partially and Totally Damages Houses Due to Typhoon "Yolanda" Guidelines for Implementation, MC No. 24, s.2014 https://www.dswd.gov.ph/issuances/MCs/MC_2014-024.pdf
Emergency Shelter Assistance Project for Families with Partially and Totally Damages Houses Due to Typhoon "Yolanda" Guidelines for Implementation, MC No. 24, s.2014 https://www.dswd.gov.ph/issuances/MCs/MC_2014-024.pdf
Hazard Zone Classification in Areas Affected by Typhoon Yolanda https://www.lawphil.net/statutes/repacts/ra2010/ra_10121_2010,html
Joint Memorandum Circular 2014-01, DENR, DILG, DND, DPWH, DOST https://server2.denr.gov.ph/uploads/rmdd/jmc-2014-01denr-dilg-dnd-dpwh-dost.pdf
Kirkby, Patrick, Casey Williams and Saleemul Huq. A Brief Overview of Community Based Adaptation. http://icccad.net/wp-content/uploads/2015/12/A-brief-overview-of-Community-Based-Adaptation.pdf
Licuanan WY, Cabriera RW, Alino PM (2019) Chapter 23 – The Philippines. In: World Seas: An Environmental Evaluation II, 2nd ed., II:515–537. Cambridge, Massachusetts, USA: Academic Press
Local Government Code R.A. 7160 www.lawphil.net/statutes/repacts/ra1991/ra_7160_1991.html
Office of the Presidential Assistant for Relief and Recovery, M.O. No. 62 https://www.officialgazette.gov.ph/2013/12/06/memorandum-order-no-62-s-2013/http://yolanda.neda.gov.ph/yolanda-rehabilitation-and-recovery-efforts
Malenab-Hornilla L (2017) Enhanced comprehensive land use plan and zoning ordinance. https://www.coursehero.com/file/46569623/1-CLUPpdf/
Mangaoang E, Montes HR, Cesar SA, Fiecas KC, p. 16, Impacts of climate change slow-onset events of sea level rise, increasing sea surface temperature, and ocean acidification. https://icsc.ngo/wp-content/uploads/2019/04/VSU_SOE_Final_Web.pdf
Perez R, Amadore LA, Feir RB (1999) Climate change impacts and responses in the philippines coastal sector. Climate Res 12:97–107
Quitalig R, Orale R (2016) Comprehensive land use planning capacity of local government units in Samar Philippines. Countryside Develop Res J 36–55
Rouse HL, Blackett P, Hume TM, Bell RG, Britton R, Dahm J (2013) Coastal adaptation to climate change: a New Zealand story. J Coastal Res 2013:1957–1962
Sales RF (2009) Vulnerability and adaptation of coastal communities to climate variability and sea-level rise: their implications for integrated coastal management in Cavite City. Philipp Ocean Coastal Manag 52(7):395–404
Transforming Our World, The 2030 Agenda for Sustainable Development, United Nations, A/RES/70/1 https://sustainabledevelopment.un.org/content/documents/21252030%20Agenda%20for%20Sustainable%20 Development%20web.pdf
U.S. Agency for International Development (USAID) (2009) Adapting to coastal climate change, a guidebook for development planners, May 2009: https://www.climatelinks.org/sites/default/files/asset/document/PNADO614.pdf, pp. 1–162
Water Code, Presidential Decree No. 1067, 1976 https://www.lawphil.net/statutes/presdecs/pd1976/pd_1067_1976.html

Websites

http://122.54.214.222/population/MunPop.asp?prov=CEB&province=Cebu http://blgf.gov.ph/wp-content/uploads/2019/04/FY-2019-SGLG_20-IRA-Utilization_Municipality-1.pdf
http://daanbantayan.gov.ph/history/ http://daanbantayan.gov.ph/vision-mission/). http://rsso07.psa.gov.ph/poverty http://rsso08.psa.gov.ph/sites/default/files/2006%20and%202009%20SAE%20Report%20FINAL.PDF
http://siteresources.worldbank.org/INTPHILIPPINES/Resources/PEM05- ii-xi.pdf

http://www.ipcc.ch/publications_and_data/ar4/wg2/en/contents.html http://www.ph.undp.org/content/philippines/en/home/countryinfo.html https://cebudailynews.inquirer.net/277524/state-of-calamity-declared-in-daanbantayan medellin#ixzz6OBkIn3sr
https://climate.nasa.gov/vital-signs/sea-level/ https://en.wikipedia.org/wiki/Daanbantayan#Islands_and_thalassic_attractions
https://en.wikipedia.org/wiki/Daanbantayan#Islands_and_thalassic_attractions
https://en.wikipedia.org/wiki/Typhoons_in_the_Philippines https://newsinfo.inquirer.net/737617/p18-4b-in-yolanda-emergency-aid-distributed-late
https://oceanservice.noaa.gov/facts/coastalthreat.html https://oceanservice.noaa.gov/hazards/natural
https://sealevel.nasa.gov/resources/103/video-rising-tides-understanding-sea-level-rise
https://unfccc.int/sites/default/files/resource/Climate_Action_Support_Trends_2019.pdf
https://www.cnnphilippines.com/news/2019/9/21/PH-climate-change-funds-paltry-.html
https://www.dutchwatersector.com/news/typhoon-haiyan-pushed-devastating-5-m-high-storm-
https://www.irishtimes.com/life-and-style/coastal-communities-face-opportunities-as-well-as-threats-1.1906730
https://www.meteoblue.com/en/weather/historyclimate/climatemodelled/daanbantayan_philippines_1716388
https://www.philatlas.com/ https://www.philatlas.com/visayas/r07/cebu/daanbantayan.html/surge-into-tacloban-city-philippines

Liza D. Corro served as the first Chancellor of the University of the Philippines Cebu and taught law courses as a faculty member of the School of Management. She spearheaded disaster mitigation projects of UP Cebu. Her research interests range from disaster risk reduction and management to business incubation and intellectual property management. She headed the UPCEBUinIT office and led the extension center, STiiTCH in UP Cebu. She is an awardee of the UP Gawad Pangulo Award in Excellence.

Part III
Multidisciplinary Responses, Prospects and Challenges

Chapter 13
Fill in the Blanks: Data Voids in Philippine Hazard Maps

Jonnifer R. Sinogaya, Chito L. Patiño, and Melanie Montaño

Abstract Communities are exposed to natural hazards such as flooding and landslides while relevant data collection and risk assessments are insufficient. If nothing is done then communities remain vulnerable to disasters. The Philippines has moved towards accurate mapping of hazards through multi-year programs (DREAM, NOAH, Phil-LiDAR) utilizing advanced surveying equipment and geoscience expertise. While the developed maps successfully provided detailed spatial information to avert disasters, the destructive Naga landslide showed gaps in our hazard information systems—people data. Sparse victim data slowed down search, rescue, and retrieval (SRR) as responders had no location information of buried houses. In the midst of a climate emergency, this chapter asserts that hazard exposure maps combined with demographic data facilitate better planning and improve disaster response and recovery. Policymakers must allocate funds for climate and disaster risk assessments and land use planning. For this purpose, they must provide plantilla items in LGUs and SUCs, ensure capacity-building, and encourage local collaborations for data collection.

Keywords Data voids · Phil-LiDAR 1 · UP Cebu CENVI · Naga landslide · CBMS

Key Points

- The destructive Naga landslide showed gaps in our hazard information systems—people data. Search, rescue, and retrieval (SRR) operation was slow as responders had no location information of buried houses and victim data were sparse.
- A rapid data collection workflow is proposed to empower barangay workers and volunteers to fill hazard maps with demographic information.

J. R. Sinogaya (✉) · C. L. Patiño · M. Montaño
University of the Philippines Cebu, Cebu, Philippines
e-mail: jrsinogaya1@up.edu.ph

K. B. Berse et al. (eds.), *Climate Emergency in the Philippines*, Disaster Risk Reduction, https://doi.org/10.1007/978-981-99-7804-5_13

- The Community-Based Monitoring System (CBMS), in tandem with open-source mobile data collection platforms, provides an opportunity to fill the gap in demographic information which is very essential for disaster risk assessments in the country.
- Policymakers must allocate funds for capacity-building for climate change adaptation and mitigation, ensure plantilla items for this purpose in LGUs and SUCs, and encourage local collaborations between LGUs and SUCs for data collection, climate and disaster risk assessments, and land use planning.

Summary for Policymakers

The problem of data voids in Philippine hazard maps exacerbates climate change vulnerabilities, particularly in flooding and landslides.

- The lack of community data in the immediate aftermath of the Naga landslide delayed the immediate search and rescue of possible survivors. Areas with similar terrain are vulnerable to the same disaster.
- About one-third of river basins nationwide still do not have detailed, probabilistic, and multi-scenario flood hazard maps thus preventing science-informed planning which could prevent the loss of lives and properties during disasters.

About one-third of river basins nationwide still do not have detailed, probabilistic, and multi-scenario flood hazard maps thus preventing science-informed planning which could prevent the loss of lives and properties during disasters.

- Several local government units (LGUs) do not have appropriate technical personnel in local disaster risk reduction and management offices (LDRRMOs) who can facilitate compliance to climate and disaster risk assessment and comprehensive land use planning. Lack of prior assessment and planning tends to expose communities to different hazards like flooding and landslides.

To address this, policymakers at the national level must ensure appropriation for capacity-building of LGUs.

- Provide permanent plantilla items for LDRRMOs.
- Provide funds for state universities and colleges to develop expertise on climate change and hazards research.
- Provide permanent plantilla items for faculty and researchers, infrastructure, and operating expenses for the conduct of climate change and hazards research. The different government agencies (national and local) involved in housing and livelihood provision must be established.

Policymakers at the local level must collaborate with the academe to address gaps in technical knowledge on climate change and its aggravating impacts on natural hazards.

- Allocate grants to the academe to provide technical assistance/consultancy in climate change and disaster risk assessments and land use planning.

- Implement regular demographic surveys on top of mandated censuses.

  Policymakers at the local level must collaborate with the academe to address gaps in technical knowledge on climate change and its aggravating impacts on natural hazards.

13.1 Introduction

The Philippines is ravaged by about 20 typhoons annually. In the span between 2000 and 2016, the Philippines suffered socioeconomic damage of about 20 billion dollars (or yearly average damages of $1.2 billion), over 23,000 deaths, and roughly 125 million people affected because of natural disasters (Jha et al. 2018).

The loss of lives and damage to property caused by natural disasters, particularly flooding, are significant concerns, which need to be addressed in order to break from this kind of tragedy.

A comprehensive assessment of climate and disaster risks provides a good basis for planning by different localities in anticipation of future scenarios. However, many local government units in the Philippines are unable to comprehensively assess their situation in relation to climate risks and population vulnerability due to an apparent lack of information. In several instances also information is available, however, there is a lack of technical expertise and/or desire to integrate the appropriate information. This is an ongoing emergency, as per the International Federation of the Red Cross (IFRC n.d.), since "poorly conducted assessment is likely to lead to poor planning decisions and an inadequate response. This often has consequences beyond the emergency phase and can affect recovery efforts too."

Science-based hazard mapping helps identify communities at risk to hazards such as landslides and flooding. This allows governments to prioritize areas where interventions can be implemented to eliminate or cushion the impact of present and future impacts of climate change, especially to the vulnerable section of disaster-prone localities.

This chapter will discuss the importance of the integration of local data and hazard maps in a climate emergency. This will also review the coverage and limitations of government hazard mapping initiatives, study a climate-smart village initiative, and identify measures, which can be taken to address current data gaps.

13.1.1 *Mapping of Hazards*

The Philippines moved towards accurate mapping of hazards throughout the country through multi-year programs (DREAM, NOAH, Phil-LiDAR, etc.) utilizing advanced surveying equipment and geoscience expertise to produce high-resolution maps of hazards over certain areas in the country.

The Department of Science and Technology (DOST)-funded Nationwide DREAM Program covered 18 major river basins of the Philippines covering a basin

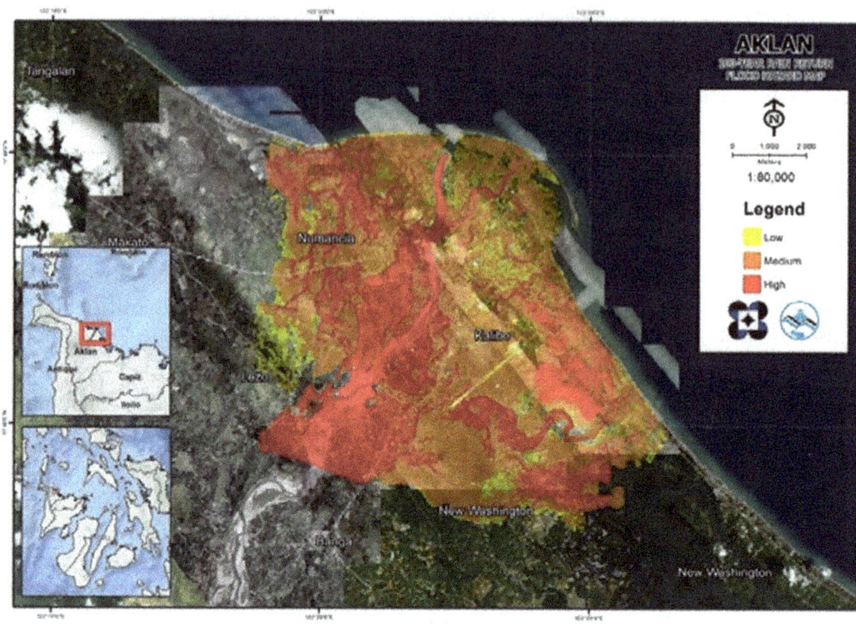

Fig. 13.1 Flood hazard map developed by the UP Cebu Phil-LiDAR 1 project covering the Aklan River Basin

area equivalent to 100,000 km^2. Meanwhile, the Phil-LiDAR 1 program was aimed to cover the other river systems in the country by "tapping state universities and colleges (SUCs) and private higher education institutions (HEIs) nationwide in the processing, validation, and modeling of the acquired LiDAR data (UP Dream 2014)."

On the other hand, Project NOAH was pursued by the DOST to "battle the threats of disasters in the country" using advanced technologies that facilitate warning against floods with a 6-h lead-time to at-risk communities and also to improve existing geohazard vulnerability maps in the country (DOST-PCIEERD 2017).

Samples of the maps produced by the abovementioned projects are shown in the figures (Figs. 13.1, 13.2, 13.3, 13.4 and 13.5).

13.1.2 Limitations

While the successive efforts of government through projects NOAH, DREAM, and LiDAR facilitated flood hazard exposure assessment of river basins through highly accurate mapping of hazards, there are still limitations. Foremost is the limited coverage particularly of the flood hazard mapping of river basins. The DREAM and LiDAR programs modeled a total of 280 (including the 18 principal river basins) or just two-thirds of the 421 river basins in the country (Palmones et al. 2017; Valencia

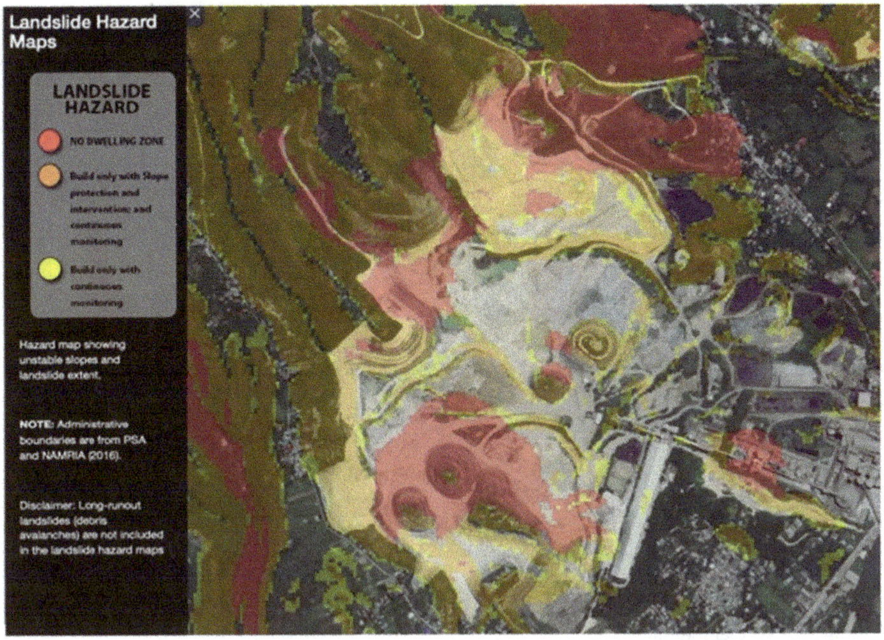

Fig. 13.2 Landslide hazard map culled from the Project NOAH web GIS platform showing the area of the City of Naga where a massive landslide occurred

Fig. 13.3 Storm surge hazard map showing the areas which will be possibly affected by a Storm Surge Advisory 4 level event

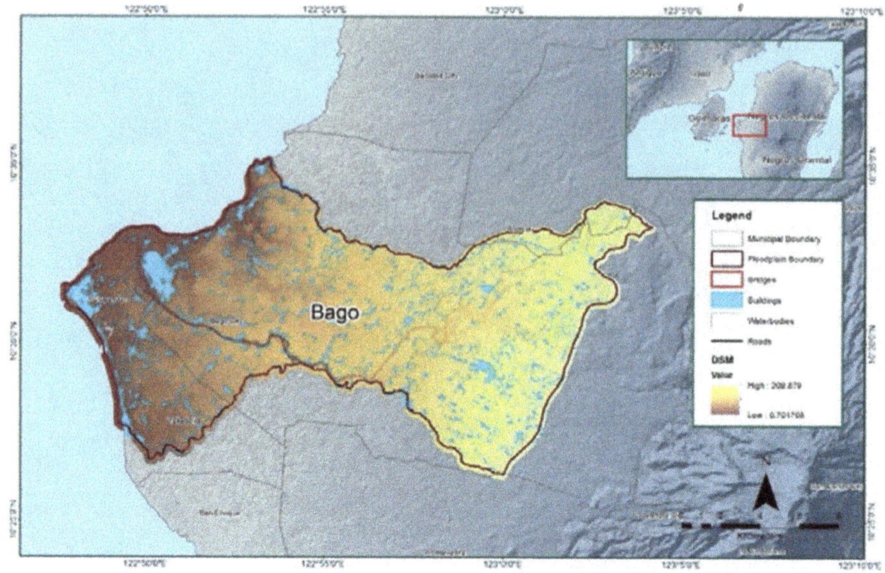

Fig. 13.4 Map of digitized features/structures (colored light blue) located within the Bago River Basin in Negros Occidental

2012). Thus, many cities and municipalities including those with highly dense residential areas still do not have flood models and maps which show multiple scenarios under different precipitation scenarios.

Second is the slow adoption of local government units in utilizing the maps for its risk assessment and development planning. The shortage of personnel in local disaster risk reduction and management offices (LDRRMO) is among the culprits. COA's (2014) findings noted the deficiencies in terms of capacity and "preventive technical DRRM mechanisms" among LGUs. The same COA report shared that only 42% of LGUs located in major river basins have complete rosters for local disaster risk reduction and management councils while only 67% of LDRRMOs have full staffing. Domingo and Manejar (2018) also observed minimal integration of climate and disaster risk management among LGUs with only less than one-third of all LGUs in the country able to accomplish their comprehensive land use plans (CLUP) in 2016. Worse, these CLUPs were said to not reflect yet the inputs from climate and disaster risk assessments (CDRA).

13.1.3 Need for Further Mainstreaming

The lack of technical personnel in local government units who can readily utilize the high-resolution hazard maps prevents the widespread adoption of these tools for disaster preparation, risk reduction, and urban planning. This corresponds to the

Fig. 13.5 Zoomed in sample of area with digitized structures. While the high-resolution maps from Phil-LiDAR have inventoried building structures, there is no info on what is inside these structures. Thus, social-economic analysis is limited

observed general mismatch between mandated responsibilities of local governments and their corresponding DRRM capacities (Commission on Audit 2014; Domingo and Manejar 2018; Domingo 2016).

Local universities are in the best position to have the personnel, which can be trained for hazard mapping, risk assessment, and urban planning. Universities can disseminate knowledge to their communities and local governments. Unlike local governments, universities experience less personnel turnover issues as long as appropriate funding is available.

13.2 Buried Information: Naga Landslide Gaps and Realizations

We take a look at the realizations from the interventions done by the University of the Philippines, through its Center for Environmental Informatics and Firecheck Project, during the search, rescue, and retrieval (SRR) operations in the aftermath of the landslide in the City of Naga, Cebu last September 2018. We make use of the lessons gleaned from the SRR operations to fill gaps/identify solutions to plan for climate and disaster resilience.

The fatal landslide in Naga, Cebu occurred around six in the morning on September 20, 2018. The soil debris spanning 80 hectares and as high as 20 m tall buried 57 houses and around 90 residents of barangays Tinaan and Naalad.

No direct trigger was found to have set off the Naga landslide (Catane et al. 2019). However, it is important to note the history of rain-induced landslides in the Philippines (Indelicato 2015). The hazard potentially puts in peril several lives and properties across the country in the context of more extreme precipitation with climate change.

The expanse of the landslide exposed limitations in SRR operations. The main challenge was to locate the victims buried in meters-high debris with little to no information of their actual locations. The houses became the primary clue to finding buried victims. However, the volume of debris removed visual traces of the structures that were mostly buried deeply causing delay and effectiveness of search and rescue efforts (see Fig. 13.6). With lives at stake, it is crucial for SRR operations to be fast and accurate and the mobilization of resources optimal (Lagmay et al. 2008).

Using LiDAR data to fast-track victim search, rescue, and retrieval.

With the landslide extent (see map in Fig. 13.7) processed and provided by Project DATOS (Briones 2018), UP Cebu CENVI was able to pinpoint the estimated locations of the buried houses. SRR teams from the Armed Forces of the Philippines, Philippine National Police, Bureau of Fire Protection, Office of Civil Defense, Cebu Provincial Disaster Risk Reduction and Management Office, etc., were each briefed on the improved operations zones thereby removing overlaps in areas of responsibilities and allowing targeted excavations of possible victim locations through the house coordinates (see Fig. 13.8).

The Pasig City Search and Rescue Team which was sent from the national capital to provide life-sensing technical guidance to local rescuers was provided coordinates (See Fig. 13.9) in order to fast-track the search for possible survivors underneath the rubbles.

While the house location information culled from LiDAR data allowed faster search operations, the limited data on household prevented SRR teams from excavating new locations faster. Had the number of household occupants been known, SRR teams would know how many individuals to search and when to move to another potential digging point. This further optimizes the operation in terms of time and effort.

Fig. 13.6 Personnel from UP Cebu CENVI and Firecheck join SRR teams to provide estimated locations of houses based on LiDAR data. No visual clues of locations of most houses in ground zero can be found

In the Naga landslide incident, LiDAR showed its utility as a disaster response tool—in fast-tracking SRR operations. However, the technology is more than that as it can facilitate planning to achieve climate-resilient communities.

13.3 The Way Forward

While residences and critical structures have been geotagged, there is no information of who, what, or how many are in the structure.

With data limited to just structures located in flood hazard areas, we cannot answer these questions:

1. How many families/individuals will be/are affected?
2. How many people with special needs (PWDs, senior citizens, pregnant, children, etc.) will be/are affected?
3. How much will implementing a municipal/city resilience program entail given the socioeconomic profile of its people?

Fig. 13.7 CENVI personnel orient search, rescue, and retrieval teams on how the victim location and zoning of operations were done. (Photo by Cebu PDRRMO)

Crampton (2004) notes that the goal should include the identification of relationships among people and physical structures along with their risks and limitations if mapping is to be utilized for governance. Governments can take advantage of the information and solutions, which can be acquired from technologies such as GIS, LiDAR, and other remote sensing products. They benefit more when local information of people and the available resources are precisely mapped—to understand vulnerabilities of the population, to know the underlying hazards, which may endanger social and economic activities, and to find capacities which enable communities to adapt to the challenges of the time including that of climate change. Thus, shifting the mindset from reactive response to disaster into preparedness and adaptation in the context of the climate emergency.

13.3.1 Community-Based Monitoring System (CBMS)

The Community-Based Monitoring System (CBMS) provides an opportunity to fill the gap in demographic information. Disaster risk assessments in the country will be more robust if filled with appropriate socioeconomic information (Aitsi-Selmi et al. 2015; Cohen et al. 2017). Efforts to integrate the CBMS with hazard maps have been sparse among local government units—usually dependent on the level of experience of local governments' DRR and GIS offices/personnel.

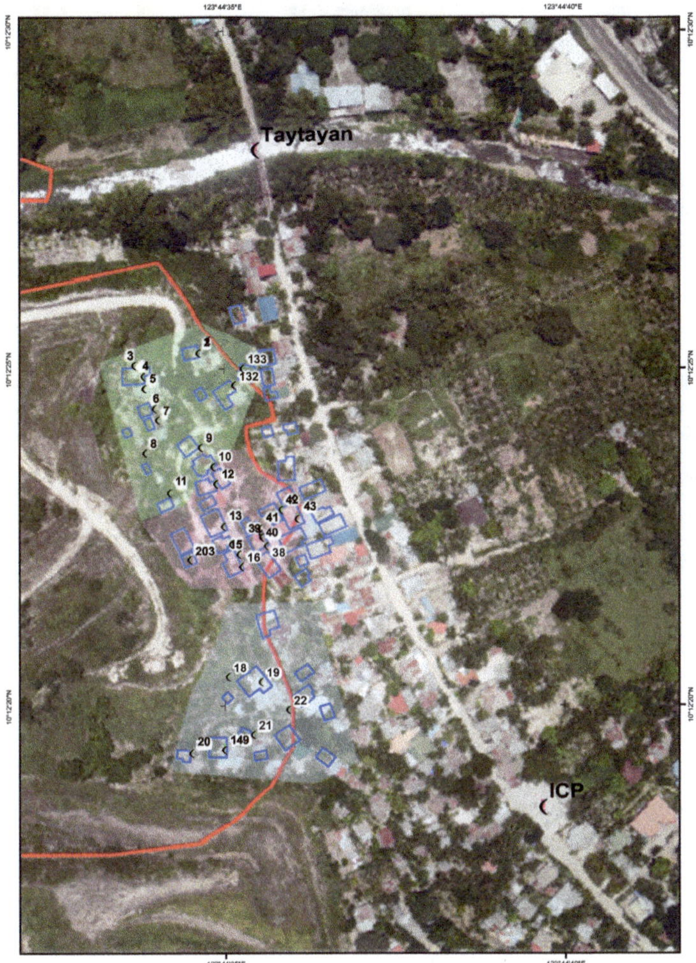

Fig. 13.8 Delineated search, rescue, and retrieval zones after possible victim locations were identified using LiDAR data (Photo by UP Cebu CENVI)

The CBMS has already been in use in the Philippines since 2000 (Partnership for Economic Policy 2019). It is a tool developed in the 1990s to evaluate the impacts of national policies and reforms on the local level. It aims to provide legislators and program implementers a reliable disaggregated data for planning, program creation, and impact monitoring by involving the local community in the data collection. The disaggregated data gives a better face of the local scenario and will allow for targeted and customized solutions in the community.

Just recently, the CBMS Act of 2019, Republic Act No. 11315, was signed to institutionalize community-based monitoring across the country.

Fig. 13.9 Pasig City Search and Rescue Team, assigned to detect survivor signs of life, inputs geotagged locations to their GPS device to speed up their search. (Photo by UP Cebu CENVI)

The Community-Based Monitoring System Act aims to create a CBMS which will improve poverty analysis in the country.

The CBMS is "a technology-based system of collecting, processing, and validating important local data for planning, program implementation, and impact monitoring while empowering communities to be involved in the process." The system will fuse the data collection methods used by national agencies, geo-tagging, and the CBMS implemented by the LGUs.

The new law mandates that the CBMS be implemented regularly (every three years) and simultaneously by all cities and municipalities. Its implementation will be led by the Philippine Statistics Authority (PSA).

The data the CBMS will generate will be the basis for the creation and implementation of poverty alleviation and development programs that respond to the basic needs of the local community.

The CBMS adopted by the LGUs and the National Anti-Poverty Commission (NAPC) uses the following indicators to measure and monitor poverty: health, nutrition, housing, water and sanitation, education, income, employment, and peace and order (National Anti-Poverty Commission n.d.).

While this presents an opportunity for a regularly updated and rich disaggregated data, the current CBMS is limited to population statistics. It is missing an important and timely indicator—disaster preparedness and resilience. This is important as communities with a high incidence rate of poverty tend to suffer more from disasters due to "hazardous location, weak disaster management, and limited coping strategies of households (Ballesteros 2010)."

The new system must then integrate with disaster preparedness and resilience mechanisms.

13.3.2 Using Open Data Kit (ODK) for Up-To-Date Data

The regularity of the CBMS surveys is mandated to be every 3 years. Such can be supplemented by surveys which make use of free and open-source software like the Open Data Kit (ODK) that accommodates the limitations of local government units in terms of budget and skilled personnel. ODK can accept text, image, and location (coordinates) data. It runs on Android platform which is pervasive among Filipino mobile phone owners. Survey questions for CBMS-related data, commercial establishments, and critical infrastructure and resources can be written into the app to digitize the survey process.

The ease of use of ODK allows surveys to be conducted without the need for high technical expertise thus barangay tanods, health workers, and/ or youth volunteers/officials can readily assume data enumeration (See Fig. 13.10 for sample ODK screen). This will speed up the process while expanding the breadth of data that is surveyed. Consequently, this adds more depth of information to existing hazard maps.

13.3.3 Integration with Probabilistic, Climate-Adjusted, Scenario-Based Hazard Maps

Recent collaboration of the UP Cebu Center for Environmental Informatics (CENVI) with the Philippine Augustinians Development Office (PADO) Inc. and Sto. Nino de Cebu Augustinian Social Development Foundation (SNAF) for the Monte Maravilla Augustinian Project in Dumanjug, Cebu provides a good template for integrating community socio-economic profiling and the integration of probabilistic and scenario-based hazard mapping. The Monte Maravilla Augustinian Project is envisioned as a climate-smart complex in a 40-hectare mountainous expanse. The project is planned to include an "integrated farming system, farming adventure camp with a learning site for organic farming, a nature park, farm products display center, and a village of the needy (Augustinian Philippines n.d.)."

The project utilized the hazard information from mandated government agencies found on the HazardHunterPH website, https://hazardhunter.georisk.gov.ph. HazardHunterPH generates hazard assessment reports for specific locations, in this case,

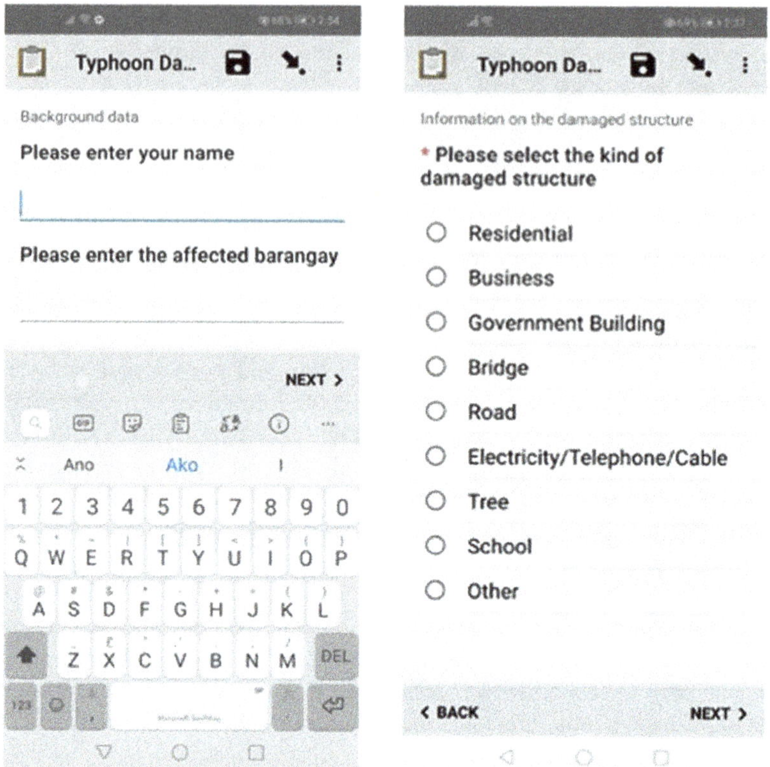

Fig. 13.10 Screenshot of a typhoon damage survey form developed using ODK

Barangay Kang-actol, Dumanjug, Cebu. Table 13.1 gives a short overview of the natural hazards in the area.

The area is safe for most natural hazards based on HazardHunterPH, thus, the project focused on acquiring probabilistic, scenario-based, and high-resolution maps of the flood and rain-induced landslide hazards. Recurrence probability based on precipitation records from PAGASA was computed to get the reference rainfall intensities which may produce flooding in the area. Based on these rainfall recurrence probabilities, 5-year, 25-year, or 100-year flood hazard maps were produced. The landslide hazard maps from Project NOAH were utilized for the landslide layer. The hazard data are joined into one map (see Fig. 13.11 for sample map) and combined into one index for possible development using the rain-induced landslide hazard map of Project NOAH and the flood hazard maps generated by CENVI considering 5-year, 25-year, or 100-year rainfall return periods.

The project then developed a matrix (see Table 13.2) which considers the different combinations of "flood and landslide hazards that could hamper development in the area and under which conditions, specific developments could be made (Pakes and Campomanes 2019)."

Table 13.1 Overview of Natural Hazards in Barangay Kang-actol, Dumanjug, Cebu, based on HazardHunterPH

Seismic hazards	Hazard level
Ground rupture	Safe, Approximately 7.8 km from the Cebu Fault System
Liquefaction	Safe
Earthquake-induced landslide	Safe
Tsunami	Safe
Volcanic hazards	
Nearest active volcano	Kanlaon (51.5 km) Safe
Lahar	Safe
Lava pyroclastic flow	Safe
Permanent danger zone	Outside
Hydro-meteorological hazards	
Flood	Most likely suitable for development, needs further validation
Rain-induced landslide	Moderate Susceptibility; Possible landslide occurrence

Field validation and property development planning in consultation with CENVI staff was done by the Augustinians for their climate-smart village in Dumanjug.

The participation of locals in the validation of the maps and ground truthing gave more depth, engagement, and ownership to the process of establishing a climate-smart village.

The Climate Smart Village project of the Augustinian Province of Cebu shows how a pocket community can adopt disaster risk assessments to identify areas suitable for varying forms of development based on different hazard projections. Areas exposed to hazards can be assessed for different levels of contingencies and/or engineering interventions to ensure a climate risk-resilient locality. The project demonstrates mainstreaming of landscape and contingency planning at the ground level. It deviates from the observed weakness of local governments in terms of availability of technical know-how on hazards projection in attaining climate-smart communities. Collaborations among different sectors including the academe could fill the void in spatial data and expertise.

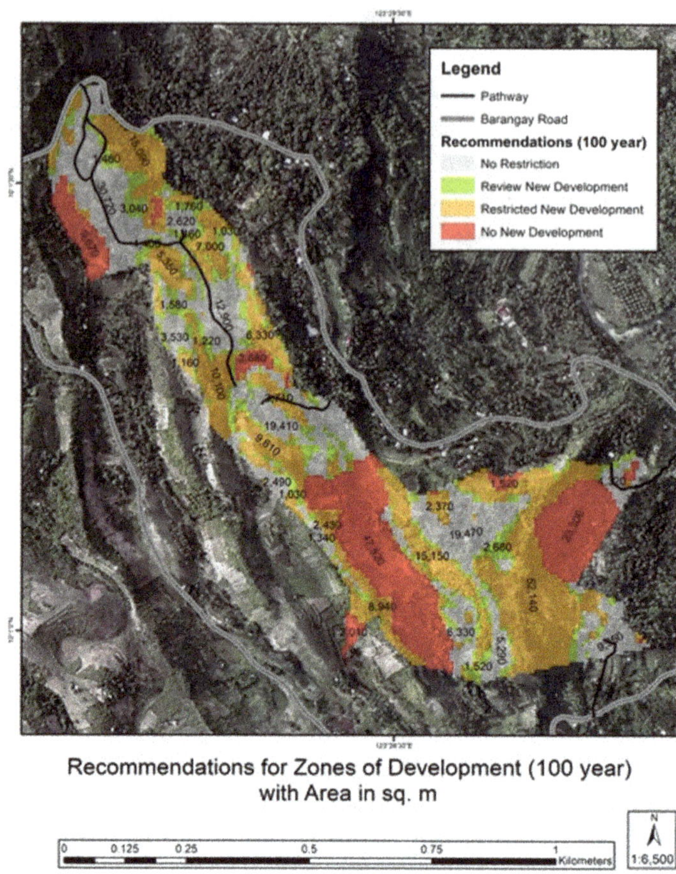

Fig. 13.11 Development suitability map integrating flood and landslide hazards. (Map produced by UP Cebu CENVI)

Table 13.2 Suitability matrix based on landslide and flood hazards

Hazard type	No landslide hazard	Low landslide hazard	Moderate landslide hazard	High landslide hazard
No flood hazard	No restriction	Review new development	Restricted new development	No new development
Low flood hazard	Review new development	Review new development	Restricted new development	No new development
Moderate flood hazard	Restricted new development	Restricted new development	Restricted new development	No new development
High flood hazard	No new development	No new development	No new development	No new development

13.4 Conclusion

Gaps in technical expertise on climate and disaster risk assessments put vulnerable communities at continuous risk as government interventions will be unable to identify the locations and sectors exposed to hazards such as flooding and landslides. CBMS in tandem with open-source mobile data collection platforms provides an opportunity to fill the gap in demographic information which is a very essential for disaster risk assessments in the country. More effort is needed to integrate the demographic and resource information with hazard maps of cities and municipalities. This can be done by continuously capacitating local governments' DRR and GIS offices/ personnel with the necessary technical knowledge while addressing the problem of personnel tenure and compensation in the long term. The academe, meanwhile, with the appropriate funding, can facilitate the knowledge transfer and data upkeep to support the requirements of local government units.

References

Aitsi-Selmi A, Egawa S, Sasaki H, Wannous C, Murray V (2015) The sendai framework for disaster risk reduction: renewing the global commitment to people's resilience, health, and well-being. Int J Disaster Risk Sci 6(2):164–176. https://doi.org/10.1007/s13753-015-0050-9

Ballesteros M (2010) Linking poverty and the environment: evidence from slums in philippines cities. In: PIDS Discussion Paper Series, No. 2010-33

Briones JB (2018) Bridging social media and geospatial technologies during disasters. Flip Science October 2018. https://www.flipscience.ph/news/features-news/bridging-social-media-and-geospatial-technologies-during-disasters/

Catane SG, Nathan AS, Veracruz JR, Flora R, Chatty MM, Go RE, Enrera, and Erizza Rose U. Santos. (2019) Mechanism of a low-angle translational block slide: evidence from the september 2018 Naga Landslide, Philippines. Landslides 16(9):1709–1719. https://doi.org/10.1007/s10346-019-01212-9

Cohen O, Goldberg A, Lahad M, Daniel L (2017) Building resilience: the relationship between information provided by municipal authorities during emergency situations and community resilience. Technol Forecast Soc Chang 121:119–125. https://doi.org/10.1016/j.techfore.2016.11.008

Commission on Audit. 2014. "Assessment of Disaster Risk Reduction and Management (DRRM) at the Local Level," 31. http://www.google.co.uk/url?sa=t&rct=j&q=&esrc=s&source=web&cd=2&cad=rja&uact=8&ve.php%2Freports%2Fdisaster-risk-reduction-and-management-reports%3Fdownload% 3D20592% 3Aassessment-of-disaster-risk- reduction-and-management-drrm-at-the-l

Crampton JW (2004) GIS and geographic governance: reconstructing the choropleth map. Cartographica 39(1):41–54. https://doi.org/10.3138/h066-3346-r941-6382

Domingo SN, Manejar AJ (2018) Disaster preparedness and local governance in the Philippines. In: PIDS Discussion Paper Series, No. 2018-52. http://hdl.handle.net/10419/211072

Domingo N, Sonny (2016). An Assessment of the sectoral and institutional implementation of the NDRRMP." Discussion Paper Series No. 2016-49

DOST-PCIEERD (2017) "DOST TURNS OVER PROJECT NOAH OUTPUTS TO PAGASA, PHIVOLCS AND OTHER AGENCIES." 2017

IFRC (n.d.) "Emergency Needs Assessment." Accessed 30 Nov 2019. https://www.ifrc.org/fr/introduction/gestion-de-catastrophes/operations-en-cours- /disaster-response-system/emergency-needs- assessment/
Indelicato A (2015) Landslides in the Philippines: assessing the role of bioengineering as an effective alternative mitigation technique. In: Proceedings of Slope 2015
Jha S, Martinez A, Quising P, Ardaniel Z, Wang L (2018) Natural disasters, public spending, and creative destruction: a case study of the Philippines. 817. Tokyo
Lagmay AMA, Tengonciang AMP, Rodolfo RS, Soria JLA, Baliatan EG, Paguican ER, Ong JBT, et al. (2008) Science guides search and rescue after the 2006 Philippine Landslide. Disasters 32(3):416–33. http://ovidsp.ovid.com/ovidweb.cgi?T=JS&PAGE=reference&D=emed8&NEWS=N&AN=18958912
Pakes U, Campomanes FV (2019) Suitability assessment for the proposed climate-smart village of the Augustinian Province of Sto. Niño de Cebu
Palmones U, Rodillas R, Cruz CD (2017) LiDAR AT THE FOREFRONT OF THE DISASTER RISK REDUCTION AND MANAGEMENT IN THE PHILIPPINES : LESSONS AND CHALLENGES The Case of Cebu Phil-LiDAR 1 Program
Partnership for Economic Policy (2019) CBMS Philippines. https://www.pep-net.org/cbms-philippines
UP DREAM (2014). "Phil-LIDAR 1." 2014. https://dream.upd.edu.ph/about/phil-lidar-1/
Valencia GR III (2012) "DOST'S DREAM Project Takes Off." 2012

Jonnifer R. Sinogaya Ph.D. is an Associate Professor of Physics at the Department of Biology and Environmental Science, College of Science, University of the Philippines Cebu. He was the Project Leader of Phil-LIDAR 1, a research project on flood hazard mapping in Western Visayas under the the Disaster Risk Exposure and Assessment for Mitigation (DREAM) Program. He was the Project Proponent of the UP Cebu Center for Environmental Informatics in Central Visayas (CENVI), a DOST-GIA funded project under the Niche Centers in the Regions for R&D (NICER) Program. Dr. Sinogaya is currently the Project Consultant of UP Cebu CENVI.

Chito L. Patiño is a Computer Science and Geographic Information System professional with experience in machine learning, flood modeling, risk analysis, and natural resource assessment. He currently works with the University of the Philippines Cebu Center for Environmental Informatics (CENVI) as its Supervising Science Research Specialist.

Melanie R. Montaño is a graduate of the Master of Science in Environmental Studies program in the University of the Philippines Cebu, where she also worked as a researcher and a lecturer. She now writes web content and copy for sustainable living brands and environmental health and safety companies.

Chapter 14
Mainstreaming Climate and Disaster Risk Assessment in the Comprehensive Land Use Plan

Alfredo Mahar A. Lagmay, Joy T. Santiago, and Jerico E. Mendoza

Abstract As the country addresses the climate emergency, it is vital for local governments to mainstream climate and disaster risk information into their local plans to be able to adapt to predicted climate change impacts. The 1987 Constitution provides for the sustainable management, development, allocation, and use of land resources, while the Local Government Code of 1991 devolves environmental responsibilities and planning to local governments through the Comprehensive Land Use Plan (CLUP). This chapter discusses the process on how to integrate climate and disaster risk assessment (CDRA) into the CLUP through the example of planning for the Science City of Muñoz (SCM) in the province of Nueva Ecija, executed by the University of the Philippines Resilience Institute (UPRI) from 2018 to 2019. It presents how CDRA is mainstreamed in every step of the CLUP formulation process. Using a multi-stakeholder participatory approach strengthens local ownership and understanding of the city's risks and vulnerabilities, specifically to floods. This chapter is intended to serve as a practical guide and resource for policymakers, practitioners, and planners. Without probabilistic climate and disaster risk assessments, local governments will not be able to anticipate future hazards that are predicted to be bigger than what is covered by local knowledge and historical records.

Keywords CCA-DRR · CLUP · Mainstreaming · Vulnerability · Risk

Key Points

- The 1987 Constitution provides for the sustainable management and development of all the country's resources, particularly the proper allocation of land and water resources.

A. M. A. Lagmay (✉)
University of the Philippines Diliman, Quezon City, Philippines
e-mail: aalagmay@up.edu.ph

A. M. A. Lagmay · J. T. Santiago · J. E. Mendoza
University of the Philippines Resilience Institute, Quezon City, Philippines

K. B. Berse et al. (eds.), *Climate Emergency in the Philippines*, Disaster Risk Reduction, https://doi.org/10.1007/978-981-99-7804-5_14

- Through the Local Government Code (LGC) of 1991 (RA 7160), local government units are mandated to create Comprehensive Land Use Plans for their respective jurisdictions to manage their land resources sustainably.
- The CLUP is a government-required document meant to determine the specific uses of land and other natural resources of a city or municipality. During the formulation of the CLUP, the assessment of the vulnerabilities and risk of the city to natural hazards are incorporated.
- Using participatory approach benefits LGUs and local stakeholders as it raises their awareness of climate and disaster risks, increases their belief and sense of ownership in the planning process, and contributes to effective implementation of the CLUP.

Summary for Policymakers

The Comprehensive Land Use Plan (CLUP) is required by law to help local government units manage their land, water, and natural resources sustainably. To effectively fulfill this mandate, it is crucial for cities and municipalities to incorporate disaster risks and climate change impacts in the planning and implementation of their respective CLUPs.

The UPRI facilitated the mainstreaming of climate and disaster risk assessment (CDRA) in the CLUP of the Science City of Muñoz (SCM), Nueva Ecija. The CDRA process sheds light on the natural hazards faced by an LGU given the climate emergency. The climate risk assessment conducted for SCM made stakeholders realize that climate change increased their risk and vulnerabilities to floods, whereas landslide risk did not increase significantly.

> Climate risk assessments shed light on the natural hazards faced by an LGU given the climate emergency. The climate risk assessment conducted for SCM made stakeholders realize that climate change increased their risk and vulnerabilities to floods.

Lessons from the SCM experience can benefit policymakers, especially those working at the local level. These are as follows:

- **Mainstream climate change and disaster risk information to help LGUs adapt to predicted climate change scenarios**. In the case of SCM, it helped them identify measures that are truly adaptive to projected climate change impacts. Without these assessments, there is no anticipatory planning for future hazards that are predicted to be bigger than those in the historical record.
- **Communicate using the local language that is used and best understood by stakeholders**. Communicating disaster-related information in a way that could be understood by the locals of the SCM encouraged stakeholders to get immersed in the planning of their community. It also capacitated them with better knowledge to become more effective leaders.
- **Apply participatory approaches using both science and local knowledge**. The mainstreaming process for the SCM was conducted in a participatory manner where science was discussed with stakeholders of the community to complement local knowledge. This way, there was ownership of response actions and

development plans for the community. This approach paves the way for effective and efficient implementation because the members of the community themselves would believe the outcomes of their preparations and that these will benefit them as individuals and as a society.

- **Partner with the academic community and other relevant agencies that can supply science-based information.** Establishing such partnerships in the mainstreaming process of CDRA and the development of CLUP can help LGUs produce evidence-based and science-based policies and plans.

Mainstreaming CDRA in the CLUP is technical and participatory. The endeavor of planning for a disaster and climate resilient future for LGUs entails a whole-of-society approach.

To produce science-based plans and policies, LGUs should establish partnerships with the academic community and other relevant agencies that can supply science-based information in the mainstreaming process.

14.1 Introduction

The Comprehensive Land Use Plan (CLUP) is a government-required document meant to determine the specific uses of land and other natural resources of a city/municipality (HLURB 2013). It represents the aspired land use patterns of barangays in cities or municipalities and prescribes the desired future growth and development of the community. This is necessary to ensure that the long-term goals of the community will prosper by means of a well-laid-out plan, just as in the execution of any successful business venture.

The CLUP is a multi-sectoral analysis of land use patterns in the city/municipality, which are influenced by internal conditions and activities, and by external factors. It requires a technically sound and evidence-based analysis using the biophysical condition of the city/municipality as the critical focal point to allocate resources (HLURB 2013). Included in the analysis during CLUP formulation are the assessment of the vulnerabilities and risk of the city to natural hazards, mainstreamed into the planning process. Furthermore, risk levels and adaptation measures to climate change impacts must be incorporated to secure the safety of the people and mitigate losses and damages in the future. With the conduct of a Climate and Disaster Risk Assessment (CDRA) as a prerequisite for the CLUP, zoning ordinances are crafted for resilient planning of the city/municipality.

The CDRA, aside from assessing the vulnerabilities and risks of barangays in a city/municipality, aims to assist the local government in formulating risk-sensitive land use plan towards more resilient development. Even though the assessment mainly provides the existing condition of the city in relation to natural hazards, the CDRA also plays a big role in every step of the CLUP planning process.

This paper documents the process of the formulation of the CLUP of the Science City of Muñoz (SCM), in the province of Nueva Ecija, and how climate and disaster risk assessment is mainstreamed into the local plan. Their CLUP was executed by the University of the Philippines Resilience Institute (UPRI) from 2018 to 2019 as part of its main goal to build resilient communities by incorporating climate change and natural hazards information into land use and development plans.

A full copy of the city's CLUP and CDRA can be requested from the local government unit (LGU) of SCM. This paper is an abridged version of the CLUP and focuses on how the CDRA was incorporated and mainstreamed in the CLUP planning process to formulate a risk-sensitive land use plan. The results of each planning step, workshop results, and CDRA are also summarized. We refer the reader to the LGU for all the detailed outputs of the entire CLUP.

14.2 Legal Mandates and Enabling Policy Environment

The 1987 Constitution provides for the sustainable management and development of all the country's resources, particularly the proper allocation of land and water resources. The law contains provisions on how the country's resources should be managed, used, developed, owned, and disposed of. It also provides that the state shall promote the right of the people to a sound and healthful ecology in accordance with the rhythm and harmony of nature.

Through the Local Government Code (LGC) of 1991 (Republic Act [RA] 7160), the LGUs are allowed to legislate and implement policies and projects which would contribute to the development of their jurisdiction. Environmental responsibilities are devolved to LGUs. Among others, the LGC enables and mandates the LGUs to create CLUPs to sustainably manage land resources subject to the adoption of the Sangguniang Panglunsod (SP) and the approval of the Housing and Land Use Regulatory Board (HLURB).

As the frontline government unit in dealing with disasters and the effects of climate change, the Climate Change Act of 2009 (RA 9729) also requires LGUs to formulate the Local Climate Change Action Plans (LCCAP). The LCCAP, which is sectoral by nature, is complementary to the spatial nature of the CLUP. The term sectoral means that planning is done across all sectors of society, including agriculture, coastal, biodiversity, infrastructure, tourism, energy, and mining, to adapt to future climate change impacts. On the other hand, the term spatial refers to planning through the delineation of zonal areas in the community, such as residential, economic, agricultural, and mining zones.

Further emphasizing an LGU's role as a frontline agency in dealing with the adverse effects of climate change is the Philippine Disaster Risk Reduction and Management Act (RA 10121). This landmark legislation in 2010 required the mainstreaming of Climate Change Assessment and Disaster Risk Reduction (CCA-DRR) concepts in the Local Disaster Risk Reduction and Management Plan (LDRRMP), which is a requirement in the formulation of a CLUP and a CDP.

In the same vein, the Climate Change Act of 2009 (RA 9219) provides that LGUs shall be the frontline agencies in the formulation, planning, and implementation of climate change action plans in their respective areas, consistent with the provisions of the Local Government Code, the Framework, and the National Climate Change Action Plan.

The creation of a CLUP is guided by laws such as the Urban Development and Housing Act (UDHA) (RA 7279), which defines parameters for a sustainable land use plan. LGUs are also guided by the standards set by the HLURB through Executive Order (EO) 648. Since 2018, the Climate Change Commission (CCC) issued a memorandum order 2019-0001 that clearly defines the Climate Risk and Management Framework (CRMF), requiring probabilistic risk assessment to protect lives and properties from future adverse impacts of climate change through anticipatory planning. Probabilistic risk assessment is also emphasized in the Philippine Development Plan (2017–2022).

Further guiding laws on the management and conservation of resources that are needed are found in RA 7586, RA 8550, RA 8435, RA 9003, RA 10066, and RA 9593. The CLUP formulation also considers RA 7942, RA 8371, PD 705, Commonwealth Act (CA) 1412, and planning guidelines from the Department of Environment and Natural Resources (DENR), the Department of Interior and Local Government (DILG), as well as the National Economic and Development Authority (NEDA) as these affect the LGU's mandate in its performance of their land use planning and management functions.

14.3 Relationship With National Plans And Programs

The formulation of CLUPs should be in line with national plans such as the Philippine Agenda 21, which supports the coordinated integration of a sound and feasible economy, responsible governance, ecological integrity, and social cohesion. The LGU, in the creation of the CLUP, must look after the support for food security, ecological integrity and environmental stability, spatial integration, equitable access to physical and natural resources, regional urban development, private–public sector partnership, market orientation, and recognition of the rights of the people. This must also be in accordance with the National Framework for Physical Planning (NFPP) of the country. Lastly, the contents of a CLUP should be in line with the National Strategic Framework for Climate Change (NSFCC) and the National Climate Change Action Plan (NCCAP), which states that the national strategy should be able to give assurance and strength to the adaptation of our human communities and natural ecosystems to climate change (HLURB 2013).

14.4 The Planning Process

The CLUP formulation requires a 12-step procedure. It begins by assembling the staff, resources, and support needed for community planning, and ends with the monitoring and evaluation of the CLUP and the formulated zoning ordinance (ZO). The ZO serves as the implementing medium of the CLUP (Fig. 14.1).

The CLUP process begins with a pre-planning phase, which is vision-oriented. The city's vision statement should be supported by locally gathered data and facts, made possible through the planning team, composed of a Planning Core Group from the LGU, and assisted by a Technical Working Group (TWG).

Once the pre-planning process is done, there is a need to identify and engage participants who play an active role in the main planning phase. Participants are referred to as stakeholders whose roles are key to the planning process. Their involvement raises the quality of the CLUP as information is shared between the LGU and the citizens of the community. Issues and concerns of the stakeholders are identified in the planning process to evaluate options to formulate strategies that are beneficial to the populace. The vision, goals and objectives, development thrusts, and spatial strategies are then identified along with the situational analysis of the city or municipality to prepare a viable land use plan.

In planning, decisions are critical. It is therefore necessary to have a set of diverse and well-represented decision-makers. This is to ensure that the concerns and biases of the stakeholders are balanced. This group of decision-makers belongs to the planning team and the TWG. The planning team and the TWG are composed mainly of the Sangguniang Panglungsod (SP) members, and heads from different departments and non-government organizations. These groups are responsible for coordinating planning activities and decision-making.

After the preparation of the land use plan, the zoning ordinance is drafted and presented in a public hearing. It is followed by the adoption and approval of the CLUP and the ZO. Afterward, the city or municipality can proceed with its implementation,

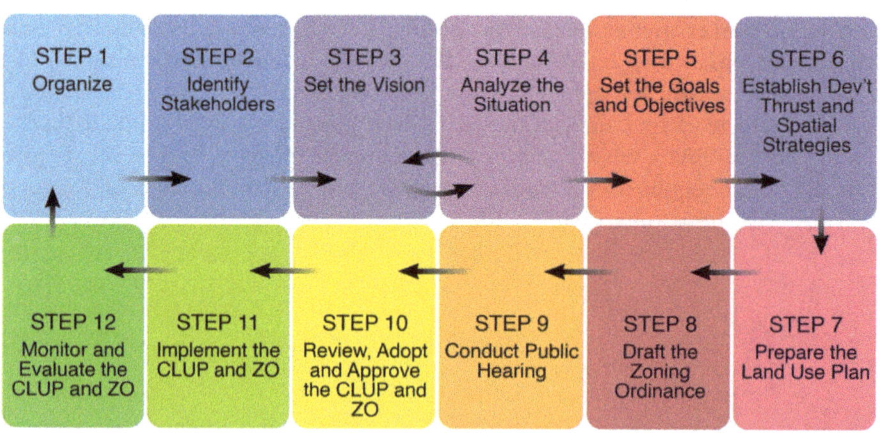

Fig. 14.1 The 12-step process in the formulation of the CLUP (HLURB 2013)

monitoring, and evaluation. The CLUP may be updated earlier than the three-year required period to revise the plan. This is done, for example, when a powerful typhoon with consequent floods and landslides quickly alters the landscape. In such cases, land use management strategies are crafted with policies including CCA-DRR and mitigation measures incorporated into the updated CLUP to address the lessons from such calamitous events.

14.5 Spatial and Temporal Planning Areas

The city's or municipality's political boundaries including the barangay and city/ municipal waters extending 15 km from the shoreline in the case of coastal LGUs is the spatial planning area as defined by (HLURB 2013). In addition, the physical coverage area for planning is the specific watershed/sub-watershed basin area within the city/municipality and other adjacent localities.

The updated CLUP lays the foundation for the city's vision for the next 9 years and should be reviewed every 3 years or as often as necessary. It starts from the planning area of the city, systematically translating the city's strategic vision, objectives, and directions into physical and spatial actions. The CLUP presents the city's current state based on the available baseline data and provides analyses about the environmental, social, economic, infrastructure, and local institutional capability of the city (HLURB 2013). The baseline studies conducted as part of the CLUP preparation provide a basis for projections of population and the determination of land supply and demand for various land use activities.

14.6 Principles and Approaches Adopted in the Clup

It is good practice to adopt several principles in the CLUP preparation process. For example, it is essential to consider watersheds for a ridge-to-reef approach. This is most significant at the provincial level since their administrative boundaries commonly encompass entire watersheds. Integration in the overall plan of the watershed ecosystem highlights the need for specific land uses and development controls within the territorial jurisdiction of an LGU and its adjacent LGUs. One way to visualize the cause and effect of natural and anthropogenic impacts in the entire watershed is through the use of a Geographic Information System (GIS). Such a system also allows simulation of processes that may have beneficial or detrimental effects on communities anywhere within a watershed.

Furthermore, land use plans and development plans are products of collaboration among the state, civil society, and the private sector. When observed in the planning process, inclusive and expansive governance is embraced. All stakeholders in the LGU who participate become part of the growth and development of their community and are also responsible for whatever outcome their plans lead to. In the formulation of the CLUP, identified stakeholders are capacitated through GIS, CDRA, and LCCAP trainings and workshops with hands-on exercises to facilitate an effective transfer of knowledge.

For protected areas, resources are co-managed by the national government. They are both responsible for managing and maintaining the ecological balance of the protected territories. Both local and national governments make sure that the integration of gender and population interrelationships in the entire planning process is applied and up to monitoring and evaluation standards. Gender mainstreaming is integrated by soliciting and processing sex-disaggregated data during the preparation of the Ecological Profile (EP) and Sectoral Studies for the city or municipality. Men, women, and representatives from the lesbian, gay, bisexual, transgender, and queer (LGBTQ) community were also actively involved in all the steps undertaken in the formulation of the CLUP to ensure that the CLUP is sensitive to these genders.

Barangays are also mandated under the LGC of 1991 to prepare their development plans. The integration of the barangay-level plans into the city's land use plan, development goals, and objectives ensures a bottom-up approach that aligns the needs at the grassroots level to the desired thrust of the city. During the formulation of the CLUP, all the barangay profiles are consolidated and integrated into the planning process. Equally important in the formulation of a CLUP is the top-to-bottom approach. In contrast to the bottom-up strategy, higher level plans such as the Provincial Land Use Plan (PLUP) or the Provincial Physical Framework Plan (PPFP) are utilized to serve as a framework for the formulation and updating of the CLUP. This ensures that approved area-specific plans in the city's vision, goals, and objectives are aligned with the provincial and other higher plans.

Last but not least is the use of the probabilistic approach, which is in contrast to the deterministic approach for risk assessment of flood, landslide, and storm surge hazards that all cities and municipalities of the country until now employed during the formulation of their CLUP. In simple terms, the deterministic approach provides a single-scenario (Fig. 14.2a) representation of a hazard, whereas the probabilistic approach (Fig. 14.2b) is a multi-scenario representation of a hazard (e.g., flood). In the Philippines, official hazard maps for floods and landslides are historical worst-case scenarios, a single-scenario depiction of a hazard, bereft of input from worsening climate change conditions. In order to capture climate change impact consequences as projected by IPCC, the multi-scenario depiction of any type of hazard, which goes beyond the historical worst-case record, is imperative. Such multi-scenario hazard maps are simulated by computers as they can not be derived from anecdotal accounts or expert opinion that is manually input into a hazard map. Probabilistic hazards assessment will allow the community to do plans in an anticipatory manner, which is synonymous with adaptation against worsening hydrometeorological hazards (e.g., floods, landslides, storm surges) due to climate Change.

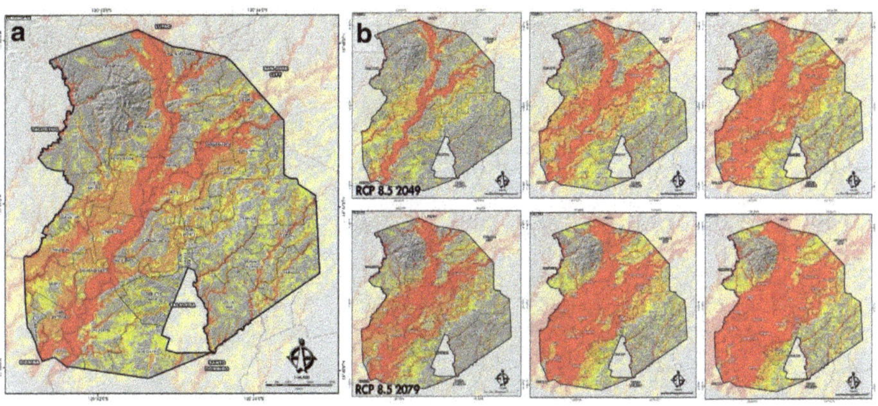

Fig. 14.2 **a** Deterministic hazard assessment. **b** Probabilistic hazards assessment of floods with climate change scenarios. Representative Concentration Pathways (RCPs) are according to the projections made in the IPCC report

14.7 Existing Land Use

The existing land use (Fig. 14.3) of the SCM in 2018 was gathered and evaluated through the use of GIS and data triangulation. After incorporation into a GIS, the land use of the city was validated by a technical working group (TWG) and barangay representatives of the city.

GIS analysis shows that the city is predominantly agricultural, with approximately 80.90% of its total area utilized for agricultural production. Institutional areas that make up 3.81% of the city's total area include the city hall, fire stations, police stations, barangay halls, schools, daycare centers, health centers, churches or places of worship, government buildings, and other facilities that provide essential social services. Likewise, a significant percentage of the institutional area is allocated for the Central Luzon State University (CLSU), the biggest state university in Central Luzon Region. Commercial areas cover approximately 0.20%.

Rivers, namely Baliuag, Cumabol, Oluulu, Panlasian, and Rio Chico form approximately 2.92% of the total area. Road networks and infrastructure, on the other hand, constitute 2.18% of the total land area. Other land uses include agri-industrial (0.80%), cemeteries (0.07%), open spaces (0.06%), agro-industrial (0.04%), tourism areas (0.03%), and parks and recreation (0.01%). The disputed area also constitutes 3.18% of the total land area.

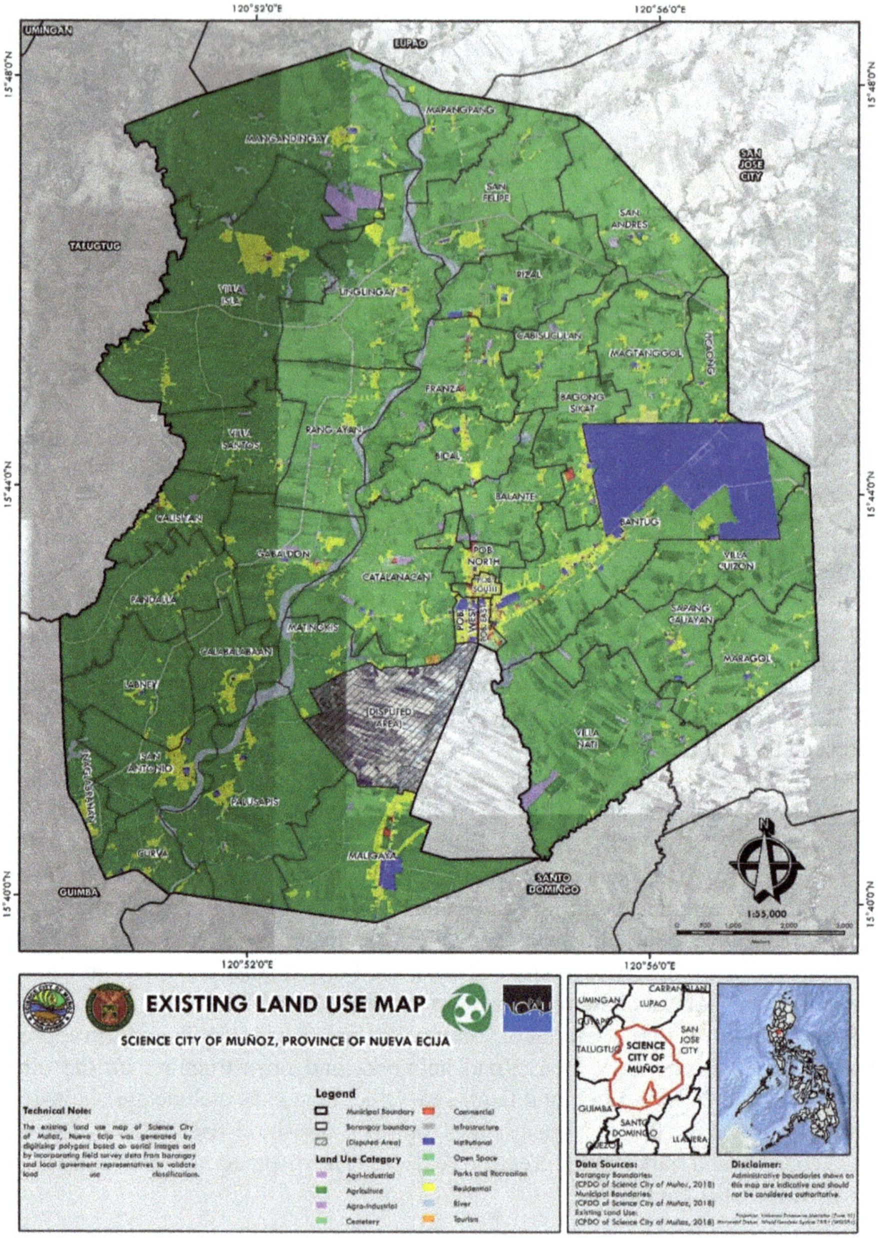

Fig. 14.3 The 2018 existing land use map of the SCM (Science City of Muñoz 2019)

14.8 Climate and Disaster Risk Assessment (CDRA)

The Climate and Disaster Risk Assessment results show the level of vulnerability and danger for flood and rain-induced landslide hazards for every barangay in the city. In addition, decision areas were identified, defining the need for priority response and risk-sensitive policy interventions. To create the CDRA for the SCM, it was necessary to conduct: (1) Climate Change and Hazard Information Collection and Organization; (2) Scoping of Hazard and Climate Change Impacts; (3) Exposure Database Development; (4) Climate Change Vulnerability and Risk Assessment; (5) Disaster Risk Assessment; and (6) Summarize Findings.

During the CDRA process, increasing risk rates were identified as climate change and flood scenarios became worse. All exposure units showed the same trend for vulnerability assessment. As the climate scenario and the return period increase, the number of barangays, facilities, infrastructure, and extent of affected areas increases (Science City of Muñoz 2019). On the other hand, rain-induced landslide vulnerability assessment did not differ significantly across climate change scenarios.

The disaster risk assessment also showed that the risk level declines as the return period of the flood hazard increases. This is due to the lower probability of occurrence of a severe but rare event compared to a small but frequent hazard event. For instance, a 100-year. return duration offers a low-risk scenario, although its severity is high because it has only 1% chance of happening annually. This is unlike a 5-year return period rainfall event, which presents a higher risk as it can occur more often with a 20% chance of occurrence annually. The number of high-risk areas therefore usually decreases as the return period increases due to its lower likelihood of occurrence.

14.9 Discussion

The planning process is divided into four phases. The preliminary phase includes the organization of the planning team, stakeholder analysis, and formulation of the vision statement. The second is the situational analysis, whereas the third step is the preparation of the plan, which includes goal setting, the establishment of development thrusts and spatial strategies, and the preparation of the proposed land use plan and the zoning ordinance. The last step is the conduct of a public hearing, review, adoption, approval of the plan, and implementation and monitoring process. These three phases are where the CCA-DRR is mainstreamed into the local plans.

14.9.1 Preliminary Phase

As part of the planning process and to ensure the involvement and participation of different government offices, the Local Chief Executive of the SCM creates a

TWG through an executive order. The organization and stakeholder analysis aims to identify and engage participants who should play an active role in the main planning phase. Participants are referred to as stakeholders whose roles are key to the planning process. Their involvement raises the quality of the CLUP as information is shared between the LGU and the citizens of the community.

There are many tools to identify the stakeholders, one of which is the Power-Interest Grid (PIG). The PIG arrays stakeholders into a quadrant categorized based on the level of interest and power to influence the planning process and the planning area (Eden and Ackernlann as cited in HLURB 2013). For the case of SCM, stakeholders composed of department heads, members of civil society organizations, barangay representatives, and local and regional government offices were identified using the PIG approach (Fig. 14.4). During this process, stakeholder concerns or those who may be affected by the impacts of natural hazards are considered.

Through a series of workshop consultations, the local government and the stakeholders put into words the collective aspirations of a city.

> The Science City of Muñoz as a center of agricultural science and technology for education, tourism, arts and culture, driven by a globally competitive and progressive economy, supported by an ecologically friendly and safe environment, with livable infrastructure nurtured by healthy and resilient Muñozonians, guided by a passionate, and reliable local leadership.

Inward-looking descriptors summarized in Table 14.1 provide a clear view of the aspirations of the SCM.

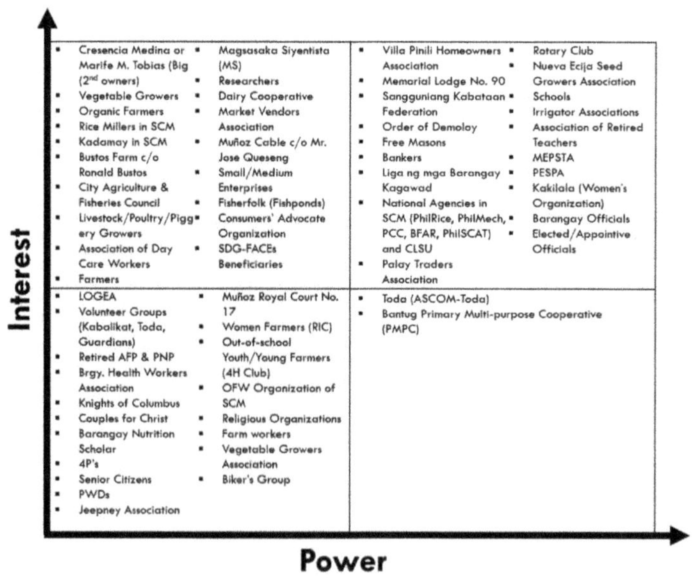

Fig. 14.4 Stakeholder analysis using power-interest grid in preparation for the CLUP formulation of the SCM (Science City of Muñoz 2019)

Table 14.1 Inward-looking indicators for the SCM which show the aspirations of Muñozonians for its city

Development sector	Inward-looking indicators
Environment	Ecologically friendly
	Safe
Social	Healthy
	Resilient
Economics	Globally competitive
	Progressive
Infrastructure	Livable
Institutional	Passionate
	Reliable

The formulation of the vision statement integrates climate change adaptation and disaster risk reduction (CCA-DRR) concepts. The process includes lectures and workshop activities on climate change and disaster risks. During the formulation of the vision statement, stakeholders of the city incorporated their aspiration to become resilient to adverse climate and natural hazard impacts.

Participants also raised questions and essential points during sectoral discussions on CCA-DRR. For example, stakeholders from the environment sector recognized that an ecologically friendly and safe environment is a place where people are far from risks, whether natural or human-induced. They noted that a globally competitive and progressive local economy would only be sustainable if the city has the capability to resist, adapt to, or bounce back at the soonest possible time after a hazard strike. Discussions also lead to descriptions of a livable built-up environment as well-constructed, withstands the adverse impacts of climate change, useable, and provides continued services to the people in times of disasters. Furthermore, the stakeholders emphasized during the consultations that in order to execute the vision in the long term, all of the CCA-DRR plans must be led by a passionate and reliable government that prioritizes the reduction of vulnerabilities and risks of its constituents and territory.

14.9.2 Situational Analysis

This phase involves analyzing and assessing the current condition of the city, as well as its physical environment, vulnerability and risks, strengths, weaknesses, opportunities, and threats. It includes the preparation of the ecological profile, preparation of the land use map, and climate and disaster risk assessment following best practices as earlier described in this article. Situational analysis requires participatory and technical methods. It serves as the baseline, presenting thematic, sectoral, and cross-sectoral assessments as well as the biophysical analysis of planning areas. It is formulated together with the vision statement and can be accomplished interchangeably. This step is carried out through participatory GIS mapping and data triangulation, which links different information sources that involve people, places, and time to characterize possible links between program interventions and their impact on populations and the community's growth and development.

During the situational analysis, the UPRI presents the results of the CDRA to the stakeholders in order to provide a more in-depth analysis of the planning environment through the identification of issues, concerns, and problems. Such analysis provided the opportunity for a more integrated approach in formulating the land use plan (HLURB 2013) and was done with a goal in mind, which is to anticipate the impacts of climate change and natural hazards. Projected climate change variables, the descriptions of different natural hazards that plague the community, and statistics of historical damage due to disasters were presented to provide a comprehensive understanding of potential adverse impacts.

Also, the workshops served as an avenue to discuss the geologic and hydrometeorological hazards assessment of the SCM with the stakeholders. For hydrometeorological hazards, baseline scenarios and Representative Concentration Pathways (RCPs) scenarios (IPCC, 2014) were presented. CDRA was done at the barangay level using intensive exposure database collection with inputs from the local government, barangays, and the academic/scientific community. Given the hazard scenarios, stakeholders identified their strengths and weaknesses, available opportunities, and possible threats to determine gaps between the city's aspirations and its existing condition.

Based on the SWOT analysis, it was determined that the SCM's strengths in the social sector lie primarily in housing and education. Barangays have access to basic services such as electricity, potable water systems, and daycare centers. Scholarships are also available to the citizens. Other notable outputs of the SWOT analysis include the need to improve protection services to track misdemeanors and to reduce violence against women and children.

Opportunities for the environmental sector involve the 'go organic' trend and "upcycling" of materials to give greater value to their products. These programs offer livelihood and waste reduction opportunities. Threats in the SWOT analysis include the lack of discipline amongst its citizens in observing regulations. There was also a lack of enforcement of environmental laws. Moreover, the city's exposure to typhoons, El Niño, La Niña phenomenon, and climate change, made stakeholders realize that these occurrences can damage agriculture. They also identified possible environmental damage from the use of unregulated chemicals and unplanned land development by private landowners as other threats to the environment. By tapping national government financial grants and funding sources and fostering partnerships with local, national, and international agencies, SCM intends to increase and improve its human resources and its capability to address these environmental concerns.

The city boasts its recognitions as a recipient of the SGLG (Seal of Good Local Governance) and Gawad Kalasag. These awards are partly attributed to SCM's effective governance of infrastructure projects. The government always ensures that buildings are constructed in safe places and follows the building code for safety. Another identified problem in the infrastructure sector is delayed delivery and completion of projects due to the high costs of raw materials. Nonetheless, the SCM assures that infrastructure projects are not compromised with strict observance and compliance with laws.

The local government ensures enough funds for economic enterprise, sufficient agricultural supply, and improvement of the wet market to increase the local revenue of the city and to promote local products. However, the situational analysis revealed the need to improve the irrigation system, information dissemination on modern agricultural techniques, and marketing strategies for tourism products and services. Although the local government strives to provide effective management and execution as reflected in their sound and transparent fiscal management and institutional planning, there are difficulties due to a lack of critical services, funding, and tax collection. The lack of a tourism department is also contradictory to the city's ambition to promote tourism.

Health-related outbreaks also affect the city. In 2019, a nationwide vaccine scare decreased the public's confidence in the immunization system which intensified possible disease outbreaks in the region. Another is the introduction of diseases from nearby towns and cities into local livestock and crops, which were identified to hamper the growth of SCM's agro-industry.

With all the discussed strengths and weaknesses present in the city, the stakeholders identified opportunities that can work to their advantage. They also identified threats that are potentially harmful to the city. The high rate of inflation was treated as a menace to the SCM business sector that could adversely affect the city's economic development. In addition, inadequate power and the lack of contact facilities decrease both population and private sector's efficiency. Changes in leadership arising from elections and the local chief executive's short term were also mentioned as a destabilizing factor. One ongoing threat found is the transfer of certain National Government Agencies and Research and Development offices to other locations due to higher real estate taxes levied on tenants of SCM land.

The discussion of the SWOT by the stakeholders was reinforced by their knowledge on CCA-DRR. SCM as an agricultural city with generally flat topography, when affected by climate change, may experience drought or intensified flooding. These factors are evidently seen on how the stakeholders analyzed their SWOT. Based on their vision, the local government of SCM, together with the stakeholders identified measures that will help them achieve their aspirations. To better equip the city and its citizens in the process, they also determined the gap of how far the city's existing condition versus their aspired goals. Using the Vision-Reality Gap (VRG) Analysis tool, some of the concerns with the highest gaps were spotted. These include (1) low priority of physical fitness activities; (2) inadequate promotion of the city's products and services; (3) lack of ample coordination and communication with the local government and the barangay government; (4) lack of high-speed internet connection and poor mobile signal; (5) poor employment conditions; and, (6) improper sewage treatment systems.

According to the stakeholders (Science City of Muñoz 2018), physical fitness is one of the important factors in determining the vulnerabilities of one community because it is connected to the health and well-being of its citizens. For them, a healthy Muñozonian is less vulnerable to climate and disaster risk. In addition, promoting the city's products and services will also provide a better adaptive capacity to the city as it provides livelihood and other sources of income, whereas better communication between the local and barangay government will provide smoother governance, especially in preparation, during, or after a disaster.

14.9.3 Preparation of the Plan

This phase represents steps 5–8 (Fig. 14.1) of the CLUP planning process. In setting the goals, objectives, targets, and strategies (GOTS) the stakeholders were grouped into five development sectors and were asked to review the vision statement to identify problems and opportunities in achieving the desired aspirations of the city. Each sector group is assigned to formulate the GOTS of the city.

The SCM aims to serve as the center of agricultural science and technology for innovation, education, and tourism and a hub for the promotion of the culture and the arts (Science City of Muñoz 2019). Sectoral goals, which intend to contribute to the realization of the city's vision, were also identified at this stage. Goals were focused on the safety of farmers, compliance to environmental laws, access to basic services, improvement of local human resources, climate-smart farming, improvement of health and well-being, drug-free communities, promotion of tourism, additional revenue, improvement of connectivity and infrastructure, better communication systems, and proper implementation and monitoring of projects.

Since the principles of CCA-DRR are already incorporated in the vision statement, the GOTS ensured that the local economy would benefit from mechanisms that adapt to the impacts of climate and disaster risks. Furthermore, strategies assimilate measures to reduce vulnerabilities and risks. Proposed developments are climate-proofed and expected to guarantee the continuous provision of social services. Most importantly, the GOTS analysis, which is anchored on CCA-DRR principles, showcases low-risk public and private investments since assets are planned in a manner to avoid natural hazards.

Once the GOTS were identified, stakeholders established the development thrust and spatial strategies that translate the city's vision, goals, and objectives into various development alternatives or scenarios. After the development thrust was identified, spatial strategies were needed to plan the development of the city. These ensure the safety of the community and avoid urban sprawl.

Three possible development options were suggested to facilitate the achievement of the city's aspirations. These are the agri-commercial, agri-tourism, and agri-industrial options. Stakeholders were divided into three groups and asked to brainstorm on their preferred direction of growth and strategies that can help them achieve the development thrust they desire. They were also asked to consider how each development thrust will help reduce their vulnerabilities and mitigate risks of residents and other exposure units. The feasibility of pursuing proposed development thrusts in relation to various internal and external factors, including natural hazards, was considered in the process.

The results of the GAM (Table 14.2) show that agri-tourism is the preferred development thrust determined through the Goal Achievement Matrix (GAM) Analysis. The agri-tourism thrust rated at 1.70, followed by agri-commercial with a rating of 1.55. The lowest rating was for the agri-industrial direction, which only garnered an average rating of 1.26. The preferred development thrust of SCM leaned towards a combined development where the agricultural sector and tourism serve as its main engine for growth.

The stakeholders were again grouped to identify the possible urban form needed to realize the preferred development thrust. Urban forms evaluated were the concentric zone, multi-nuclei, linear, and sector development models. During the evaluation, the stakeholders were asked to refer to the hazard maps and CDRA results to identify options for development. This was made specifically to reduce and prevent future risks, ensure uninterrupted delivery of goods and services, deliver basic social support services, and put in place risk mitigation measures.

The multiple nuclei model ranked at the top based on the GAM (Table 14.3), with an average rating of 1.97. According to the stakeholders (Science City of Muñoz

Table 14.2 Goals Achievement Matrix (GAM) of the identified development thrusts for the SCM (Science City of Muñoz 2019)

Sector	Indicator	Weight (%)	Alternative development 1	Alternative development 2	Alternative development 2
			Agri-Commercial	Agri-Tourism	Agri-Industrial
Social	Healthy	8.3	2.16	2.09	1.25
	Resilient	8.3	2.18	2.09	2.11
Institutional	Passionate	8.3	1.93	2.36	1.91
	Reliable	8.3	2.07	2.22	1.98
Economic	Progressive	8.3	2.49	2.53	2.41
	Globally competitive	8.3	2.25	2.38	1.87
Environment	Safe	8.3	1.66	2.23	1.28
	Ecologically friendly	8.3	1.72	2.23	0.63
Infrastructure	Livable	8.3	2.17	2.28	1.74
AVERAGE			**1.55**	**1.70**	**1.26**

Table 14.3 GAM of the evaluated spatial strategies of SCM (Science City of Muñoz 2019)

Sector	Indicator	Weight	Linear model	Sector model	Multiple nuclei model	Concentric zone model
Social	Healthy	8.3%	2.58	2.42	2.65	2.38
	Resilient	8.3%	2.58	2.19	2.58	2.38
Institutional	Passionate	8.3%	2.62	2.46	2.62	2.42
	Reliable	8.3%	2.65	2.42	2.58	2.38
Economic	Progressive	8.3%	2.85	2.50	2.65	2.46
	Globally Competitive	8.3%	2.65	2.42	2.56	2.31
Environment	Safe	8.3%	2.38	2.65	2.77	2.08
	Ecologically Friendly	8.3%	2.19	2.50	2.60	2.31
Infrastructure	Livable	8.3%	2.65	2.52	2.62	2.35
AVERAGE			**1.93**	**1.84**	**1.97**	**1.76**

2018), it was the winning spatial strategy because the multiple nuclei model promoted a safe environment, a progressive economy, and citizen health with ratings of 2.77, 2.65, and 2.65, respectively. The principles of this spatial strategy became the basis for the proposed development of the SCM.

A structure plan was then prepared after the spatial and other planned strategies for development were selected. The structure plan map is a schematic representation of the winning spatial strategy containing the approximate location of areas planned for development (HLURB 2013). Overlain with multi-scenario hazard maps, the stakeholders are able to ensure that future development are not located in hazard-prone areas. The structure plan map was also used to make sure that priorities are given to risk mitigation infrastructure, especially in areas with high risk. For example, if existing built-up areas are located in hazard-prone areas, engineering interventions or other mitigating measures are prioritized to immediately reduce the possible impacts of climate and disaster risks (Figs. 14.5 and 14.6).

14.9.4 Land Use Plan

Land is a resource that is finite. When the population increases, the built environment expands to provide space for housing, living space, and space for supporting infrastructure (DILG-BLGD 2008). In time, the unbuilt environment is reduced. This predicament requires a proper balance of the built and unbuilt environments if it were to be consistent with the vision, goals, and development thrusts of the city. To do this, a supply–demand analysis is required to determine whether there is an adequate supply of land given the plans of the city for urban development. Using GIS,

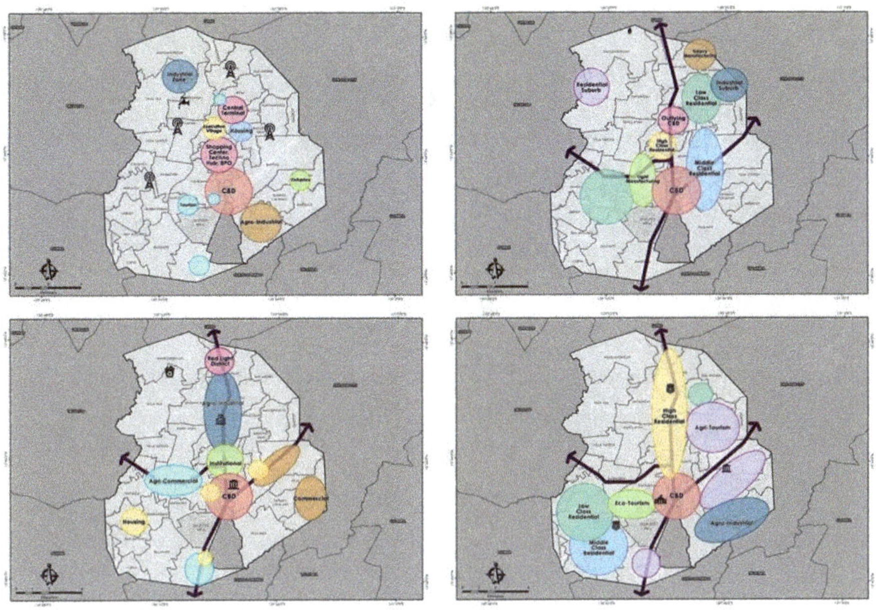

Fig. 14.5 Spatial strategies for agri-tourism development thrust from left to right and top to bottom: concentric zone, multiple nuclei, linear, and sector model (Science City of Muñoz 2018)

spatial requirements were mapped while considering the socio-economic and physical constraints of the planning area. More importantly, existing laws and regulations were considered while identifying developable areas. Last but not least, the CDRA results were incorporated into the GIS to assess the suitability of areas for expansion in relation to their hazard exposure. Land supply and demand were assessed to translate the city's vision into a land use plan.

To assess the land supply for future development of SCM, the total land area was subtracted from protected areas, urban uses areas, and special land uses (HLURB, 2013). Areas with a slope of over 18% were removed to ensure that these places are not intended for urban use. After this was done for the entire city of SCM, around 11,849.51 ha of agricultural land was left available for use in urban expansion. When the 10% reclassification cap provision of Section 20 of RA 7160 was applied, an area of 1,184.95 ha of agricultural land that can be reclassified was identified.

To compute the demand for land, population increase and use of social services and facilities were projected for the next 10 years. The projected land requirement to meet the demand of SCM in the year 2028 showed different land categories with various requirements for space. On top of the categories requiring land are residences and commercial establishments (Table 14.4).

Through the TWG, the proposed sites and construction areas up to the year 2028 were identified by digitizing potential site polygons and resulted in the creation of

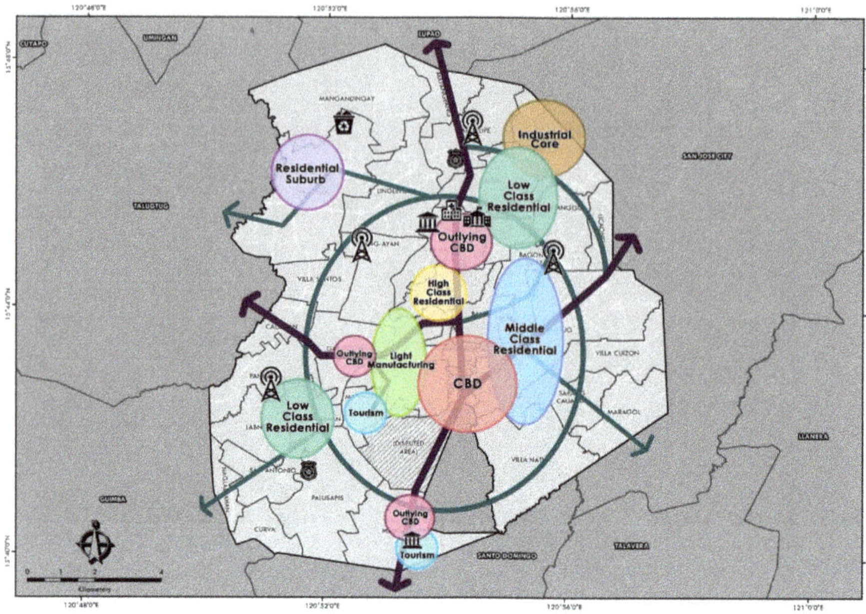

Fig. 14.6 Structure plan for the SCM using the principles from the multi-nuclei model which aims to translate the vision, goals, and development thrust of the city to spatial plan (Science City of Muñoz 2019)

Table 14.4 Projected land area requirement, Year 2028 (Science City of Muñoz 2019)

Category	Requirement (ha)
Residential	35.14
Health	1.5
Protective services	0.12
Education	0.83
Commercial	6.29–41.75
Parks and recreation	3.04
Cemetery	1.02
Solid waste	3.14
Urban roads	35.67 km

a proposed SCM land use map. In order to meet the needs of the growing population of SCM, 576.67 ha of agricultural lands were reclassified for other urban uses (Table 14.5). The creation of open spaces and buffer zones added a large percentage to the reclassification. It was also noticeable that some of the areas of the proposed land use were greater than the expected land demand because the digitized polygons followed the delineation of land in the cadastral maps.

Table 14.5 Existing and proposed land use change, Year 2018–2028 (Science City of Muñoz 2019)

Land use	Existing		Proposed		Change
	Area (ha)	Total (%)	Area (ha)	Total (%)	
Agri-industrial	118.18	0.80	127.13	0.86	8.95
Agricultural production	11,967.80	80.90	11,391.13	77.00	−576.67
Agro-industrial	6.64	0.04	6.42	0.04	−0.22
Cemetery	10.54	0.07	11.74	0.08	1.20
Commercial	29.38	0.20	55.33	0.37	25.96
Infrastructure	322.22	2.18	397.00	2.68	74.77
Institutional	564.13	3.81	565.31	3.82	1.17
Open space	8.97	0.06	469.66	3.17	460.69
Parks and recreation	2.01	0.01	5.86	0.04	3.85
Residential	856.39	5.79	852.69	5.76	−3.70
River	432.4	2.92	432.25	2.92	−0.15
Tourism	4.92	0.03	9.06	0.06	4.15
Disputed area	469.84	3.18	469.84	3.18	0.00
Total	**14,793.42**	**100.00**	**14,793.42**	**100.00**	**0.00**

As the implementing tool of the land use plan, the zoning ordinance was then formulated. The zoning ordinance considers existing local and national laws. To ensure proper implementation at the barangay level, maps for all the barangays were also made. The results of the climate and disaster risk assessment were incorporated into the zoning ordinance through hazard overlays to ensure safety of future development. Areas for Priority Actions such as built-up areas located in hazard-prone areas were also designated in the zoning map.

The land use plan and zoning ordinance were presented to the city through a series of public hearings and revised as necessary before undergoing review, adoption, and approval by the city. The implementation of the land use plan will be carried out for the next 9–10 years with corresponding monitoring and evaluation and must be reviewed regularly to ensure its applicability and may be revised as necessary (Fig. 14.7 and 14.8).

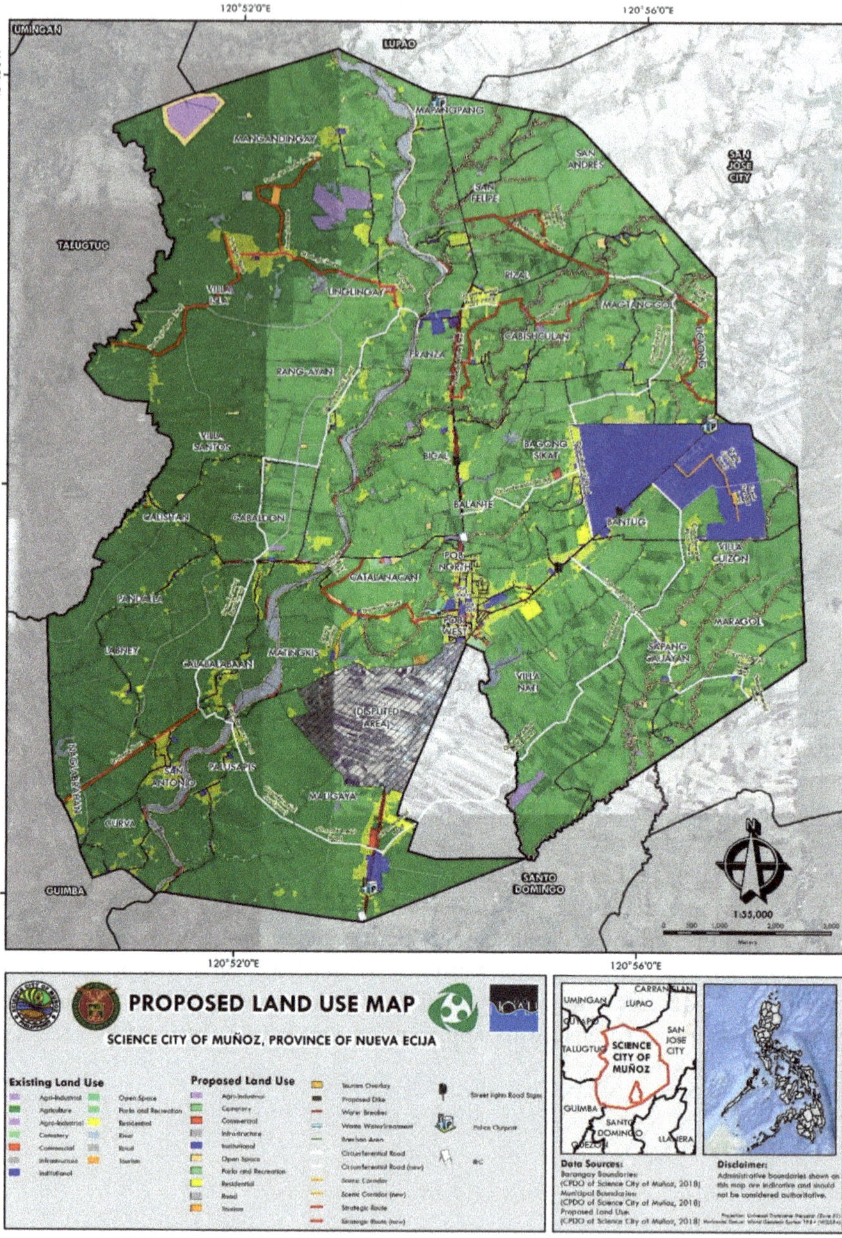

Fig. 14.7 Proposed land use map of SCM, Year 2028 (Science City of Muñoz 2019)

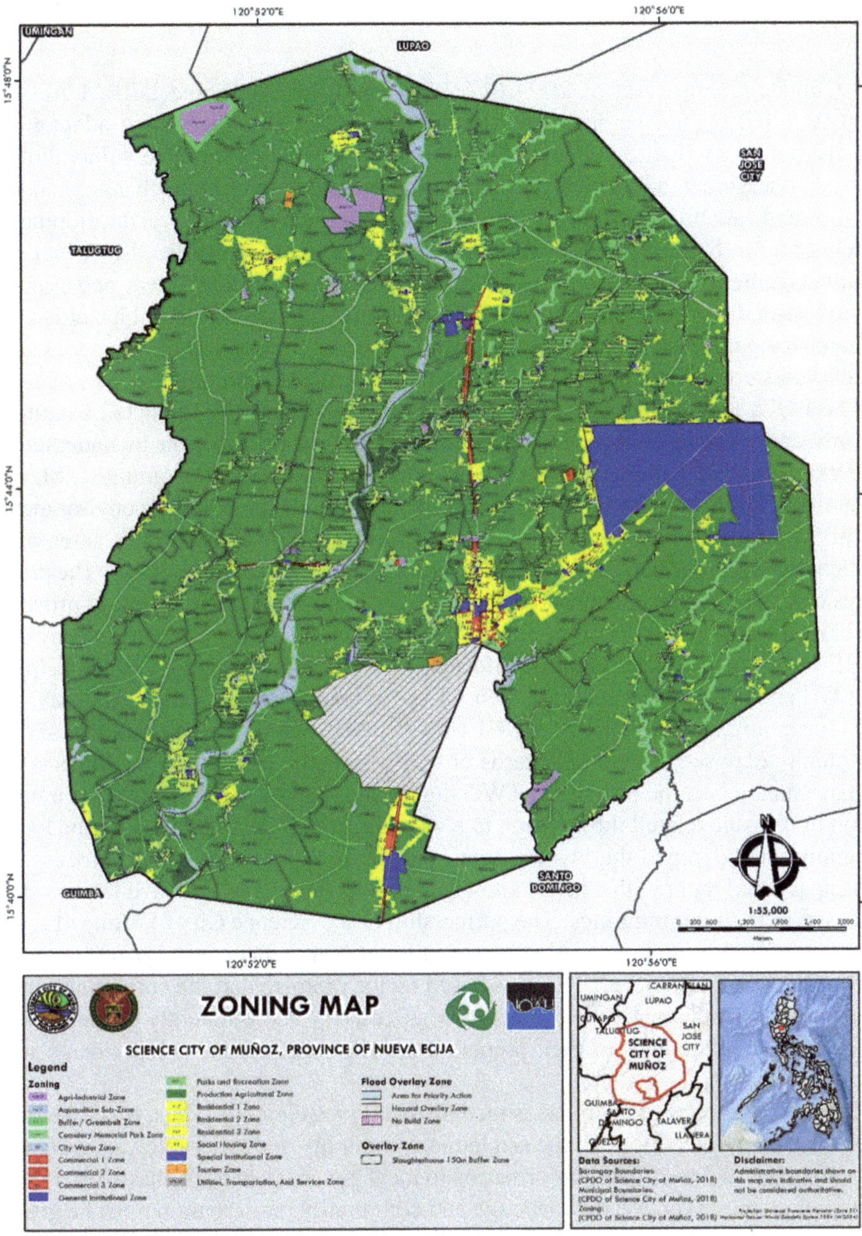

Fig. 14.8 Zoning map of SCM (Science City of Muñoz 2019)

14.10 Conclusion

The Climate Change Act of 2009 (RA 9729) and the Philippine DRRM Act of 2010 (RA 10121) are landmark legislative acts in the climate change adaptation and disaster risk reduction efforts of the country. These laws stipulate vulnerability and risk assessment, adaptation and mitigation measures to climate change, gender equality, and capability building, which must be integrated into the local development plan in order to build resilient communities. They promote a disaster risk reduction and management approach that is holistic, comprehensive, integrated, and proactive to lessen the socio-economic and environmental impacts of natural hazards and climate change. It also specifies the involvement and participation of all sectors and stakeholders concerned, at all levels, especially the local community.

The LGU, as the frontline agency mandated to address the impacts of natural hazards and climate change, must have the necessary information to understand their exposure and vulnerabilities to reduce disaster risk in their community. Mainstreaming CDRA into the CLUP enriches the analysis of the planning environment and allows stakeholders and decision-makers to formulate risk-informed, safer, and resilient local plans (HLURB- CCC-UNDP-Australian Government 2015). The principles of CCA-DRR are incorporated at the very beginning of the planning process and embedded in the vision of the city. In the case of SCM, the assistance given by the UPRI was able to provide scientific information and technical support, which helped the LGU and stakeholders prepare their CLUP to become a resilient community.

Mainstreaming CCA-DRR in the CLUP of SCM ensured that stakeholders of the community exposed to natural hazards or will be greatly affected by the impacts of climate change became part of the TWG during the preparation process. Communication of disaster-related information in a way that could be understood by the local community encouraged the involvement of participants in the planning process. It also capacitated them with a better knowledge to become more effective leaders and members of their communities. The partnership of the Science City of Munoz LGU, local communities, and the academe, was tailored in a manner that fostered ownership at the community level. This was based on the premise that the community that owns the plan implements their plans more effectively and efficiently because they believe that the outcomes of their preparations will benefit them as individuals and as a society.

The formulation of local plans is technical and participatory. Thus, it is essential to establish a partnership with the academic community and other relevant agencies that can supply science-based information to local government as the basis for policy formulation. However, local knowledge and community participation must be given equal importance to ensure a whole-of-society approach.

Acknowledgements The authors of this paper would like to acknowledge the contribution of the researchers of the UP Nationwide Operational Assessment of Hazards (NOAH) Center, Atty. Mark Anthony M. Gamboa, and En.P. Charlie Cabotake in their valuable contribution in the completion of the CLUP of the SCM. The Local Chief Executive and the members of Sangguniang Panglungsod together with the City Planning and Development Office (CPDO) and other departments of the LGU of SCM, as well as the members of TWG, barangay representatives, stakeholders, and participants, who made the successful completion of the CLUP with expertise, dedication, and participation. The UPRI would also like to acknowledge the support of the entire UP community as the institute continues to Serve the People.

References

And Program On Climate Change, Creating For This Purpose The Climate Change Commission, And For Other Purposes. Climate Change Act of 2009. Official Gazette of the Republic of the Philippines, October 23. https://www.officialgazette.gov.ph/2009/10/23/republic-act-no-9729/

Commission Resolution No. 2019-001(2019) Resolution Adopting A National Climate Risk Management Framework To Address the Intensifying Adverse Impacts of Climate Change. National Climate Risk Management Framework. Climate Change Commission, March 11. https://climate.gov.ph/public/ckfinder/userfiles/files/Resolutions/Resolution%202019-001.pdf

DILG-BLGD (2008) Rationalizing the local planning system: a soucebook. Department of the Interior and Local Government - Bureau of Local Government Development, Quezon City

EO 648, s. 1981 (1981) Reorganizing the human settlements regulatory commission. Charter of the Human Settlements Regulatory Commission. Official Gazette of the Republic of the Philippines, February 7. https://www.officialgazette.gov.ph/1981/02/07/executive-order-no-648-s-1981/

HLURB (2013) Comprehensive land use plan guidebook Volume 1: The planning process. Vol. 1. Housing and Land Use Regulatory Board, Quezon City

HLURB-CCC-UNDP-Australian Government (2015) Comprehensive land use plan guidebook: supplemental guidelines on mainstreaming climate change and disaster risks in the comprehensive land use plan. Housing and Land Use Regulatory Board, Quezon City

RA 10121 (2010) An act strengthening the philippine disaster risk reduction and management system, providing for the national disaster risk reduction and management framework and institutionalizing the national disaster risk. Philippine Disaster Risk Management Act of 2010. Official Gazette of the Republic of the Philippines, May 27. https://www.officialgazette.gov.ph/2010/05/27/republic-act-no-10121/.

RA 7160 (1991) An act providin for a local government code of 1991. Local Government Code of 1991. Official Gazetter of the Republic of the Philippines, October 10. https://www.officialgazette.gov.ph/1991/10/10/republic-act-no-7160/

RA 7279 (1992) An act to provide for a comprehensive and continuing urban development and housing program, establish the mechanism for its implementation, and for other purposes

RA 9729 (2009) An act mainstreaming climate change into government policy formulations, establishing the framework strategy

Urban Development and Housing Act of 1992. Official Gazette of the Republic of the Philippines, March 24. https://www.officialgazette.gov.ph/1992/03/24/republic-act-no-7279/

Science City of Muñoz (2018) Focus group discussion. Science City of Muñoz

Science City of Muñoz (2019) The comprehensive land use plan for the science city of Muñoz: Volume 1. Science city of Muñoz, Nueva Ecija. Science City of Muñoz, Nueva Ecija

The 1987 Constitution (1987) The constitution of the Republic of the Philippines. Official Gazette. Republic of the Philippines, February 02.https://www.officialgazette.gov.ph/constitutions/1987-constitution/

Alfredo Mahar A. Lagmay Ph.D. is a Professor at the National Institute of Geological Sciences, University of the Philippines Diliman, and concurrently Executive Director of the UP Resilience Institute. In 2013 he received the Professional Regulation Commission (PRC) Outstanding Professional of the Year Award in the field of Geology and Outstanding Filipino (TOFIL) award. He was also the first Asian to receive the Plinius Medal by the European Geosciences Union in 2015.

Joy T. Santiago is a Chief Science Research Specialist at the University of the Philippines Resilience Institute. She received the Outstanding Student Poster and PICO (OSSP) Award from the European Geosciences Union in 2016, and the International Visitor Leadership Program on Disaster Preparedness and Emergency in 2017.

Jerico E. Mendoza is a Chief Science Research Specialist at the University of the Philippines Resilience Institute. He was part of the Young Southeast Asian Leaders Initiative professional fellowship program on Environmental Sustainability in 2017.

Chapter 15
Financing in the Time of Climate Emergency

Erwin A. Alampay and Dennis G. dela Torre

Abstract The spate of typhoons in the last decade, and the El Niño that brought with it drought and forest fires across Mindanao, are watershed events demonstrating the accelerated impacts of climate change in the Philippines. The increased frequency and severity of climate impacts increases the burden of financing to address wide ranging loss and damage the country suffers. This chapter asks whether the Philippines' climate and disaster financing policies are efficiently and effectively addressing climate impacts in the country, especially to its poor municipalities, and communities. This chapter examines such policies by integrating outcomes of climate finance-related projects of the Center for Local and Regional Governance (CLRG) in the past five years. This includes cases of El Niňo 2015–2016 and the problem of funding slow-onset climate disasters, as well as assessments in the aftermath of Typhoons Ondoy, Pepeng, and Yolanda. It also looks at the issues and challenges hounding the implementation of climate and disaster risk reduction (DRR) funds such as the People's Survival Fund (PSF), risk transfer mechanisms (RTMs), and national DRR funds, especially with respect to budget execution. This chapter recommends urgent climate financing reforms as part of "transformative governance" to address the climate emergency, including the use of a climate finance policy framework.

Keywords Climate emergency · Disaster risk financing · Climate finance

Key Points

- The government incurs opportunity costs as disaster and climate risk financing predominantly depend on public funds. The prioritization of allocation towards disaster risk reduction (DRR) and climate change adaptation (CCA) means

E. A. Alampay (✉)
University of the Philippines Diliman, Quezon City, Philippines
e-mail: eaalampay@up.edu.ph

D. G. Torre
University of the Philippines Resilience Institute, Quezon City, Philippines

K. B. Berse et al. (eds.), *Climate Emergency in the Philippines*, Disaster Risk Reduction, https://doi.org/10.1007/978-981-99-7804-5_15

that funds can be diverted from other fundamental public services and public investments for development.

- Innovative measures on disaster and climate risk financing must be developed to properly address the demand for funding recovery and rehabilitation expenditure.
- There is a clear and urgent necessity for a climate finance policy framework for the Philippines where resilience needs are efficiently and effectively identified, based on a comprehensive climate and disaster risk assessment and vulnerability-based loss and damage assessment at the national/sectoral and local levels.

Summary for Policymakers

Several key lessons can be drawn from the case studies on financing in the time of climate emergency.

International commitments to finance climate change adaptation are severely lacking as most initiatives are still shouldered by the country. Less than two percent of the financing is in the form of grants, demonstrating how the rest of the world does not share in this burden.

The case of Climate Change Expenditure Tagging (CCET) in the Philippines shows that international commitments to finance climate change adaptation initiatives, especially commitments from the Global North, are severely lacking as most initiatives are still shouldered by the country.

- Only less than 2% of the financing is in the form of grants, demonstrating how the rest of the world does not share in this burden.
- An important aspect of transformative governance is to focus on the vulnerable, as there is greater impact on poorer LGUs, and the livelihoods of the poor (e.g., farmers, fisherfolk).

In recent years, insurance coverage has emerged as an important means for ensuring protection from disasters enabling communities to get back on their feet after a devastating loss.

- Despite the enactment of many laws mandating the insurance of LGU assets against disasters, public assets are still either not insured or are under-insured.
- There is a need for a working and effective insurance model for funding LGUs and these must be developed and scaled.
- While there are local sources that can be leveraged in the meantime (e.g., DRR Funds), these may not be enough to cover all critical facilities of vulnerable LGUs. National Funds such as the People's Survival Fund (and related public funds such as subsidies for state insurance institutions) must be augmented.

During the El Niño in 2015 to 2016, interventions were late despite clear evidence of agricultural communities likely to be affected in terms of projected losses and damages.

- Existing models for relief and financing do not address new dimensions of the climate emergency.

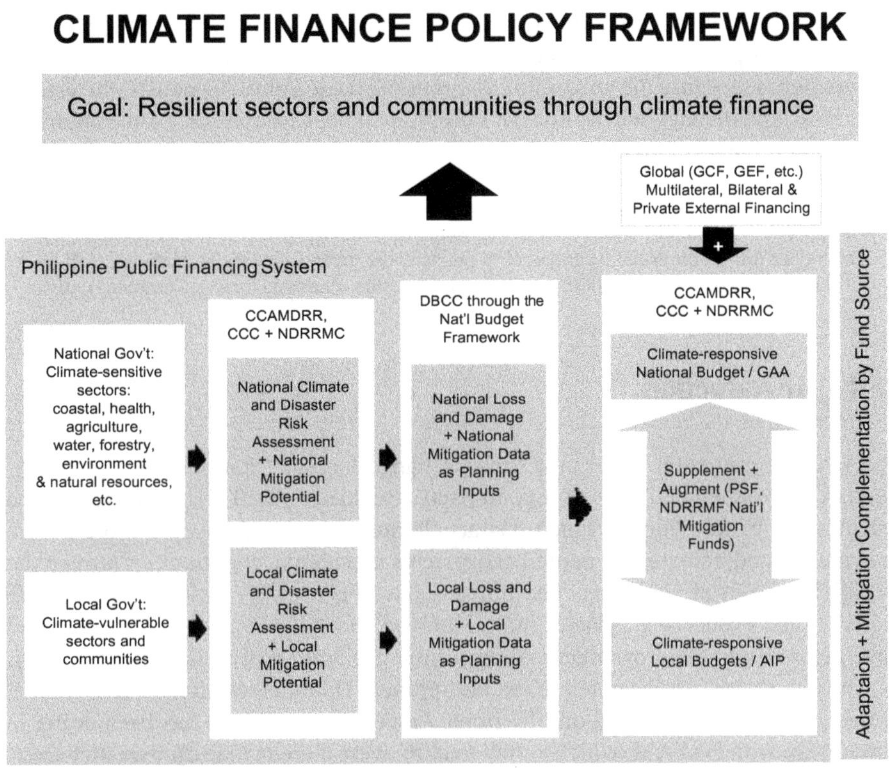

Fig. 15.1 Proposed "Climate Finance Policy Framework"

- The timely delivery of assistance to El Niño affected areas is also affected by the slow release of funds intended for disaster response covering El Niño.
- Some funds had to go through bureaucratic stages of approval, and under strict parameters which affect the timely disbursement and mobilization of these funds.

To address the problem of inadequate climate financing, the following specific policy guides are recommended (Fig. 15.1):

- A climate finance policy framework for the Philippines is necessary where resilience needs are efficiently and effectively identified. The framework has to be based on a comprehensive climate and disaster risk assessment, loss and damage assessment both at the national/ sectorial and local levels based on their vulnerability;
- The Philippines, being a highly vulnerable developing country, must have a climate finance policy. This is for the purpose of achieving climate resilience for its people and also to systematically access climate funds (both for adaptation and mitigation) provided under the Paris Agreement as a nationally intended strategic move; and,

- The policy challenge of addressing institutional fragmentation on climate finance in the Philippines is urgent because of the projected impacts under a climate emergency, because the longer this institutional fragmentation persists, the greater the loss and damage will be for the Filipino people, especially poor communities in vulnerable local government units.

The policy challenge of addressing institutional fragmentation on climate finance in the Philippines is extremely urgent due to the worsening impacts under a climate emergency. The longer this institutional fragmentation persists, the greater the loss and damage will be for the Filipino people, especially poor communities in vulnerable local government units.

15.1 Introduction

The Intergovernmental Panel on Climate Change has said that there is high likelihood of unavoidable climate change impacts (Benzie et al. 2018). There have been recent signs of this impact which has accelerated faster and become more severe than anticipated, threatening natural ecosystems and human communities around the world (Ripple et al. 2020, 8). The impacts have been more evident and felt with recent global events (e.g., glacier melt, Amazon and Australian wildfires, etc.) and have given rise to the movement calling for the declaration of a climate emergency.

The impact has also been felt in the Philippines. This was reflected in the series of projects the Center for Local and Regional Governance (CLRG) has been doing in partnership with OXFAM, which coincidentally were all related to climate financing.

This chapter will present a synthesis of three examples of CLRG's work on climate financing while also briefly touching on watershed climate events in the Philippines. The first case is El Niño in 2015, which is meant to illustrate the problem of financing in atypical disasters. The second case is devoted to climate change expenditure tagging (CCET) in the country. This will highlight how much is being financed and how much of this is shouldered by the country. Finally, it then discusses the challenge of climate financing for local government units, based on both the CCET and the Meso-scale Insurance for Disaster Readiness and Recovery (MINDER) projects.

Prior to presenting the cases, a short background on climate financing will be discussed. This will be followed by a short summary of policies related to climate financing. The lessons from the three projects serve as a take-off point for this policy review which asks: Are our climate and disaster financing policy efficient and effective in addressing climate impacts in the Philippines?

15.1.1 Background

The Paris Agreement, which the Philippines ratified, contains actions on addressing climate impacts. Article 8 emphasizes that vulnerable countries must avert, minimize, and address "loss and damage associated with the adverse effects of climate change, including extreme weather events and slow onset events." Based on Article 8, Section 4 of the Paris Agreement, "risk insurance facilities, climate risk pooling, and other insurance solutions" should be prioritized. Articles related to climate finance are discussed in the section on the Paris Agreement below. However, while the Philippines needs a lot of adaptation financing to address loss and damage due to climate and disaster impacts, a significant amount of globally available climate financing is allocated for mitigation.

The Philippines is among the most disaster-prone and climate-vulnerable countries in the world. The country's high vulnerability to natural disasters may be attributed to factors such as lack of land barriers, accelerating environmental deterioration, unsustainable development practices, and growing population.[1] Between 1985 and 2015, the country experienced 410 natural disasters, which led to over 40,000 deaths and 23 billion dollars worth of damages (CLRG 2018).

More recent data from 2010 to 2015 show a total of 96 climate-related disasters struck the country. This consisted of 44 floods, 51 storms, and one drought.[2] Eleven million people per year, were affected by these disasters during that period, with damage to property, crops, and livestock amounting to 750 billion pesos[3] and took up almost 9% of the country's annual national budget (Alampay et al. 2017, 2).

It is in this light that climate finance becomes crucial. Previous studies have reported, however, that the primary challenge in tracking international climate finance in developing countries is the absence of proper climate finance definitions and a criteria and method of for tagging (Tirpak et al. 2014).[4]

In the Philippines, climate finance is defined in RA 10174 (the Act establishing a People's Survival Fund (PSF)) as "resources that have been allocated or may be utilized towards climate change adaptation and mitigation requirements of the country and its vulnerable communities" (Sec 2e).

15.1.2 Of Climate Risks and Risk Transfers

Adopting a transboundary view of climate risk explicitly recognizes the interconnection between people, ecosystems, and economies in a globalized world (Benzie et al. 2018). Hence, while the most affluent countries have been known to contribute most for the historical greenhouse gas (GHG) emissions and have the greatest per capita emissions (Rippleet al. 2020, 9), the biggest rise in natural disasters have been in low-income countries, which have twice the global rate in disaster incidences

[1] This refers to the National Strategic Framework on Climate Change promulgated by the CCC in 2010. Note that RA 9729 previously referred to this document as the National Framework Strategy on Climate Change, which was corrected under its amendatory law, RA 10174.

[2] See Section 9, paragraphs d, e, and f.

[3] GOVPH. (n.d.). Retrieved November 27, 2016, from http://psf.climate.gov.ph/about/.

[4] See https://www.lloyds.com/.

(Barrett et al. 2007, 1). As a result, climate shocks devastate poor communities with distressing frequency. It is for this reason, that wealthier countries are also being called to support poorer nations in this fight (Ripple et al. 2020, 11), based on the principle of climate justice. While risk transfer is a market-driven strategy advocated by developed countries and multilateral institutions, its operationalization requires the formulation of clear climate finance policies in recipient countries, and policy responses to address some constraints.

The demand for risk transfer instruments in emerging markets is constrained by market gaps, lack of regulatory frameworks, lack of data on disaster risk, a lack of a culture of risk financing, and the reluctance of large reinsurance market players to invest in the development of small risk markets (Arnold 2008, 8). This dearth compounds the problem of ineffective and inefficient *ex ante* and *ex post* strategies to manage and cope with climate risks and shocks in developing countries (Barret et al. 2007, 1), as will be discussed in the Philippines' experiences later.

RA10174 also amended Sec 9f of RA 9729 that says it will "create an enabling environment for the design of relevant and appropriate risk-sharing and risk transfer instruments." It mentions renumbering the part of the PSF (RA 9729s 18) whereby the PSF is established as a special fund in the national treasury for the financing of adaptation programs and projects based on the national strategic framework.[5] One billion pesos was appropriated as an opening balance in the General Appropriations Act (GAA) for this. One of the uses of the fund included risk transfer (Sec 20(f)) to serve as a guarantee for risk insurance needs for farmers, agricultural workers, and other stakeholders. The PSF is meant to be supplementary and encourages counterpart funding among local governments, community organizations, private sector, and other entities. While it is hoped that insurance can become an adaptation and risk management tool, the climate emergency will present new challenges, as disaster impacts also grow, and affect the insurability of assets (Arnold 2008, 8).

15.2 Policies On Climate And Disaster Financing

International and domestic laws on climate and disaster financing already abound and are discussed here separately.

15.2.1 Bases in Domestic Law

This section examines the policies on financing climate change and disasters based on assessments and institutional reviews made in the aftermath of Typhoons Ondoy, Pepeng, and Yolanda. The laws are particular to the financing policies and strategies arising out of Republic Act (RA) 9729, as amended, or the Climate Change Act of

[5] See footnote 1.

2009, and RA 10121, or the Philippine Disaster Risk Reduction and Management Act of 2010. These include national climate and DRR funds and financing mechanisms created. Executive enactments formulated by government shall be discussed with respect to institutional mechanisms established on climate and DRR financing in a separate section on climate finance policy framework.

Legal mandates on climate finance is based on RA 9729, as amended, and RA 10121. In RA 9729, the Climate Change Commission (CCC) has the following climate finance-related powers: (1) Recommend legislation, policies, strategies, programs on and appropriations for climate change adaptation and mitigation and other related activities; and, (2) Recommend key development investments in climate-sensitive sectors such as water resources, agriculture, forestry, coastal and marine resources, health, and infrastructure to ensure the achievement of national sustainable development goals (underscoring ours). (Section 9). It is also empowered to "create an enabling environment for the design of relevant and appropriate risk-sharing and risk- transfer instruments."[6] However, in RA 10121 (Section 6) the National Disaster Risk Reduction and Management Council (NDRMMC) is the one mandated to "develop appropriate risk transfer mechanisms that shall guarantee social and economic protection and increase resiliency in the face of disaster."

Nonetheless, despite these national policies, their implementation at the local level still remains problematic (dela Torre and Alampay 2019, 1). RA 9729, as amended by RA 10174, and RA 10121 mandate not only the creation of an enabling environment for the development of risk transfer mechanisms (RTMs) but also appropriations and key development investments on climate finance to guarantee social protection and resiliency in the aftermath of climate change and disasters. However, the creation of an enabling environment on climate finance and risk transfer have barely progressed since the passage of these landmark legislations from a decade ago.

15.2.2 The Paris Agreement

On 23 March 2017, the Philippines ratified and acceded to the Paris Agreement. This is a landmark international legal regime outlining the global efforts on ambitiously addressing the climate change problem, based on each country-signatory's effort to formulate short- to long-term national strategies and actions. The Agreement recognizes "the urgent need to enhance the provision of finance, technology and capacity-building support by developed country Parties, in a predictable manner, to enable enhanced pre-2020 action by developing country Parties." Developed countries are expected to provide climate financing to developing countries like the Philippines, to fund national climate actions. In particular, the following text identifies the existence of climate funds that are to be mobilized for the purposes of this Agreement:

---

[6] See footnote 2.

1. *Decides* that the Green Climate Fund and the Global Environment Facility, the entities entrusted with the operation of the Financial Mechanism of the Convention, as well as the Least Developed Countries Fund and the Special Climate Change Fund, administered by the Global Environment Facility, shall serve the Agreement;
2. *Recognizes* that the Adaptation Fund may serve the Agreement, subject to relevant decisions by the Conference of the Parties serving as the meeting of the Parties to the Kyoto Protocol and the Conference of the Parties serving as the meeting of the Parties to the Paris Agreement; (underscoring ours)

Selected key provisions of the Paris Agreement relevant to climate finance are summarized in Table 15.1. Article 8(4f) provides for risk transfer, Article 9 covers climate finance mechanisms, Article 10 is on technology development and transfer, and Article 11 is on capacity development. Climate financing here is made available based on the general mechanisms provided by the Paris Agreement. Specific global funds such as the GCF may be availed of upon compliance with specific guidelines governing the access of each funding facility. Central to the availment by the Philippines of these funds identified by the Agreement is the formulation of a national comprehensive strategy on accessing them. In the case of the Green Climate Fund (GCF), the accreditation of the Land Bank of the Philippines (LBP) as the first direct access entity (DAE), and the endorsement by the Climate Change Commission (CCC) of the Development Bank of the Philippines (DBP) to be accredited as well, are welcome developments as the country seeks greater access to the said Fund. While the CCC had received readiness grants from the GCF amounting to 992,000 dollars, both were under review by the Department of Finance (DOF). As will be discussed later, the Philippines still does not have clear institutional mechanisms on how to systematically and successfully tap the said funds.

Coordinated efforts between the national and local governments are needed to address disaster and climate-related concerns; hence, the role of NDRRMC as an oversight is crucial. The council gives advice on the allocation of funds from the National Disaster Risk Reduction and Management Fund (NDRRMF), which serves as an important financial resource in addressing disaster-related needs. However, this cannot be accessed unless the agencies have fully depleted their Quick Response Fund (QRF), and require the presidential approval prior to disbursement.

Aside from the NDRRMF, the government also maintains a calamity fund which is utilized for the provision of humanitarian aid and infrastructure support in the affected areas. Likewise, agencies, such as the Department of Agriculture (DA) and the Department of Social Welfare and Development (DSWD), maintain their own QRF for the provision of emergency and immediate needs during disasters. Furthermore, the QRF can also be accessed without the recommendation and approval of the NDRRMC and the Office of the President, in contrast to the prerequisites in accessing the calamity fund (Alampay and dela Torre 2019).

The People's Survival Fund (PSF) is another funding source intended to address the impacts of climate change. It was created as an annual fund accessible to LGUs and accredited local/community organizations for the implementation of projects

Table 15.1 Paris Agreement Provisions on Climate Finance

Specific Provisions	Text
Article 7(13)	Continuous and enhanced international support shall be provided to developing country Parties for the implementation of paragraphs 7, 9, 10, and 11 of this Article, in accordance with the provisions of Articles 9, 10, and 11
Article 8(4f)	Accordingly, areas of cooperation and facilitation to enhance understanding, action, and support may include (f) Risk insurance facilities, climate risk pooling, and other insurance solutions;
Article 9(1)	Developed country Parties shall provide financial resources to assist developing country Parties with respect to both mitigation and adaptation in continuation of their existing obligations under the Convention
Article 9(3)	As part of a global effort, developed country Parties should continue to take the lead in mobilizing climate finance from a wide variety of sources, instruments, and channels, noting the significant role of public funds, through a variety of actions, including supporting country-driven strategies, and taking into account the needs and priorities of developing country Parties. Such mobilization of climate finance should represent a progression beyond previous efforts
Article 9(4)	The provision of scaled-up financial resources should aim to achieve a balance between adaptation and mitigation, taking into account country-driven strategies, and the priorities and needs of developing country Parties, especially those that are particularly vulnerable to the adverse effects of climate change and have significant capacity constraints, such as the least developed countries and small island developing States, considering the need for public and grant-based resources for adaptation
Article 10(6)	Support, including financial support, shall be provided to developing country Parties for the implementation of this Article, including for strengthening cooperative action on technology development and transfer at different stages of the technology cycle, with a view to achieving a balance between support for mitigation and adaptation
Article 11(1)	Capacity-building under this Agreement should enhance the capacity and ability of developing country Parties, in particular, countries with the least capacity, such as the least developed countries, and those that are particularly vulnerable to the adverse effects of climate change, such as small island developing States, to take effective climate change action, including, inter alia, to implement adaptation and mitigation actions, and should facilitate technology development, dissemination and deployment, access to climate finance, relevant aspects of education, training, and public awareness, and the transparent, timely and accurate communication of information
Article 11(3)	All Parties should cooperate to enhance the capacity of developing country Parties to implement this Agreement. Developed country Parties should enhance support for capacity-building actions in developing country Parties

that will better equip vulnerable communities to deal with the impacts of climate change.[7] However, it has not been widely accessed despite being available since 2015. This can be partly explained by the limited capabilities among LGUs and

[7] See footnote 3.

vulnerable communities in understanding how to comply with the requirements of the PSF (Alampay and dela Torre 2019).

15.3 Climate Financing Cases in the Philippines

This section briefly describes three recent CLRG projects that have implications on climate financing. They are presented starting from a "macro level" of financing, down to a "micro-level" financing. As such, they are not chronologically presented. The first case involves climate change tagging. This tries to situate financing on a global and national scale. The second case involves financing among local governments, and the third discusses financing impacts on vulnerable communities and individuals (e.g., farmers and fisherfolk).

15.3.1 Case 1: Climate Change Expenditure Tagging (CCET)

The Climate Public Expenditure and Institutional Review (CPEIR) is "*a starting point to mainstream climate change into the budgeting and planning process*"[8] among 20 countries, including the Philippines. The CPEIR recommended the adoption of "*Climate Budget Tagging*" as a tool to "*identify and measure the climate-relevant expenditure in the budget system.*"

Climate Change Expenditure Tagging (CCET) was implemented by the government in 2015 through Joint Memorandum Circular No. 2015-01 (JMC 2015-01). [9]Implemented It is a tool for tracking climate change expenditures and was implemented by the Department of Budget and Management (DBM). CCET can facilitate the country's readiness for accessing, administering, and coordinating flows of domestic and international climate finance. It can also be used to assess how the Philippines responds to climate change and can help promote coherence and convergence of policies and programs among government agencies and LGUs.

The National Economic Development Authority (NEDA) and the Climate Change Commission (CCC) also have roles in climate budget tagging to ensure a responsive national budget for rehabilitation in the aftermath of El Niño (Alampay and dela Torre 2019). The development of CCET is indicative of the priority being given by the Philippine Government to address the issue of Climate Change. According to JMC 2015–01, the purpose of CCET is "to identify, prioritize, and tag climate change P/A/

[8] See http://unfccc.int/files/cooperation_and_support/financial_mechanism/standing_committee/application/pdf/takeshi_kuwabara_biography.pdf.

[9] See https://www.dbm.gov.ph/index.php/dbcc-matters/about-the-dbcc.

Ps..." CCET was also applied in the succeeding budgets (FY 2016–2017) following FY 2015.[10]

CLRG's review found increasing local and international investments for climate change adaptation initiatives in the Philippines, particularly for the infrastructure and agriculture sectors. There was also an equal mix in local and international (ODA) funding. However, less than two percent (2%) of the financing was in the form of grants. Hence, much of the financing for climate change adaptation initiatives has been shouldered by the Philippines (Alampay et al. 2017, 1).

15.3.2 Case 2: Meso-scale Insurance for Disaster Readiness and Recovery (MINDER)

The Philippine government shoulders majority of the burden in recovering from losses and in rehabilitating impacted communities during disasters.

Most of the work is performed at the local level. It is more difficult for poorer LGUs to do this because they possess less financial resources (e.g. lower Internal Revenue allotments; smaller tax base), and yet, they are also more vulnerable to disaster and climate impacts.

Fiscal vulnerability to climate impacts, however, can be reduced by means of disaster risk financing and ex ante or preventive measures like insurance. In fact, all government offices, except second to sixth-income class municipalities, are mandated by the Property Insurance Law (RA 656) to avail of insurance from the mandated government insurance provider, the Government Service Insurance System (GSIS). This exception leaves the availment of second to sixth-class municipalities to their own discretion, such that fourth- to sixth-income-class municipalities have been found to have lower insurance take-up. Hence, poorer LGUs are less likely to be insured.

OXFAM's MINDER Project was intended to design a viable insurance program for third to sixth-income-class municipalities by looking at feasible insurance options. Findings from the project suggest that among the factors that result in low insurance availment are low appreciation and the low awareness among LGUs regarding its importance, benefits, and cost, as well as perceived unaffordability. These contribute to insurance being less prioritized as it competes with other programs and projects that need to be locally financed (CLRG 2018).

Higher insurance availment rates, however, do not automatically translate to optimum coverage. What must be insured are the infrastructure critical to the operations of the municipalities. Insurance coverage should also consider that the perils covered coincide with the municipalities' actual local exposure and vulnerabilities. Insured municipalities often insure their municipal halls and market buildings. Municipal halls are considered a critical facility since most of the local government

[10] See https://www.greenclimate.fund/about/resource-mobilisation.

operations and transactions originate there. Since the minimum coverage for peril insured is for fire and lightning, all properties that the municipalities have insured are assumed to be covered against this peril. However, based on the National Disaster Risk Reduction and Management Plan 2011–2028, the most frequently occurring hazards in the country are actually due to typhoons and floods. Hence, there is an apparent disconnect between the minimum insurance coverage and the disasters experienced by LGUs. A large proportion of LGU properties have been found to be uninsured against typhoons and floods (CLRG 2018).

15.3.3 Implications on Climate Financing

In traditional insurance, the cost of insuring facilities is dependent on the number and kind of facilities they indemnify, the perils they insure against, and the level of risk that may affect the possibility of payout. The premium cost considers the assessed replacement cost of a property in case of damage and its exposure and vulnerability to certain risks, among others.

The insurance data on second- to sixth-class municipalities provided by GSIS shows the total amount that LGUs could claim as payout for the certain amount of premium that they have paid covering varying types of perils. Based on this, the average percentage of premium over TSI shows the portion of amount that an LGU would shoulder for an equivalent amount of expected payout. Coverage for additional perils entails an additional 0.06% for typhoon and 0.19% for both typhoon and flood. This means that with additional perils covered, the average premium increases, while the average TSI decreases. But considering the increased likelihood of the additional perils due to climate change, the added expenditure would be justifiable.

When recovering from disasters, one OXFAM partner LGU specified that the funds are sourced out internally (from their own DRRM fund) and externally (from the provincial and national government and international donors). These LGUs also indicated typhoon and flooding as their most commonly experienced disasters and expressed interest in insuring their critical facilities, but cited their limited budget and other priority programs as hindrance for financing it. LGUs can normally charge their insurance coverage from their general fund, maintenance and other operating expenses (MOOE), insurance expenses, and their DRRM Fund. However, few LGUs have charged insurance to their DRRM fund, even though it is an allowable expense under this fund (CLRG 2018).

Local DRRM funds constitute 5% of the total budget of LGUs, 30% of which could be utilized as QRF and 70% for the other DRRM items. According to the 2016 Commission on Audit (COA) Reports, some LGUs have unspent balances under their LDRRM funds. Unspent QRF balances must be secured in a trust fund that could be used for the same purpose for 5 years. These findings suggest that although low-income class municipalities may have less budgets, proper utilization of their existing funds and prioritization of significant projects with beneficial outcomes is still possible. These, however, do not guarantee that the local DRRM funds of municipalities would be adequate to cover the entire premium expenditure for disaster insurance. Other possible sources of financing or external funding might be needed and the amount would still depend on the number of vulnerable facilities LGUs intend to insure and the other DRRM projects that need to be executed.

15.3.4 Slow Onset El Niño

In 2015 and 2016, the Philippines encountered one of the worst El Niños ever recorded. The timing of the phenomenon coincided with a national election; hence, it posed challenges in terms of timely allocating funds required to mitigate its impacts.

To respond to the impending crisis, President Aquino instructed the formulation of the Roadmap to Address the Impacts of El Niño, also known as RAIN, with a budget of nineteen billion pesos (PHP 19B). RAIN was spearheaded by the National Economic and Development Authority (NEDA) as part of the Philippine El Niño Task Force (ENTF). This was established in August 2015 with a goal of having a comprehensive approach in terms of coordinating efforts with various line agencies and their corresponding regional units.

Key areas that were most affected by the 2015 El Niño include (1) food security, (2) energy security, (3) health concerns, and (4) safety concerns. Food security became a concern due to reduced food production and decreased farm income, which resulted in fluctuations in food prices. Meanwhile, energy security through the hydroelectric-powered generation was affected due to limited water availability. Health security, on the other hand, was threatened by an increased prevalence of communicable diseases as water, which is mainly used for hygienic purposes, became scarce. Safety was also identified as an affected area, mainly owing to higher occurrence of fire outbreaks. Lastly, livelihoods involving farming and fishing experienced disruptions due to El Niño.

Furthermore, the effects of the phenomenon extended beyond the Philippine borders. According to UN-OCHA (2016)7, it also affected eleven other countries in the region, with approximately 50 million individuals experiencing scarcity in food and water, coupled with diseases. With this, urgent interventions (i.e., provision of food assistance and health necessities, livelihood support) were carried out within 3–6 months.

Three factors were seen to be imperative to address the El Niño phenomenon: (1) livelihood recovery, (2) poverty alleviation measures, and (3) food security. In the Philippines, the Department of Agriculture (DA) estimated losses amounting to twelve billion pesos8.

Many local government units (LGU) down to the barangay level have declared a state of calamity due to the effects of El Niño. In May 2016, Region XII (SOCC-SKSARGEN) registered the highest number of LGUs with 34 declarations. At the provincial level, Cotabato recorded the highest number of declarations at twelve.

According to the DA's report on El Niño damages, Region VI—Western Visayas (85, 529 individuals), Region X—Northern Mindanao (78,752 individuals), and

Region XII—SOCCSKSARGEN (72,755 individuals) recorded the most number of affected farmers. Aside from its livelihood effects, El Niño highly impacted the food and water availability in the communities, leading them to reduce their meals to one or two per day (from normally three meals). To ease the burden of El Niño, many families were reported to have resorted to cash loans and livestock selling. Worst of all, some of them turned to petty crimes as a desperate means to ease hunger (Alampay and dela Torre 2019).

15.3.5 The Problem of Financing During El Niño

Climate change, including the occurrence of El Niño as one of its impacts, has compelled the government to make significant allocations to address its worsening effects. In 2016, the government allocated 6% of the government budget to climate change expenditure, as an indication of their commitment to the mitigation and adaptation to climate change. The biggest chunk of the budget went to water sufficiency projects (42%) and sustainable energy initiatives (38%). The agencies that were funded the most out of the total climate change expenditure budget, aside from the Department of Environment and Natural Resources (DENR), include the Department of Public Works and Highways (DPWH) and the Department of Agriculture (DA).

Despite having funds to address CCA and DRR concerns, there remain limitations in its use, especially for El Niño. As a slow-onset disaster, it cannot easily tap into the Quick Response Fund, since it can only be accessed in a post-disaster scenario. In fact, the sluggish release of funds was manifested when the RAIN project only utilized 1.1 billion pesos out of the 6.6 billion pesos funded under the GAA9.

Another limitation on the use of funds was exhibited when the Department of Social Welfare and Development (DSWD) was tagged with adverse Commission on Audit (COA) findings for extending "cash for work" programs to the El Niño-affected areas.10 Shoreline cleaning, food production, and planting trees were some of the activities under the program. Based on protocols, DSWD is only authorized to intervene when an LGU declares a state of calamity. As a result, they are implicated when they decide to use funds to support affected areas that have not officially declared a state of calamity.

Through the Philippine Crop Insurance Corporation (PCIC), the government has established another intervention—crop insurance. This mechanism supposedly protects the farmers from the aftermath of disasters that ruin their crops. On the other hand, the PCIC remains risk averse in covering crop losses situated in places with peace and order dilemmas like Maguindanao (Alampay and dela Torre 2019).

As a resolution, the National Development and Economic Authority (NEDA) together with the NDRRMC should review the parameters in financing response and recovery. Aside from that, there should also be reassessment on the usage of the Quick Response Fund, most especially for slow-onset disasters like El Niño (Alampay and dela Torre 2019).

15.4 Analysis

This section analyzes the cases by applying the lens of "transformative governance" as recommended by the International Panel on Climate Change (IPCC). Transformative governance here will discuss a policy framework applicable for the Philippine setting. The package of reforms is discussed, such as the need for conducting national climate and disaster risk assessments to guide financing for climate change, making national and local budgets climate-responsive, with options on addressing institutional issues.

Transformative governance is "an approach to environmental governance that has the capacity to respond, manage, and trigger regime shifts in coupled social-ecological systems (SESs) on multiple scales. Its objective is to actively shift degraded SESs to alternative, more desirable, or more functional regimes by altering the structures and processes that define the system (Chaffin et al. 2016). Applying this lens to the Paris Agreement, St. Claire and Aalbu (2016, 41–46) mention four transformative governance shifts emanating from the Paris Agreement: (1) shifting of responsibility towards subnational and non-state actors; (2) placing adaptation on par with mitigation in terms of importance to climate change; (3) the need to develop measures for transparency and accountability; and, (4) special attention to the most vulnerable to climate change. In this analysis, we try to extract key lessons from the climate finance cases above, then critique and expound on its applicability to climate finance governance, ending this with a climate finance policy framework the authors have formulated.

15.4.1 Transforming Climate Financing

Disaster and climate risk financing, at present, rely mostly on public funds, especially for recovery and rehabilitation. As a result, government incurs opportunity costs. This means that large amounts of public resources go to DRR and CCA rather than other basic public services or public investments for development. There is a need, in terms of transformative governance, to develop innovative measures on disaster and climate risk financing to properly address the demand for funding recovery and rehabilitation expenditure. In this regard, the three cases highlight some issues in financing arising from the climate emergency.

The case of CCET in the Philippines shows that international commitments to finance addressing climate impacts, especially commitments from the North are severely lacking. Even the Philippines has adequately met square on the needs to finance the needed climate change adaptation initiatives, most are still shouldered by the country. Noting that less than 2% of the financing is in the form of grants, clearly the rest of the world does not share in this burden. While St. Claire and Aalbu (2016) argue the need to shift responsibility towards subnational and non-state actors, consistent with the principle of subsidiarity, these are not necessarily applicable to financing, given that local communities and vulnerable LGUs, often do not have the needed funds for it. To that extent, the need to focus on the vulnerable is an important aspect of transformative governance, as the impact on poorer LGUs, and the livelihoods of the poor (e.g., farmers, fisherfolk) have been apparent in the cases.

The second case shows the importance of insurance as a proactive financing measure for alleviating and controlling disaster and climate risks in the Philippines. In recent years, insurance coverage has emerged as an important means for ensuring protection from disasters enabling communities to get back on their feet after a devastating loss. According to Lloyd's[11] an increase of 1% on insurance coverage (penetration) saves public spending for rehabilitation and recovery by 22%, or reduces the tax burden on the people by 22%. Despite the enactment of many laws mandating the insurance of LGU assets against disasters, public assets are still either not insured or are under-insured. [12]With the climate emergency, and as natural disasters in the Philippines worsen and intensify, the need for a working and effective insurance model becomes more compelling, especially for vulnerable LGUs. Models for funding LGUs have to be developed and scaled. While there are local sources that can be leveraged in the meantime (e.g., DRR Funds), these may not be enough to cover all critical facilities of vulnerable LGUs. This is where augmentation from national funds such as the PSF (and related public funds such as subsidies for state insurance institutions) can come in.

The last case demonstrated the impact of the climate emergency on livelihoods, especially among the most vulnerable being farmers and fisherfolk. In the 2015/16 El Niño crisis, the likely impacts, and areas likely to be affected were already known. There were already projections on how it will inevitably lead to massive loss and damage in agriculture. Despite this, interventions were not done with urgency, nor in a timely manner. Agencies were slow in mounting anticipatory actions for slow-onset events. It was contrary to the intent of both the NDRRM Law and the Climate Change Act.

When a taskforce is created, it consequently excercises the policy and planning functions on climate change and disaster risk reduction from the very bodies Congress mandated. This resulted in questions about mandates and led to problems in coordinating and implementing government's El Niño response.

The slow release of funds intended for Disaster and Emergency Response during El Niño, in turn, delays the delivery of assistance affected areas.

Some funds had to go through bureaucratic stages of approval, and strict guidelines, which affected the timely disbursement and mobilization of these funds. It highlights the problems with existing models for relief and financing that do not address new dimensions of the climate emergency.

[11] See footnote 4.

[12] See footnote 15.

15.4.2 *Critique of Transformative Governance*

15.4.2.1 Shifting of Responsibility Towards Subnational and Non-state Actors

This shift towards subnational actors apparently is an application of the subsidiarity principle or primacy of local governments in frontline implementation, which is a long-held governance principle now finding its relevance in climate governance. In the Philippines, this is also consistent with the tenets of local autonomy and devolution. The non-state actors can be construed as referring to civil society organizations (e.g., OXFAM), in recognition of the prominent role they play in Philippine democracy, especially on empowerment, transparency, accountability, and whole-of-society mobilization, with respect to climate governance, in addition to external development financing they can mobilize to flow to vulnerable communities.

15.4.2.2 Placing Adaptation on Par with Mitigation in Terms of Importance to Climate Change

Developing countries such as the Philippines tend to have lower adaptive capacity compared to developed countries. As a result, adaptation actions are prioritized over mitigation actions. This approach is reflected in budget prioritization where mitigation is only viewed as a support to adaptation efforts. In fact, this approach is explicitly stated in the National Framework Strategy on Climate Change (NFSCC). It says that the national priorities should be anchored with respect to adaptation, with mitigation only serving as a "function of adaptation" (CCC 2010, 5).

15.4.2.3 The Need to Develop Measures for Transparency and Accountability

Transparency and accountability are already regarded as key principles in Philippine governance. These are especially applicable in climate finance in terms of tracking fund flows (e.g., CCET), and ensuring delivery of funds to climate-vulnerable sectors, presuming that the said climate funds are adequate to address climate impacts, in the first place.

15.4.2.4 Special Attention to the Most Vulnerable to Climate Change

This special attention approach at the global level means that climate-vulnerable countries must be given priority with respect to climate fund flows. In the Philippine context, the Climate Change Act acknowledges the differential impacts of climate change and recognizes that vulnerable sectors, including the poor, bear the most burden on loss and damage due to climate impacts. Institutional mechanisms establishing fund flows in targeting and prioritizing the vulnerable must be integrated purposively into the country planning and budgeting frameworks both at the national and local levels.

Parenthetically, while this paper focused on the initial step for climate finance policy, the authors recognize that further research is needed on institutional issues on disaster financing with respect to its convergence areas with climate financing, considering that RA 10121 created the NDRRMC for this. In addition, further discussions are needed on gaps and barriers on the entire institutional processes on climate finance readiness in the Philippines, from tracking global fund flows into the Philippine public financial system, down to the subnational flows, and the verifying and monitoring of such.

Distilling the lessons on the climate finance cases by the authors using the transformative lens, certain recurring issues have surfaced. They can be summarized as such:

1. Lack of an institutional framework with respect to national agencies mandated to perform various aspects of climate finance; and,
2. Lack of an institutional process on climate finance within the public financial system to guide governance efforts from climate risk assessment until planning and budgeting both at the national and local levels.

To reiterate, there must be a climate finance group in the executive department of the Philippine government which purposively takes on the task of determining the adequacy, efficiency, and effectiveness of public funding for climate actions, both at the national and subnational or local levels. The Philippines is not lacking in legislative and executive enactments allowing the opportunity to formulate enabling policies on climate finance. The question is, who is willing to take the lead, and what institutional mechanism will be used.

15.5 Towards A Climate Finance Policy Framework

Given the key lessons above, a policy framework is recommended as a starting point in making the Philippine public climate governance more efficient and effective in addressing issues on financial resilience, focusing on the planning and budgeting aspects. This policy framework seeks to address institutional fragmentation by unentangling the climate finance budget prioritization aspect from the cabinet cluster system, and instead tasks the Development Budget Coordination Committee (DBCC)[13] created under EO No. 292, series of 1987, as the lead climate finance group. After all, it is the DBCC that formulates and issues the National Budget Framework guiding the planning and annual budgeting of NGAs and LGUs. Note that it is the DBCC that finalizes budget ceilings, based on revenue sources. The

[13] See footnote 15.

Table 15.2 Approved GCF funding proposals of selected Southeast Asian nations[16]

Rank	Country	Number of projects	Approved funding (in US$)	2018 global climate risk index ranking
1	Indonesia	5	213.3M	64
2	Viet Nam	3	146M	16
3	Cambodia	3	44M	38
4	Timor-Leste	1	22.4M	135
5	Philippines	2	15.6M	2

Sources https://www.greenclimate.fund/countries, https://germanwatch.org/en/17307

DBCC an consider the viability of blending climate financing from external entities as additional and steady sources, given the pronouncements under the Paris Agreement. Presumably, it is at the DBCC where policy discussions are conducted on development setbacks due to climate change and disaster impacts in the form of loss and damage, within the context of development financing. The task of finalizing climate finance budget prioritization presumes however that a comprehensive national climate and disaster financing strategy exists in the first place.

An emerging climate finance strategy is the drafting by the CCC of the GCF country program for the Philippines, to access this 100 billion dollars global fund.[14] This country program informs the GCF Board of the adaptation needs of the Philippines to guide it on which funding proposals should be prioritized. To date, the GCF has funded a total of 129 projects worldwide, 27 of which are with the private sector.[15] As of this writing however, this country program is still to be finalized, which may have been one of the reasons why the Philippines has not been able to substantially access the GCF, compared to other Southeast Asian countries. As Table 15.2 shows, the Philippines lags behind Southeast Asian countries despite being ranked as the second most climate-vulnerable country in the world based on the 2018 Climate Risk Index (CRI). Indonesia lords it over among Southeast Asian nations despite a low CRI rank of 64th.

Granting that a country program is finalized, it strategically bodes well for the Philippines to ensure that a comprehensive climate finance strategy that is truly national in scope and covers all external climate and disaster fund flows, and not just the GCF. In addition, it must have the proper climate finance architecture—with clear institutional mechanisms, and guided by country needs based on our national climate and disaster risks and vulnerabilities. In addition, climate and disaster financing must comply with the principles of climate justice, including additionality. **For these purposes, the "Climate Finance Policy Framework" shown below is suggested**.

---

[14] See footnote 10.

[15] See https://www.greenclimate.fund/sectors.

[16] Other Southeast Asian countries such as Thailand and Malaysia don't have approved projects yet (aside from funding for readiness support, which are not counted here), while Singapore and Laos don't have data.

To emphasize, there might be wisdom in conducting a Climate & Disaster Risk Assessment (CDRA) through the National Panel of Technical Experts created under RA 9729, and in partnership with the NDRRMC. A true national assessment can be formulated to provide a clear picture of both the country's climate and disaster profiles, instead of having such assessments separately conducted as contained in the NCCAP and the NDRRMP. RA 9729, as amended, and RA 10121 commonly mandate that CCA should be integrated with DRR. While both laws provide the institutional mechanisms for the CCC and the NDRRMC to coordinate with each other on these aspects, we cannot ignore that the policy mechanisms chosen by the Duterte administration are those specified under EO No. 24, series of 2017. The said EO mandates policy and program coordination and participatory governance on climate change and disasters through the Cabinet Cluster on Climate Change Adaptation, Mitigation, and Disaster Risk Reduction. Here, policy guidance should be made on the national and local loss and damage framework/assessment or in short, the climate and disaster analytics that will become the bases for planning inputs and budget estimates for annual national and local plans, as specific actions/interventions (programs, projects, and activities) are identified.[17] Hence, a comprehensive CDRA is a necessary step in determining the financing needs of the country. Note that the above framework includes the determination of the mitigation potential (both national and local levels) so that this can also be leveraged against external and domestic sources of mitigation funding. Finally, the CCC, NDRRMC, and the DBCC, after coordinating with each other to determine the magnitude of financing and funding gap on climate change and disasters, can make policy decisions on how to complement external sources of adaptation and mitigation multilaterally (Paris Agreement funds), bilaterally, and domestically.

15.6 Conclusion

This chapter showed gaps, issues, and challenges on budget execution of the national climate and DRR funds and financing mechanisms created by RAs 9729 and 10121. In presenting three cases that CLRG has undertaken with Oxfam, it has highlighted various institutional mechanisms on climate and disaster financing that illustrate the complexity and lack of coherence with respect to fund administration and lack of efficiency and effectivity of delivery mechanisms especially with respect to vulnerable LGUs.

Transformative governance involves a broad set of governance components, but requires additional capacity to foster new social-ecological regimes including increased risk tolerance, significant systemic investment, and restructured economies and power relations (Chaffin et al. 2016). Among climate risks, droughts, and floods impact more people around the world than other types of disasters (Barret et al. 2007, 3). In the Philippines, this was evidenced by El Niño in 2015–16 which caused

[17] See footnote 15.

droughts (Alampay and dela Torre 2019), while, on the other hand, local government partners of OXFAM indicated typhoon and flooding as their most commonly experienced disasters (CLRG 2018).

Recent experiences in the Philippines as illustrated in the cases here have emphasized the need for transformative governance as urgent policy responses are required to effectively address financing the climate emergency.

Climate finance governance must be deeply integrated in broad policy arenas (St. Claire and Aalbu 2016). It is transformative because the arena spans the global, national, and local:

- At the global level—nations have to be called out for their pledges.
- At the national—need to support capacities for lGUs to be insured, and to revisit processes for CCET and releasing of emergency funds for disasters.
- At the local—consider using DRRM fund for insurance/RTM.

Hence, the following specific policy guides are recommended:

1. There is clear and urgent necessity for a climate finance policy framework for the Philippines where resilience needs are efficiently and effectively identified, based on a comprehensive climate and disaster risk assessment, loss and damage assessment both at the national/sectoral and local levels based on vulnerability;
2. The Philippines, being a highly vulnerable developing country, must have a climate finance policy not only for the purpose of achieving climate resilience for its own people but also to systematically access climate funds (both for adaptation and mitigation) provided under the Paris Agreement as a nationally intended strategic move; and,
3. The policy challenge of addressing institutional fragmentation on climate finance in the Philippines is extremely urgent due to the worsening impacts under a climate emergency, because the longer this institutional fragmentation persists, the greater the loss and damage will be for the Filipino people, especially poor communities in vulnerable local governments.

References

Agreements, Government Policies/Laws: Paris Agreement on Climate Change, December 9, 2015 Climate Change Act, RA 9729, October 23, 2009

Alampay EA, dela Torre D (2019) Addressing slow onset disasters: lessons from the 2015–2016 El Niño in the Philippines. In: Leal Filho W (eds) Handbook of climate change resilience. Springer, Cham

Alampay EA, Dela Torre D, Asuncion XV, Eguia GT (2017) The state of climate finance in the Philippines—with a focus on adaptation and agricultural investments. The State of Climate Finance in the Philippines—with a Focus on Adaptation and Agricultural Investments', presented at the NCPAG international conference—changing times, new norms and sustainable development, Marco Polo Hotel, Pasig City, August 22–26 2017

Arnold M (2008) The role of risk transfer and insurance in disaster risk reduction and climate change adaptation. The Commission on Climate Change and Development (March 2008)
Barrett CB, Barnett BJ, Carter MR, Chantarat S, Hansen JW, Mude AG, Osgood DE, Skees JR, Turvey CG, Ward MN (2007) Poverty Traps and climate risk: limitation and opportunities of index-based risk financing (Working Paper). Technical Report Number 07-02
Benzie M , Adams K, Roberts E, Magnan A, Persson A, Nadin R, Kelin R, Harris K, Treyer S, Kirbyshire A (2018) Meeting the global challenge of adaptation by addressing transboundary climate risk. Stockholm Environment Institute (SEI) Discussion Brief- April 2018
Center for Local and Regional Governance (CLRG) (2018) Meso-scale Insurance for Disaster Readiness and Recovery (MINDER) Project (Final Report), March 2018
Chaffin B, Garmestani AS, Gunderson L, Benson M, Angeler DG, Arnold C, Anthony T, Cosens BA, Craig RK, Ruhl JB, Allen C (2016) Transformative environmental governance (October 2016). Annu Rev Environ Resour 41:399–423. SSRN: https://ssrn.com/abstract=2859472 or https://doi.org/10.1146/annurev-environ-110615-085817
dela Torre D, Alampay EA (2019) Risk transfer mechanisms: charting a strategy for local insurance. UPCIDS Policy Brief 2019–13.
Executive Order No. 292, Instituting the Administrative Code of 1987, July 25, 1987
Executive Order No. 24, Reorganizing the cabinet clusters system by integrating good governance and anti-corruption in the policy frameworks of all the clusters and creating the infrastructure cluster and participatory governance cluster, May 16, 2017
People's Survival Fund Act, RA 10174, August 16, 2012
Property Insurance Law, RA 656, June 16, 1951
National Disaster Risk Reduction and Management Act, RA 10124, May 27, 2010
Ripple WJ, Newsome TM, Baranrd P, Moomaw WR (2020) World Scientists' warning of a Climate Emergency. BioScience, January 2020/Vol. 70 No. 1.
St. Claire AL, Aalbu K (2016) The four transformative governance shifts emerging from COP21. In Wilhite H, Hansen A (Eds) (2016) Will the Paris Agreement save the world? An analysis and critique of the governance roadmap set out in COP 21. Oslo Academy of Global Governance Working Paper 2016.1. Oslo: University of Oslo, Centre for Development and the Environment, pp 41–46
Tirpak D, Brown L, Ronquillo-Ballesteros A (2014) Monitoring climate finance in developing countries: challenges and next steps. Working paper. World Resources Institute, Washington, DC. http://www.wri.org/publication/monitoring-climate-finance-developing-countries-challengesand-next-steps

Erwin A. Alampay, Ph.D. is a Professor of Public Administration at the National College of Public Administration and Governance at the University of the Philippines. He teaches organizational studies and management, research methodologies, voluntary sector management, and Managing ICTs and Knowledge. He has served as Director of the Center for Local and Regional Governance (CLRG) and Director of the Center for Leadership, Citizenship and Democracy (CLCD).

Dennis G. dela Torre serves as Chief Science Research Specialist of the UP Resilience Institute Research and Creative Work. He served as the National Policy Expert of the Philippines for the ASEAN Disaster Risk Financing and Insurance Program in 2017, and co-authored several climate finance papers for Oxfam in the Philippines. He served as Research Fellow of the UP Center for Local and Regional Governance (CLRG) and was formerly the Chief for Research and Policy Development of the Climate Change Commission.

Chapter 16
Reframing Climate Change: An Ecocritical Frame

Flaudette May V. Datuin

Abstract This chapter addresses the question, "if art historians, artists, or scholars and educators in the arts and humanities were to respond to the serious and urgent issue of climate change, what options do we have for confronting this crisis?" This chapter's response focuses on surfacing the deeply ingrained and largely unconscious positions and tropes that shape our relationship with the physical and natural elements that make up our ecosystem—our home. In this chapter, I argue that there is no innocent eye, only lenses through which we interpret the so-called "objective" phenomena like climate change. To be aware of these lenses is to develop a critical eye, one that will help us understand—reflexively and critically—climate change in its multiple dimensions. I will demonstrate these insights by sharing my experiences as part of an interdisciplinary team that teaches the General Education course "Disaster Risk Mitigation, Adaptation and Preparedness Strategies" (DRMAPS) in the University of the Philippines. In the process, I hope that the pedagogical frame and course delivery I will discuss in this paper will benefit policymakers and other teachers in the field of Education.

Keywords Arts and humanities · Interdisciplinary pedagogy · Ecocritical framework

Key Points

- The climate emergency must be understood in all its multiple and interdisciplinary dimensions.
- Understanding and responding to climate change involves going beyond the "binary" of climate change as a scientific and technological phenomenon (Problem of Ecology), on the one hand, and as a cultural Ecological Problem, on the other hand. An eco-critical method suggests a more well-rounded frame.

F. M. V. Datuin (✉)
University of the Philippines Diliman, Quezon City, Philippines
e-mail: fddatuin@up.edu.ph

K. B. Berse et al. (eds.), *Climate Emergency in the Philippines*, Disaster Risk Reduction, https://doi.org/10.1007/978-981-99-7804-5_16

- The General Education course, Disaster Risk Mitigation, Adaptation and Preparedness Strategies (DRMAPS) exemplifies the integration and collaboration of multiple disciplines in learning and teaching the problem of climate change and the climate emergency.

Summary for Policymakers

The most recent report from the IPCC predicts that some of the most severe social and economic damages from the rise in global temperatures could come as soon as 2040. If nothing is done then, 300 million people will be affected by sea level rise in 2050.

- Geographically, the Philippines is one of the countries in the Global South that is most exposed to risks from climate change. It is located in the western Pacific Ocean, surrounded by naturally warm waters that will likely get even warmer as average sea surface temperatures continue to rise.
- To some extent, according to the Climate Reality Project, "this is a normal pattern: the ocean surface warms as it absorbs sunlight. The ocean then releases some of its heat into the atmosphere, creating wind and rain clouds. However, as the ocean's surface temperature increases over time from the effects of climate change, **more and more heat is released into the atmosphere.** This additional heat in the ocean and air can lead to stronger and more frequent storms—which is exactly what we've seen in the Philippines over the last decade."

The climate emergency must be understood in all its multiple and interdisciplinary dimensions. it is not only a scientific problem, but also a cultural one.

On Framing the Problem of the Climate Emergency

The climate emergency must be understood in all its multiple and interdisciplinary dimensions. It is not only a scientific problem but also a cultural one.

- Lessons can be drawn from understanding the aesthetic, social, and cultural nature of climate change, and the role of deeply ingrained and widely held beliefs that shape our relationship to our ecology.
- Tropes, or widely shared cultural scripts, contain narrative stereotypes, that are so commonplace and deeply ingrained they have become "natural," "innocent", and largely unconscious.
- It is important for policymakers to understand and appreciate the tropes by which we frame and approach the climate emergency problem.
- The discipline of Art studies focuses on art, broadly conceived and defined. "Art," not only refers to the usual suspects: the "fine arts" of painting and sculpture, for example, but also to a broad array of creative products and environments—from photographs in social media, to street theater, rap music, weaving, movie billboards, malls, tabloids, realist drama, installations, montage, epics, urban planning, ballet, pictures, television, public monuments, and so on.

In the humanities, an important paradigm shift is the shift in the concept of art—from one of object (art work) to process (work of art) and from one of fixed and prescriptive outcomes to one of openness to many possibilities within the "noise" and routines of the everyday.

- Understanding and responding to climate change involves crossing the divide between climate change as a scientific and technological phenomenon (Problem of Ecology), on the one hand, and on the other hand, as a cultural Ecological Problem.

In the humanities, an important paradigm shift is the shift in concept of art—from one of object (art work) to process (work of art) and from one of fixed and prescriptive outcomes to one of openness to many possibilities within the "noise" and routines of the everyday.

- An eco-critical method and frame goes beyond these binaries, suggesting a more well-rounded encounter between, not only professionals and experts in the natural and physical sciences but also experts in the social sciences and the humanities.

On Teaching Disaster Risk Management

Policymakers can provide ways, platforms, and venues, not only for experts to talk to each other but also to fishers, poets, farmers, weavers, carpenters, and so on, so that we can work together through collaborations that are attuned, not only with the "science" of climate change but also keenly attentive to the manifold textures and shapes of the non-human and human worlds.

The General Education course, Disaster Risk Mitigation, Adaptation, and Preparedness Strategies (DRMAPS) targets students at the tertiary level at the University of the Philippines, but may also benefit all educators, including those at the basic education level.

- The humanities perspective and the disciplines of art studies contribute to the teaching of the interdisciplinary General Education course called DRMAPS that operates on an interdisciplinary framework from course content to course delivery.
- Based at the UP College of Engineering, DRMAPS is taught by an interdisciplinary team originally composed of two engineers (who conceptualized and originated the course), a psychologist, an educator, the then Dean of the UP College of Fine Arts, who specializes in visual communications, and myself.
- This team reviews the syllabus every semester, constantly finding the right mix or balance between the technical and scientific domains, with the cultural, political, and social areas of inquiry. We aim to teach students—themselves coming from diverse backgrounds and disciplines—to be firmly grounded on scientific data, facts, backstories, and findings creatively and imaginatively, in ways that go beyond stock knowledge and clichés.

DRMAPS operates on an interdisciplinary framework from course content to delivery, including the humanities and the art studies disciplines. We aim to teach students to be firmly grounded on scientific data, facts, backstories, and findings creatively and imaginatively, in ways that go beyond stock knowledge and clichés.

- For policymakers and educators, there should be a **training of trainers** who will initiate multidisciplinary conversations across disciplines, at national and local levels, similar to DRMAPS.
- The most concrete outputs will be **curricular and education modules** and strategies that can guide teachers, not only in explaining the science of climate change but also the significance of the arts and humanities as a springboard for action and transformation.

16.1 Introduction

16.1.1 Climate Change as an Ecological Problem

One of the most striking images I saw over the internet is that of an installation of tiny sculptures by Spanish street artist Isaac Cordal. It caught my attention, not only because the scene reminds us of "our own decline," as the artist puts it, but also because of how social media has appropriated the artwork with a new headline that warns us of an apocalyptic future, and it is one that will be upon us soon, if not already: "This is what politicians debating global warming will look like soon." As it turns out, this installation does not directly address global warming or climate change. Meant to draw attention to the faceless businessmen who run our capitalist global order, this 2013 piece of miniaturized figures—around 5–9 inches tall—submerged in a Berlin puddle are part of a larger street art project called "Cement Eclipses," and this particular piece is actually titled: "Follow the Leaders."[11] This mismatch between the artist's intention, that is as a commentary on electoral politics, on the one hand, and that of social media remediation, on the other, could be an indication of how art can generate new ways of seeing and understanding, directly or indirectly through images that provoke us to confront one—if not the most pressing issues of our time.

The most recent report from the United Nations Intergovernmental Panel on Climate Change predicts that some of the most severe social and economic damage from the rise in global temperatures could come as soon as 2040. Citing the result of a study, UN Secretary-General Antonio Gutierres, in a briefing after the Association of Southeast Asian Nations (ASEAN)—UN Summit in Thailand said that an analysis of the speed of the rise of sea level showed that it is faster than what was known earlier. The report foresees that "if we are not able to defeat climate change we will have, in 2050, an impact on the sea level rise over 300 million people," he said.[22]

Another study (released 10 Oct 2019) by Climate Central and published through the Nature Communications journal forecasts that at the rate sea levels are rising,

---

[1] *For more details on the works of Isaac Cordal, please see* http://cementeclipses.com/Works/follow-the-leaders.com *(Accessed 22 December 2022).*

[2] https://www.pna.gov.ph/articles/1084901 *(accessed 25 November 2019).*

several Asian countries will be under water by 2050, including the Philippines, Thailand, Bangladesh, Indonesia, India, China, Japan, and Vietnam. In the Philippines, low-lying areas of Manila, Navotas, Malabon, Pasay, and Bulacan in Luzon (Northern Philippines), as well as coastal areas in Visayas in Central Philippines (Kalibo, Aklan, and Roxas City in Capiz) and Mindanao in Southern Philippines (Cotabato City, along Datu Piang and Northern Kabuntan in Maguindanao), are likely to be affected by coastal flooding.[3]

People are getting worried about climate change, and yes, we do have a problem. Greg Garrard in his book Ecocriticism[4] suggests, via John Passmore, that this problem is perceived and addressed at two levels: as a scientific one, which he refers to as falling under the rubric of "Problems in Ecology" and a cultural one—as an Ecological Problem—and with it, the general feeling that something is not right somewhere in our planet, and that such awareness becomes widely perceived and disseminated, and in the case of Cordal's viral image, in social media.

The question I would like to explore in this paper is: how is this Ecological Problem framed? When the tiny, almost invisible figures representing electoral politics turn into politicians debating climate change amidst an apocalyptic scenario, I detect a deep-seated fear triggered by a widely shared cultural script or trope, perhaps hitherto unarticulated, until a compelling picture brought it to the surface of social media imagination. These tropes contain narrative stereotypes, that are so commonplace and deeply ingrained that they have become "natural," "innocent" and largely unconscious. My aim in this paper is to surface these tropes, particularly those concerned with how we view our relationship with the natural and physical elements that make up our ecosystem—our home, the English equivalent of "Eco," which comes from the root word "oikos" (ancient Greek) or home. I argue that there is no innocent eye, only lenses or positions through which we interpret the so-called "objective" phenomena—Problems of Ecology—like climate change.

I will outline the positions and tropes from a humanities framework, and the discipline of art studies, which studies art, broadly conceived and defined. "Art," not only refers to the usual suspects: the "fine arts" of painting, sculpture, for example, but also to a broad array of creative products and environments from photographs in social media, to street theater, rap music, weaving, movie billboards, malls, tabloids, realist drama, installations, montage, epics, urban planning, ballet, pictures, television, public monuments, and so on. When I study these art forms, I draw on the methods of various disciplines that frame our programs at the Department of Art Studies (AB and MA Art Studies), including art history, art criticism, curation, art theory and aesthetics, and art production.

---

[3] https://technology.inquirer.net/92179/rising-sea-levels-likely-to-sink-metro-manila-cities-by-2050-study#ixzz66EmRi1pR *(accessed 23 November 2019).*

[4] *First published by Routledge in 2012.*

16.2 Drmaps and the Arts

In the following account, I will demonstrate how the humanities perspective and the disciplines of art studies contribute to the teaching of an interdisciplinary General Education course called DRMAPS (Disaster Risk Mitigation, Adaptation, and Preparedness Strategies). I hope that the experiences I will share will demonstrate how the interdisciplinary framework can become operational, from course content to course delivery. In the process, I hope that what I share will benefit educators and policy makers, particularly in the field of Education.

Based at the UP College of Engineering, DRMAPS is taught by an interdisciplinary team assembled 7–8 years ago (and counting). The team is composed of two engineers (who conceptualized and originated the course), a psychologist, an educator, a then fine arts Dean, who specializes in visual communications and myself. This team-which has expanded to include faculty members from geology, mass communication, among others - reviews the syllabus every semester, constantly finding the right mix or balance between the technical and scientific domains, with the cultural, political, and social areas of inquiry. We aim to teach students—themselves coming from diverse backgrounds and disciplines—to be firmly grounded on scientific data, facts, backstories, and findings creatively and imaginatively, in ways that go beyond stock knowledge and clichés. In retelling the story of the forest, for example, we remind them to communicate, not only the correct science of forests. The ultimate aim is to change the way we THINK about forests.

We ask our students to PROBE their surrounding, learn where to search for information, which ideas to pay attention to, in this case, the key terms: risk, vulnerability, hazards, exposure, and capacity. We learn to approach these terms from different points of view and different disciplines. In art, we call this PERSPECTIVE. For example, from a structural perspective, the UP Catholic Chapel or Church of the Holy Sacrifice[5] may be described as a modern church made of reinforced concrete that is formed in a thin circular shell, inside very spacious and free from posts or columns. It is airy and maximizes available light, although worshippers can also get wet during rainy and stormy weather, because of its open structure.

From an arts perspective, we can also point out that the church has multiple entry points and we can approach and view the center altar from different vantage points. Culturally, this multiple perspective might also indicate that worship has become more intimate, more flexible.

We can also say that it is possible to have a sturdy building—one that is strong enough against earthquake and fire—that is approachable and not necessarily massive, like the old church buildings, which crumbled during earthquakes in Bohol and Cebu (2013, 2017) and Pampanga (2019). Combining structural and cultural factors may yield certain questions and conclusions: A building's massive size does not necessarily mean it is strong enough for earthquakes. Why then, do buildings fail? This may be a structural issue, but the question could also well be: Why were these

---

[5] https://www.parishoftheholysacrifice.ph/

buildings built? Are structures built for comfort? To live in? Was beauty considered in building structures? What is beautiful? What is comfortable?

The cultural, psychological, and aesthetic reasons complicate the structural and safety concerns. When we study art history, we could see that such issues preoccupy the art historian, as much as it does the engineers. The history of art could well be a history of finding creative solutions to multi-dimensional problems. And this is why, when we talk of structures, the structural engineer and the art historian can work together as a team.

There are also moments in this class when we looked at the earth from an aerial view, as in for example, the maps in a new app launched in July 2019 called the Hazard Hunter developed by a multi-agency team of scientists and communications specialists. Upon opening the app by clicking https://hazardhunter.georisk.gov.ph, we can find the location we are interested in down to the household level. Upon clicking a particular location on the map, the app provides information on the hazards you are exposed to in more general, technical, and scientific terms: a bird's eye view, so to speak. During the course of their final projects and case studies, the students realize that we also have to come down to earth, and focus on human beings at ground level, at the level of the body, and its capacity, not only to make sense of the world but also to SENSE it: to hear more, feel more, see more.

In the arts, we refer to the study of sensory experience at ground level as AESTHETICS. To be aesthetic is to become fully aware, fully alive, our senses activated at their peak, as Dr. Ken Robinson defined it in a TED Talk on Changing Paradigms in Education.[6] We open up to the world and become fully conscious that we are part of a larger world.

Through research, we expand our understanding beyond what we know and even challenge the correctness of that knowledge. We are called upon to go beyond the "default" of tropes towards generating concepts and courses of action that are novel and useful. Learning from arts education, we call this ability to go beyond the default CREATIVITY, which according to Bloom's taxonomy[7],[7] is the highest form of learning. It is in and through creativity that we gather "data" and "facts," and put them all together in accessible and engaging narratives through informed and nuanced analysis. We cannot be creative if we do not know how to ask questions, if we do not gather relevant information, and if we do not connect this knowledge together through a process that balances logic, intuition, and affect—critically and systematically.

We translate these principles into activities that make the students go through collaborative creative processes. We turn the classroom—and the world outside the classroom as a space for *pakikipagkapwa (roughly translates to "fellowship." I will discuss this term in more detail in another section)* through the creative experience—*pagdanas* or *danas*—and make space for our hearts and minds to activate and process feelings—*dama/pagdama*—that are taken for granted or unacknowledged as we go

[6] https://youtu.be/zDZFcDGpL4U (accessed 29 November 2019).

[7] See for example: https://cft.vanderbilt.edu/guides-sub-pages/blooms-taxonomy/.

through routines of everyday life, including our academic lives. To demonstrate, I will describe one such activity in the following section.

Going Beyond Our Defaults: The Nuclear Cow

One of the activities that we have been conducting at the start of each semester is dubbed "Nuclear Cow" (also known as the Radioactive Cow).

The Instructions and Mechanics by Way of an Activity Guide Follows:

A spaceship is leaving earth soon (we do not tell them why), and only ten passengers can be accommodated. In 30 minutes, your group should decide which of the following will be on the spaceship:

1. *36-year-old male filmmaker (director)*
2. *59-year-old female concert pianist*
3. *48-year-old male congressman*
4. *25-year-old male meteorologist*
5. *40-year-old male artist (painter)*
6. *35-year-old male political scientist*
7. *30-year-old female exotic dancer*
8. *52-year-old male architect*
9. *19-year-old male actor*
10. *72-year-old female teacher (secondary or "high school")*
11. *50-year-old male biologist*
12. *42-year-old female transgender cosmetologist*
13. *60-year-old female psychologist*
14. *42-year-old male doctor (GP)*
15. *57-year-old male physicist*
16. *The nuclear cow*

Mechanics:

1. *Each member of the group will pick at random a piece of sealed paper on which is written one of the "characters" or dramatist personae listed above, except for the nuclear cow. (Take note that, this is a big class—around 80–100 students. Each group has an average of four to five members.)*
2. *Each member will play the role of the character that they picked out of the raffle, and persuade the group why she or he has to be in that spaceship. A male student might pick a female character, and so on, but he/she has to assume that role.*
3. *After the "defense," the group forms a consensus. They can reject or accept a member according to their agreed criteria.*
4. *The groups will present their list in the "plenary" session, and post the group report to our University Virtual Learning Environment (UVLe) platform.*

- End of Study Guide -

We have been conducting this activity for at least four terms now, for the purposes of group dynamics, introduction of key terms related to disaster risk reduction, and the nature, character, and objectives of the subject as a General Education requirement

Semester after semester, most of the groups swallow the bait, hook, line, and sinker and with a few exceptions, chose those who will board the spaceship according to how they can contribute to the project of building the new world or planet. The exotic dancer is in the list of most of the groups and is chosen for her reproductive and "entertainment" value, which, as I always point out during the plenary processing smacks of sexism; the scientists, the psychologist, the architect, the political scientist, and the doctor are chosen for their expertise and their contribution to building a new society and new world. The 72-year-old secondary teacher is often left behind because, unlike the exotic dancer, she is perceived to be much too old to be of any use. In a similar vein, the groups decide to leave most of the artists because they probably agree with Plato: "Artists have no place in the ideal republic," and their skills may not have any practical application. If artists do get included, such as the filmmaker or the painter, they are chosen for their skills in documentation; the pianist, like the exotic dancer, can also provide recreation and entertainment.

Most of the groups did not raise questions before they rushed to fill up the spaceship, probably because they were given very little time (as in any disaster, "quick response" is crucial), or probably because they bear deeply ingrained habits of thought—or what I referred to earlier as tropes—that they bring with them to the table. They worked on the usual assumptions: The earth is now unlivable and so we have to leave it; we are going to another planet to rebuild; and we have limited space.

They chose the passengers according to how they were listed (a pedagogical trick): age, gender, and specialization and proceeded to make their decisions according to stereotypes. In their haste, the students were unable to reflect and ask such questions: "Who are these people, beyond their profiles?" Perhaps the secondary teacher is more fit than any of them and that, although she has retired from active service, she continues to learn and update herself about twenty-first-century knowledge and modes of delivery. Perhaps the politician is not your ordinary corrupt leader, but is God-fearing, and is initiating projects that benefit grassroots communities.

As the semester progresses, we learn that it is risky to approach a problem from one default point of view, which, as the nuclear cow exercise has demonstrated, contains unexamined stereotypes, tropes, and positions that have become unquestioned precisely because they have become common sense, knee-jerk and "natural," so much so that we are unable to critically examine the premises of the exercise. What was the problem of the exercise, and what is the problem with that problem? How is the problem framed and why is that framing problematic? Why do we have to leave earth? If it has become unlivable, how did we come to this situation? Beyond age, gender, and profession, what else do we need to know about each character? Come to think of it: Why leave anybody at all? Aren't all human lives important? Why can't we all stay and figure this out?

At least one or two groups came to ask these questions and have come to the decision that nobody leaves except for the "nuclear cow," which they discovered after a quick Google search is no "cow" at all but a medical laboratory device. Perhaps the pilot and the cow can come back—if there is any earth or humans to come back to—with pertinent data. In other words, at least a few groups were willing to take risks, explore and imagine other options and other possibilities, get out of the box, their defaults, and their comfort zones.

In a related exercise, the realization that they could have approached the problem from a more nuanced humanist perspective is captured in a meme from one of the students, which she submitted after we asked them to put a caption in one of the photographs from another exercise, entitled: Role-Play a Disaster Through a Tableau. The photo in Fig. 16.1 depicts one group preparing for their final pose, the final tableau output of which aimed to portray a group of people who are preparing to evacuate following early warnings of a volcanic eruption. After we posted this "stolen" photo on the Students Bulletin Board of UVLe with the instruction "Caption this," one student re-purposed and captioned it to refer to how most of the groups responded to the Nuclear Cow exercise, and how they have since rethought this response.

The Nuclear Cow case shows that students tend to think, feel, and behave according to what my psychologist colleague refers to as "cultural scripts," and which I refer to—drawing on the ecocritical frame—as "tropes," which underline our habits of thinking towards our home, our ecosystem, as well as our relationship with non-human world within that ecosystem. Tropes are boxes that limit the possibilities for generating novel insights and actions. In the following sections, I outline some of the main characteristics of the tropes and the positions, using the terms from Garrard (Fig. 16.1).[8]

16.2.1 Going Beyond the Apocalyptic Trope: The Unlivable Earth in Popular Culture

From the Nuclear Cow exercise, we learn that students tend to immediately assume that a group of people have to leave earth because it has become unlivable. This **apocalyptic**, doomsday trope is retold in many different ways in popular culture, particularly in movies, such as *Bird Box, Day After Tomorrow, the Avengers: Endgame, the Book of Eli, and Mad Max***,** among many others. The Nuclear Cow scenario recalls such films as *Elysium* (2013), which talks about a new world hovering above earth available only to a privileged few who have the means to pay for the passage. A spaceship flies in out of the ravaged earth, inhabited by the wretched poor and a luxurious habitat shielded by a force field, where the inhabitants live free from disease, hunger and the many pains and discomforts of life on earth—until rebels from the lower classes demand to be included and thereby disturb the perfect system.

[8] Garrard lists the tropes and positions as follows:

Tropes:

Pastoral Wilderness Apocalyptic Dwelling **Positions:**

Cornucopia: the dangers are deemed illusory or exaggerated.

Shallow ecology: anthropocentric view of nature, mechanistic Deep Ecology: biocentric, organismic view of nature.

Social Ecology or Eco-Marxist Ecofeminist.

Eco-phenomenology.

Fig. 16.1 Screenshot of a Meme posted on the UVLe Bulletin Board in response to the instruction "Caption this" (Photo credit: Leonardo Rosete; Meme by Jyllean Marye Tupaz)

From an **Eco-Marxist** position, the film comments on contemporary political and sociological themes such as immigration, overpopulation, transhumanism, health care, and worker exploitation. This position takes diligent note of the complex aspects of identities, and the inequalities that arise out of uneven structures and relations. The position becomes a springboard for asking questions, such as Who are most exposed to risks? Who are most vulnerable? How do we develop our capacities to reduce these risks? These questions lead us to conclude that ecological problems are also structural problems intricately connected with poverty, lack of water, food, shelter, and education and that these lack make certain sectors of the population more vulnerable and exposed to hazards and risks than others.

Geographically, the Philippines is one of the countries in the Global South that is most exposed to risks from climate change. It is located in the western Pacific Ocean, surrounded by naturally warm waters that will likely get even warmer as average sea surface temperatures continue to rise. Warm waters, along with wind regional patterns or currents can affect the Philippines in various ways.[9] To some extent, according to the Climate Reality Project, "this is a normal pattern: the ocean surface warms as it absorbs sunlight. The ocean then releases some of its heat into the atmosphere, creating wind and rain clouds. However, as the ocean's surface temperature increases over time from the effects of climate change, **more and more heat is released into the atmosphere.** This additional heat in the ocean and air can lead to stronger and more frequent storms—which is exactly what we've seen in the Philippines over the last decade (emphasis mine)."

---

[9] https://www.climaterealityproject.org/blog/how-climate-change-affecting%20%20%20-%20philippines.

Despite being one of—if not the—most vulnerable to climate change,[10] The Philippines is one of the low carbon emitters compared to the top emitters like the US and China.[11] This is probably one of the reasons why activists from the Philippines are among the most articulate and passionate in the climate justice movement. "Owing to our long history of natural disasters and inadequate government support, Filipinos are known to forge international solidarity among highly vulnerable countries," according to Alanah Torralba, who produced a series entitled "Daluyong: Sights and sites of climate change in the Philippines held in Quezon City, September 2019."[12] But while Filipinos are at the forefront of demanding accountability from rich nations and polluting industries, the question is: who will punish/reward those who increase or decrease their carbon emissions, especially when it is difficult, not only to establish jurisdiction but also cause and effect. As Elaine Kamarck (2019) asks, "if coal plants in China and cattle ranching in Australia increase their outputs of greenhouse gases in one year and there are droughts in Africa and floods in Europe, who is responsible?" And if responsibility and the cause and effect are tracked down, who will prosecute, who will be prosecuted, and how? If the United Nations tells us that climate change is very serious, can we rely upon them to enforce or execute the international agreements aiming to address their catastrophic forecasts?

Calls for collective action, instead of inertia, as Cordal eloquently expresses through his art are important and necessary. But without awareness of structural inequities and without the ability to hold corporations, countries, and individuals accountable, no major change can occur. The apocalyptic trope paints a frightening scenario, one that tends to overshadow problems of social justice at the local, national, and international levels, and one which cannot proceed without an understanding of the structural, social, and political connections between a warmer planet and dangerous climate changes.

16.2.2 Beyond Anthropocentrism: The Dutch Boy

While *Elysium* hints at these political and social concerns, its solution is drawn from a **cornucopian** position, which puts so much faith in technological solutions, or in other words—the "cornucopia" of wealth, growth, and scientific knowledge. *Elysium* asserts that such cornucopia must be equitably distributed. The cornucopian position is also apparent in another example, *Geostorm* (released 2019), which shows the world's leaders and scientists coming together to create an intricate network of satellites to control the global climate and keep the world safe from the natural disasters that threatened the planet. The satellite network was nicknamed "Dutch

[10] The **Global Climate Risk Index 2015** listed the Philippines as the number one most affected country by climate change, using 2013's data. Ibid.

[11] See for example: https://www.visualcapitalist.com/all-the-worlds-carbon-emissions-in-one-chart/

[12] https://www.rappler.com/nation/244524-those-power-long-abused-filipino-resiliency/.

Boy," referring to the story of a boy who plugged a hole in the dike with his finger, in order to hold off a coming deluge that was threatening to sink his village.[13] The Dutch Boy bought his village time to escape to safety and to make the necessary repairs. In *Geostorm*, however, the Dutch Boy assured the world that the devastation can be held off. It is a technological fix—a way of controlling nature and, as one of the characters put it in a crucial scene, a way of playing God. With satellites plugging the damage, the world's populations do not have to change their ways, lifestyles, and attitudes; they do not have to understand why the weather is behaving the way it does. But they are safe. After a few years, the satellites malfunction due to a sinister plot, and Dutch Boy could no longer keep at bay the series of freak disasters building up to a giant Geostorm—until the hero steps in just in time to save the world from total destruction. With the bad guys eliminated, it is assumed that a much improved and failsafe Dutch Boy will be resurrected to keep the world safe from the Geostorms of the future. With the power of technology, the problems of global warming can be "plugged." In fact, the problem, from the cornucopian perspective, might not be that serious, after all.

Despite mounting scientific evidence on the effects of climate change, there remains a belief or perception that the seriousness of global warming is largely exaggerated. For example, in the US, a Gallup poll conducted for the past two decades shows that about a third to almost half of the public holds this belief.[14] The persistence of such perceptions can be due to many factors: the cornucopian belief in the power of technology, the inability to grasp the complexity of the issue, and the difficulty of establishing cause and effect, between, for example, coal plants and carbon emissions in one part of the world and Super typhoons like Yolanda (international codename Haiyan) in the Philippines. In contrast, when dead fish float on top of Manila Bay, it is not hard to understand the link between what is happening in the Bay and the garbage being dumped into it.

---

[13] https://en.wikipedia.org/wiki/dutch_boy

[14] https://www.brookings.edu/articles/the-challenging-politics-of-climate-change/

Those who do understand the connection are indifferent[15]; others are probably in denial. Still others may be so overwhelmed with helplessness and inertia, they could take the fatalistic response—"we are all going to die anyway." Like Cordal's submerged, huddled figures, the apocalyptic scenario of rising sea levels can leave us paralyzed and immobilized with inaction, confusion, and dread.[16]

The apocalyptic trope and the cornucopian positions may at first glance appear as opposite assumptions clustered around two extremes: one is grounded in action, based on technological solutions; and the other is based on a resigned, fatalist, inactive response. What they have in common is an **anthropocentric** view of the environment, which is considered of no inherent value and accorded importance only in so far as it is of interest to human wealth, health, and welfare. Nature is separate from the human realm; it exists out there in the external world. At times, it can be mindlessly and randomly powerful, unpredictable, and destructive; at other times, it can be managed, known, and controlled. Swinging between extremes, depending on the situation, humans view the environment **mechanistically**—nature and the rest of the nonhuman world is an inert, inanimate resource that can be exploited for human interest.

---

[15] In a report on the Changing Politics of Climate Change, Elaine Kamarck, Founding Director of the Center for Public Management of the Brookings Institution, expressed concern about the public's lukewarm response to climate change. One of the most telling pieces of evidence she cites on public indifference to the problem comes from the results of 30 years of open-ended polling conducted by Gallup. She reports that between 1989 and 2019, "Gallup has asked "What do you think is the most important problem facing this country today?" Jobs, the economy, and health care are often at the top of the list. "Environment/pollution" is not often mentioned. In fact, over a 30-year period, it was mentioned by anywhere from less than 0.5%–8% of the public. In the most recent 2019 poll (August), "the government/poor leadership" was mentioned by 22% of the public, and "immigration" by 18%. "Environment/pollution/climate change" garnered only 3% of the public. And in some earlier polls, climate change is not even mentioned by a significant portion of the public (although people could be including that within the term environment.)" She noted that, "over time, the most worrisome environmental problems are visible pollution problems. Water, soil, and ocean and beach pollution are at the top. These are things average people can see and smell. Global warming or climate change is toward the bottom…" https://www.brookings.edu/research/the-challenging-politics-of-climate-change/ (accessed 25 November 2019).

[16] In the lesson on psychosocial framework, students learn that attitudes to risk are personal and cultural. The apocalyptic, Pastoral, and Wilderness tropes can lead to attitudes towards risk, and the corresponding behavior. Using the terms of the Psychosocial framework, some of them are:

- dreadful risk playing on fear, which could lead to fatalism,
- unknown risk and indifference and therefore no change in lifestyle,
- acceptable risk (we know the risk but we find it acceptable due to the sheer beauty of the countryside, depicted in, for example, in the Amorsolo paintings).

16.2.3 The Pastoral Landscape in the Wilderness Mode: Female and Femaleness as Objects of Sight

The nonhuman world is not just a source of food, shelter, clothing, and other basic needs; it is also a source of emotional retreat, leisure, and aesthetic pleasure. We go to beaches and mountains and other places seemingly left "unspoiled," "pristine," or untouched by human interventions, at least on the surface. In concrete and actual terms, in the age of capitalism, we know that virtually nothing is left untouched. Resorts are carved out of landscapes, hotels are built for our comfort as we go snorkeling, trekking, or birdwatching. We go to Tagaytay or Baguio to picnic, enjoy the view, with some of us opting to live there on a long-term or periodic basis. When we reach the peak of the mountain, we take a picture as proof of having been there and conquering, not only the landscape but also ourselves and our weaknesses. Nature and its beauty become a photo op, a painting, a film, and a romantic song or dance in the pastoral or wilderness mode, where the former highlights the bucolic calm of the countryside.

The wilderness trope, on the other hand, finds solace in the primeval—at times terrifying and exotic beauty of the wild—a stable, unchanging Heart of Darkness which lies at the core of human existence stripped of the veneer of civilization. We wish to protect, preserve, and save such images and instances of wilderness, not for the environment's own sake, but for our own so that at the end of the day, we can seek refuge in that space and recharge.

The anthropocentric and mechanistic view of the environment is anchored on a hierarchical relationship between elements that make up an ecosystem, along class and gender lines. The **Eco-feminist** position draws attention to, among others, the language of patriarchy and its system of representation. There are certain terms associated with being female: the nature, the earth, and the land, which as we have earlier discussed are resources that sustain man's survival. In the Nuclear Cow exercise, femaleness is associated with reproduction and entertainment, and this is why the young exotic dancer is at the top of the list of those who are going to board the spaceship to escape the unlivable earth.

We can also detect this meaning system in the use of terms: Nature is always the nurturing **Mother** Nature, which according to the apocalyptic trope, will be lost to us, if we do not do something about it. In various images, woman is associated with the beautiful, pastoral landscape, and functions as a poster girl for the bucolic, as seen for example in National Artist Fernando Amorsolo's comely barrio lass/dalagang Pilipina. As the Marxist critic John Berger[17] famously and succinctly stated: Men Act, Women Appear. The woman is the object of the gaze, they look at themselves as objects of sight. In classic advertisements of the Marlboro Man, for example, men are seen as active doers, as pioneers who conquer and tame wild lands and animals through masculine power and skills, much akin to colonizing and taming another planet or ecosystem, in case the earth ends or dies.

---

[17] John Berger, Ways of Seeing, 1972. BBC, London.

16.2.4 Beyond Surface Appearances: Deep Ecology and the Reenchanted World

The mechanistic and anthropocentric positions and tropes can also be described as instances of **shallow ecology**. Saving or leaving the earth, as we have seen in apocalyptic films, is important for the sake of saving our skins. The strategies are often short-term, quick fixes. Recycle, reuse, and create substitutes. Use technology creatively and sustainably. There is no accompanying overhaul in lifestyle and consciousness. On the other hand, **deep ecology** posits a change in paradigm. Home is not just a place where humans dwell, separate from the non-human world; home includes that world, which is not an inert resource, but is alive, with an agency or will of its own. Morris Berman describes that connectedness as one of enchantment, where rocks, trees, rivers, and clouds were all seen—millions of years ago down to the eve of the scientific revolution—as "wondrous, alive, and human beings felt at home in this environment. The cosmos, in short, was a place of belonging." Today, at least at the level of the mind, the story of the modern world is that of progressive disenchantment. "From the sixteenth century on, mind has been expunged from the natural world."[18]

For deep ecologists, the call is for a reenchantment of the world, for reconnection with nature, which we will protect and save for its intrinsic value. The process of reenchantment is long-term, rooted in magic, beyond logic, beyond surface appearances. The relationship between humans and non-humans is egalitarian, not one based on hierarchy and domination. The self is one with, instead of separate from the rest of creation.

In the Philippines, this sense of interconnectedness beyond the self is concretely manifested in the practice of "Pakikipag-kapwa"[19] which roughly translates into treating your kapwa (fellow beings) as equals. Kapwa cannot be framed or subsumed under the I/they, self/others, I/non-I binarist Western schema; as kapwa, one is not separated from "the others" but rather the self, the "One" is inextricably part of others. Pakikipag-kapwa translates into helping out, or providing "tulong," not as a matter of "charity" or simple volunteerism (as in "giving back," for example, through volunteer work for relief operations) but should be seen as addressing a need (kailangan), and that it is but the humane (makatao) and proper or right thing to do (dapat, karapatdapat) for a person to retain and reclaim a sense of dignity (karangalan, pagkatao, amor propio) and humanity (pagiging tao). Pantay pantay na karapatan or equal rights are not just about addressing discrepancies in what is delivered and what is promised in terms of material and basic needs, but rather they are about not being scorned and treated as less than human just because one lives in poverty or are elderly (like the 72-year-old schoolteacher in the Nuclear Cow exercise), or ill and incapacitated, or because they are "mere" animals like our pets or our livestock. To be deprived of resources is equal to being deprived of the means, not

[18] Morris Berman, **The Reenchantment of the World.** Cornell University Press, 1981.

[19] Benedict Kervliet, **Class and Status Relations in a Central Luzon Village,** Rowman & Littlefield Publishers; Updated edition (March 1, 2002). Cited in Datuin 2014.

just to survive or address basic needs, but to live decently and to be treated decently as equals. In deep ecological terms, our kapwa can include not only humans but also other creatures in an ecosystem where Man does not have dominion over others.

Deep ecology is exacting and requires a lot of time and effort; consciousness and lifestyle overhauls do not happen overnight, especially since—as previously pointed out—positions and tropes are deeply ingrained and are not that easy to unlearn. But perhaps we can find and learn from traces of deep ecological practices that have survived in traditional societies through time. Among the Aytas of Pinatubo, land is life, not a commodity; it is a legacy handed down to them by Apo Namalyari. To lose their land is to lose their lives and their claim to dignity and humanity. The essay by Cynthia Zayas may prove instructive, at the very least, for understanding why the Aytas were resistant to evacuating despite Pinatubo's then impending eruption.[20] My essay on the Pasig River and a ritual in Batanes, where I conducted a short fieldwork for a documentary also represents an effort to understand how a community can preserve a watershed from a deep ecology position.[21]

16.2.5 Ecophenomenology and The Sensory World of Art

Deep ecology also intersects in significant ways with **ecophenomenology,** which draws on some of the principles outlined by Heideggerian philosophy. The first concept—that of **intermodality (as formulated by Paolo Knill)**—is anchored on the sharpening of our senses, and being more attuned to our environment, to feel and experience (danas, dama) the world around us beyond the literal and beyond our usual ideas of what home is. Art, with its powerful resources, can open up to the world and make it come alive, thus expanding our initial contexts, which becomes consequently molded by this constant exchange between the body and the environment. As we attune ourselves to the living, breathing world around us, we develop a heightened sense of being that makes us more active participants to even the most subtle, mundane moments of our everyday life."[22] In the process, we are reminded that home is contiguous with a larger world. As the Robin Williams character in *August Rush* tells us, Music (and perhaps all the other arts) "is God's little reminder that there is something else besides us in the universe and there is harmonic connection between all living beings everywhere, even the stars."

---

[20] Cynthia Zayas, "Land is Life and the Life is Land: Development, Resilience, Family and Knowledge for the Ayta of Mt. Pinatubo," in **THE CONSEQUENCES OF DISASTERS:** Demographic, Planning, and Policy Implications, eds. Helen James and Douglas Paton. Ebook: https://www.ebooks.com/en-af/2483769/the%20%20%20-consequences-of-disasters/helen-james-douglas-paton/

[21] Flaudette May Datuin, "For the Birds," **The River Project. Exhibition Catalogue.** Campbelltown Arts Center, Australia, 2015.

[22] I appropriate the ideas on intermodality and poiesis from my mentee and thesis advisee Zoila Gerente, who wrote about these concepts at length in her undergraduate thesis "Resyncing Capacities: Exploring The Role Of Art Practice In Post-Disaster Recovery." (BA Art Studies, 2016).

Art and artmaking are thus more than expressions, reflections, or representations of emotions, events, or ideas. Nor is it a finished product or object—"a perspective that puts more emphasis on the outcome (artwork) instead of the process (work of art). The art as object paradigm sees the arts as a tool for producing an expected, often profitable, outcome"[23]—as collection, for example; as object of display; as status symbol, etc. On the other hand, art is a process of **poiesis—**of bringing forth through creative revealing, responsiveness, and interconnectedness or pakikipagkapwa, thus opening up many other forms and opportunities for thinking, acting, and feeling outside the box.

Poiesis is about imaging and imagining other worlds—not to escape this one, like what we see in apocalyptic movies—but by making present and bursting open a broader range of experiences, interactions, and realizations. It does not mean we ignore the overwhelming feeling of grief and pain or wish away the death and destruction resulting from the impact of climate change and pretend they do not and will not happen. We acknowledge the feeling and the event and put it in perspective and a larger context—our bodies, other bodies, and the world around us. When we open ourselves to the rhythm of the cosmos and the pulses of the larger life world, we are able to sync the beats of our heart and of our breaths with that larger world. In effect, we draw strength from that process of fully sensing and interacting with the material, spatial, and temporal environment.

To be intermodal and poietic is to realize that at the very heart of our defaults, we look at nature from an anthropocentric view. We think of nature as resource to keep us alive, and now that resource is dwindling, we think up of technological ways to control or reduce this deterioration. We save the planet for our sakes. We save it so that we won't perish. But what if we think beyond ourselves, beyond these defaults, our cultural scripts about being in control? Anything can happen. Buildings may fail, no matter how hard we try to make them as safe as possible. Mountains will shed their skins, the ground beneath us will move, and volcanoes will explode whether we like it or not, and not according to OUR timetable. We could shift perspectives—from being man-centered, with the view that we humans are the center of the universe, and that nature is inert, resource to a perspective that makes us a small part of an ecosystem: much like musical notes. We can only understand ourselves in relation to other notes, either as part of a melody or a chord.

16.3 Grounded Creativity: A Suggested Study Guide

The preceding discussion of tropes and positions demonstrate that climate change is an Ecological Problem that cannot be understood by one discipline alone. My discussion is a contribution from the arts and humanities, which highlights the aesthetic, social, and cultural nature of climate change and the role of deeply ingrained and widely held beliefs that shape our relationship to our ecology. The sources of the

[23] Gerente, *ibid.*

examples I have cited are drawn from our lessons in DRMAPS, which targets students at the tertiary level, but it is also my hope that these lessons and case studies can also benefit all educators, including those at the basic education level. The following summary, which I cull from my Study Guide on the Eco-critical framework might prove instructive:

The Summary by Way of a Study Guide Follows:

1. *Tropes can underline and shape our attitudes and behavior towards home, as well as our relationship with the non-human world.*
2. *The pastoral, apocalyptic, and wilderness tropes are anthropocentric—focused on the perception of the non-human world as resources that are inert and without agency (mechanistic), rather than inherently valuable (biocentric), with an agency and will of their own (organismic). The mechanistic worldview is said to belong to the sphere of shallow ecology.*
3. *The human-centered and mechanistic view of the non-human world can lead to minimal change in habits, should this resource be endangered or threatened. A default solution is demonstrated by post-apocalyptic films, which show much faith in technology and capitalism's capability to correct its damages (cornucopia). In everyday life, we see this cornucopian view through such solutions as "green capitalism," modernization of jeeps, and sustainable mining, which are input and capital-intensive, and may in the long run, prove to be short-term solutions to problems that are long-term and global, as we have seen in the way the ideal worlds of Elysium and Geostorm can backfire and malfunction after a period of perceived safety and comfort.*
4. *The mechanistic and anthropocentric worldview are also hinged on deeply ingrained tropes or cultural scripts about maleness and femaleness. From an eco-feminist position, women are associated with nature (the land) and culture and civilization with men. Such configurations translate into stereotypes and assigned roles that limit the range of what is possible in real life, as we have seen in our Nuclear Cow exercise.*
5. *On the other hand, the organismic and biocentric positions call for, not only short-term solutions and superficial changes in lifestyle (e.g., recycling and segregation of waste) but also a change in consciousness. It is a consciousness that fosters a sense of belonging, a sense of place, and a sense of identification and respect for the non-human world. This deep ecological sense acknowledges the importance of scientific literacy, involving technical and "objective" solutions, e.g., early warning systems, and structural adjustments. On the other hand, deep ecology also pays attention to those areas of human life that are imaginative, beyond the logical and literal, and deeply affective. Bridging the gap between the domains of rationality and the domain of feelings and perceptions is one important step in understanding such issues as climate change.*
6. *The tropes are deeply ingrained and internalized narratives and scripts that naturalize the way things are. They are boxes that limit the possibilities for generating novel insights and actions. The pastoral, apocalyptic, and wilderness tropes efface gender, racial, ethnic, religious, and class differences and inequalities. An*

Eco-Marxist position takes diligent note of these complex aspects of identities, and the inequalities that arise out of uneven structures and relations, at local, national, and international levels. An Eco-Marxist position examines problems of accountability, social justice, and other socioeconomic and political concerns that may lead to concrete political actions that are informed, multidisciplinary, and multidimensional.

7. *One way we can contribute to the urgent and serious problem of climate change is through communication and education strategies that can explain the complexities of the "science" of climate change, in a way that goes beyond stock knowledge, clichés, and tropes. We look for backstories, focusing not on the EVENTS themselves but on yet-unexplored areas of inquiry, yet-to-be uncovered blind spots, and uncritically received defaults (for example, in media coverage). In this lesson, we call for the sharpening of the senses (intermodality) and poiesis, going beyond fixed outcomes, towards the processual, figurative, imaginative, and intuitive. We are called upon to go beyond the "default" of tropes towards generating concepts and courses of action that are novel and useful but also grounded on diligent research. In DRMAPS, we call this GROUNDED CREATIVITY.*

- End of Study Guide -

16.4 Conclusion

16.4.1 From Object to Process, from Art Work to Work of ART

In sum, several paradigm shifts have been forwarded in this paper, as shown in the following pairs:

1. Anthropocentrism, viz. biocentrism.
2. Mechanistic, viz. organismic.
3. Shallow ecology, viz. deep ecology and eco-phenomenological frame.
4. Cornucopia, viz. eco-Marxist and eco-feminist frame.

But the most important paradigm shifts, I would like to stress from a perspective honed in the humanities is the shift in concept of art—from one of **object** (art work) to **process** (work of art) and from one of fixed and prescriptive outcomes to one of openness to many possibilities within the "noise" and routines of the everyday. This shift suggests that understanding and responding to climate change involves crossing the divide between climate change as a scientific and technological phenomenon (Problem of Ecology), on the one hand, and on the other hand, as a cultural Ecological Problem. An eco-critical method and frame goes beyond these binaries, suggesting a more well-rounded encounter between, not only professionals and experts in the natural and physical sciences but also experts in the social sciences and the humanities as well. Policymakers can provide and find ways, platforms, and venues, not only for experts to talk to each other but also to fishers, poets, farmers, weavers, carpenters, and so on so that we can work together through collaborations that are attuned, not only with the "science" of climate change but also keenly attentive to the manifold textures and shapes of the non-human and human worlds. In the humanities, we believe in the power of the intermodal and poietic imagination, and it is this power that we all need to recover in these dire times. From such transformative and processual viewpoint, we may find a springboard for action, instead of collective inertia depicted in the tiny, almost invisible figures in Cordal's powerful artworks.

As disarmingly expressed by Nitz Sezcon of Iligan, a wise woman whom I have encountered in my travels giving art workshops to traumatized communities with the artist Alma Quinto: *Healing the earth means healing, not just poverty in the material and economic sense, but the poverty of the imagination and the spirit.*

References

Armstrong P (2010) Bloom's taxonomy. Vanderbilt university center for teaching. Retrieved [11 January 2024] from https://cft.vanderbilt.edu/guides-sub-pages/blooms-taxonomy/.

Berger J (2008) Ways of seeing. Penguin Classic. (International paperback edition)

Berman M (1981) The Reenchantment of the World. Cornell University Press, Ithaca and London

Conde M (2019) Those in power have long abused the Filipinos' resiliency. Rappler. https://www.rappler.com/nation/244524-those-power-long-abused-filipino-resiliency/)

Cordal I (2023) Follow the leaders. Cement Eclipses. www.cementeclipses.com/portfolio/follow-the-leaders/

Datuin FM (2004) Interrupting the roll call of suffering: bearing wit(h)ness through art. In: Bong S (ed) Trauma, memory and transformation: southeast asian experience. Strategic Information and Research Development Center, Malaysia, pp 33–49

Datuin FM (2015) For the birds. In: The river project. Exhibition catalogue. Campbelltown Arts Center, Australia, pp. 26–41

Dougherty S (2014) This is what politicians debating global warming will look like soon. The World. https://theworld.org/stories/2014-03-26/what-politicians-debating-global-warming-will-look-soon

Garrard G (2004) Ecocriticism. London and New York: Routledge (Taylor and Francis Group)

Gerente Z (2016) Resyncing capacities: exploring the role of art in post-disaster recovery. MA Thesis, Department of Art Studies, University of the Philippines, Diliman

Ghosh, In (2019). All the World's carbon emissions in one chart. Visual Capitalist. https://www.visualcapitalist.com/all-the-worlds-carbon-emissions-in-one-chart/

Inquirer.net (2019) Rising sea levels likely to Sink Metro Manila by 2050 – study. https://technology.inquirer.net/92179/rising-sea-levels-likely-to-sink-metro-manila-cities-by-2050-study#ixzz66EmRi1pR

Kamark E (2019) The challenging politics of climate change. Brookings. https://www.brookings.edu/articles/the-challenging-politics-of-climate-change/

Kerkvliet BJ (1990) Everyday politics in the Philippines: class and status relations in a Central Luzon Village. University of California Press, Berkeley

Robinson K (2010) RSA animate: changing paradigns in education. YouTube. https://youtu.be/zDZFcDGpL4U?si=rzVikT0UF6NG6Ysu)

The Climate Reality Project (2016) How is climate change affecting the Philippines. https://www.climaterealityproject.org/blog/how-climate-change-affecting-philippines)

Villanueva J (2019) UN chief calls for strengthened efforts vs. climate change. Philippine News Agency www.pna.gov.ph/articles/1084901

Zayas C (2015) Land is life and life is land: development, resilience, family and knowledge for the Ayta of Mt. Pinatubo, in James, Helen and Douglas Paton (Eds). The consequences of disasters. demographic, planning, and policy implications, pp. 85–94. Charles C. Thomas Pub, Ltd.

Flaudette May V. Datuin is a Professor at the Department of Art Studies, College of Arts and Letters, University of the Philippines, Diliman, and teaches an interdisciplinary course on Disaster Risk Mitigation, Adaptation and Preparedness Strategies based at the Institute of Civil Engineering (ICE), College of Engineering, University of the Philippines Diliman.

Chapter 17
Using Creative Approaches in Communicating Risks to Populations Exposed to Natural Hazards

Leonardo C. Rosete and Aurora Odette Mendoza

Abstract What is so special about risk communication? In a country beset with natural hazards and disasters that are projected to worsen due to climate change, there is an urgency for policymakers, practitioners, and educators to understand their role and become better skilled risk communicators. Communication is considered an art and a science. Risk communication comes with greater burden in the sense that it seeks to deliver correct information that the viewer cannot interpret in any other way, elicit the desired response, and incite action with urgency. It is also imperative to communicate messages within the context and 'realm' of the audience, otherwise it will be lost. Capturing the attention of the viewer with technical information is a creative challenge. This chapter presents an overview of theoretical foundations of communication, discusses risk communication standards in texts and graphic requirements, accepted conventions, language and symbols for specific groups, and provides concrete applications of communication strategies and new approaches. Finally, this chapter gives concrete recommendations on how to design and create risk communication materials to help and equip the reader who is also a communicator.

Keywords Disaster risk communication · Risk reduction strategies · Creative risk communication · Subjective risk communication · Risk communication strategies

Key Points

- The climate emergency is an urgent issue affecting everyone in the world, regardless of how differently people perceive, interpret, and communicate "climate change."
- Anyone has the potential to become an effective risk communicator. Following the principles of risk communication, we are better able to call for urgent measures to address climate change in a more holistic manner.
- Risk communication is a strategy in disaster risk reduction that contributes to more comprehensive risk assessment by integrating information from hazard analysis

L. C. Rosete · A. O. Mendoza (✉)
University of the Philippines Diliman, Quezon, Philippines
e-mail: acmendoza1@up.edu.ph

K. B. Berse et al. (eds.), *Climate Emergency in the Philippines*, Disaster Risk Reduction, https://doi.org/10.1007/978-981-99-7804-5_17

mapping and forecasting conducted by specialized scientists, with information from the analysis of vulnerabilities, exposures, and possible damages as validated by people in their own contexts. It is the communication of real climate risk scenarios that makes it crucial and effective.

Summary for Policymakers

Anyone has the potential to become an effective risk communicator, but for policymakers and educators, practicing the art and science of risk communication is crucial given the climate emergency. To hone our skills, we must embed in ourselves certain traits that will help us become effective risk communicators.

Study the audience. Collect information, observe audience attributes, and involve the audience as much as possible in crafting the communication material. Use local narratives, color, and shapes that the audience can relate to, and avoid the ones that they dislike.

The following is a proposed set of guides in crafting communication for an audience:

- **Study the audience**. Collect information, observe audience attributes, and involve the audience as much as possible in crafting the communication material. Use local narratives, colors, and shapes, that the audience can relate to, and avoid use of the same to the ones that they dislike. Translate technical terms to words that relate to the audience's experience. Check what information channels are available to them, and what they enjoy using the most.
- **Adhere to standards when it is required**.
  The set of standards for signs and signals exist for public order. Where standards are binding, observe them. But beware, some accepted standards and conventions are only influenced by industry and are not established by law. In this case, the so-called standards can be modified and translated to symbols that are closer to the audience's realm of experience.
- **Use technologies that can work in the area**. While the current level of technologies facilitates the authoring of creative media, the communication process can only work if the audience can receive and use the material as designed. Check communication delivery systems such as available transmission and projection systems, access to the internet, space, facilities, etc., to verify that the communication material will not be distorted by way of delivery. However, note that the creation of communication materials is not totally dependent on delivery.
- **Allow the audience to interact with media**. Stimulation through different senses, including movement and interaction with the media and content, facilitates learning. For instance, using digital media, engaging the audience in appreciating the communication media to the point where they may inject new information materials establishes their "stakeholder" posture and allow multiple points of data entry.

- Plan for a communication campaign. Planning campaigns that are audience-focused and time-sensitive ensures the appropriate use of resources. Campaigns are tactically short and focused but strategically protracted and extensive. Latch onto current trends, but explore original creations. Disrupting the mainstream but keeping on track of delivering the correct science behind the message is a creative challenge.

Plan audience focused and time-sensitive campaigns. Latch onto current trends, but explore original creations. Disrupt the mainstream but keep on track that the science behind the message is delivered.

For Example:

As a policy implementer, did you ever experience sounding the alarm of an impending flood but some residents refuse to leave their houses despite the risk to their lives? How about informing residents of the threat of an imminent volcanic eruption but some residents choose to secure their dwellings, harvest, and livestock first? You may conclude that these people are just too stubborn to understand what's good for them, that they fail to understand that the protocols set in place for these kinds of situations are for their own good.

However, when people choose to secure their houses and belongings rather than evacuate, *maybe* they are really so mindful of the danger but the investment and sacrifices made just to acquire their properties, and the thought of leaving them behind as legacies for their children—are just as valuable as a life. We need to communicate to them then that preserving their lives preserves their legacy. In saying that, we understand their predicament but that their safety and their lives will lead to nurturing these legacies even more, could *maybe*, lead them to evacuate with less reluctance.

"Maybe" is used because just as communication is an art, there may be some discrepancies between the *expected* and the *actual* results. On the other hand, science tells us that in observed conditions, the use of language, tone, and inflection, even gestures in face-to-face communication say a lot about our messages.

- In graphic forms, color, texture, shape, composition, and layout—and duration of movements in moving images—convey a profusion of meanings.
- In choice of media, going personal or face-to-face, through the airwaves or the internet, on paper, the choice of tone and story, are decisions that must be made by risk communicators.

Educating ourselves on how to use tools in communication and knowing how the population will relate to these will most likely reduce the discrepancies between the *expected* and the *actual* results in our communication initiative.

Educating ourselves on how to use tools in communication and knowing how the population will relate to these will most likely reduce the discrepancies between the expected and the actual result in our communication initiative.

17.1 Introduction

The Philippines is a hotspot of hazards. It sits in the "Pacific Ring of Fire," and is prone to earthquakes and typhoons, and with it, floods and landslides—even fires when heatwaves occur during the hot months of summer. Inculcating safety procedures, training instinctive responses, and securing environmental structures—these are tasks that are embedded in the principles of Disaster Risk Reduction that are being applied to the level of municipalities and communities. The increasing incidence of disasters due to natural hazards worldwide has been—by a large segment of scientists—attributed to "climate change." Climate change in the Philippines is still a label not so clearly critical to many, being perceived as a naturally occurring phenomenon, perhaps even with religious undertones, and therefore, often we deal with climate risks when they already turn into disasters. Still, many now believe we ourselves are the main culprits in hastening the advance of climate change due to our uncontrolled use of earth's resources. We burn fossil fuels—coal, oil, natural gas—and cut down more forests, raise methane-emitting livestock—practices shown to produce "greenhouse" gases that are causing global warming which in turn is believed to drive climate change and its dangerous impacts.

We are possibly at the tipping point of the "climate change" crisis and yet, this is not being visibly addressed with urgency. Maybe in part it is because we live in a country with so many hazards. However, we do notice the manifestations of climate change in headline-hogging disasters that strike us. We can relate to more heatwaves and stronger typhoons in the recent years, brought about by changing weather systems as we come across more announcements by PAGASA. We can easily relate to the polluted air that we breath and the contaminated water that we drink. If we can view each of these real hazards as the many "frames" that form the bigger "window" called climate change, we can better understand it, become more aware of our personal risks with these hazards, and learn how to cushion its advance in specific ways, and develop adaptive lifestyles that mitigate its effects. We can do this using the principles of risk communication in calling for urgent measures to address climate change in a more holistic manner. Risk communication is a strategy in DRR as it creates a more comprehensive risk assessment by integrating information from hazard analysis mapping and forecasting conducted by specialized scientists, with information from the analysis of vulnerabilities, exposures, and possible damages as validated by people in their own contexts. It is the communication of real climate risk scenarios that makes it crucial and effective.

Technology today allows us to communicate in many formats—text, graphics, animations, emoticons, videos—actively in real time or passively in own time. With so much information coming our way, and in so many formats, the communicator has to be armed with information, tools, and devices that will capture the attention of the intended audience, discriminate on the content, and act in desired fashion. Advertisers are challenged in the same way, goading consumers to buy products, choose a person over another, or work on a task. This form of communication often uses metaphors with veiled messages in language and visual formats, with the aim of quickly capturing the attention of the viewer and delivering a short and direct message in a burst of seconds. Risk communication comes with greater burden in the sense that risk communication seeks to deliver correct information that the viewer cannot interpret in any other way, elicit the desired response, and incite action with urgency. Capturing the attention of the viewer with technical information is a creative challenge.

The early part of the chapter discusses the emergence of interdisciplinary research into natural hazards and disasters along with the theoretical foundations of communication in the way that it is formed, enunciated, and received. Scientific studies are cited so that the reader may be more discriminating of information, or study the concepts in greater detail. Suffice it to say that communication has a robust scientific background, and studies on the subject either drive the development of technology associated with it or are driven by technological advances. Today, the media for communication has punched through virtual space, further enriching the realm of the communicators' domain. Still, even armed with rich resources, the communicator must play within the bounds of the audience's realm of experience or risk losing their interest and their participation in the thread of things.

Risk communication has been studied and used to develop materials that can be applied worldwide. These are made using standardized templates, many of which are regulated by law or prescribed by industries. Regulated signs should be used when called for. The meanings are established and documented and should be applied without subjecting it to reinterpretration. However, they are limited and designed with utmost simplicity so that they are visually distinct and steer clear of cultural biases. We see most of these standard signs in traffic and institutional environments, but where no rules apply, policymakers, and designers can develop signs and symbols that are formed along thematic lines, often cultural appropriations or functional derivations that are closely associated with the lifestyle of the population. Studying how the population responds to images then is a prerequisite to developing these communication materials. Better still, involve the community itself in crafting these materials so that they will own the messages and their embodiments in text, shapes, and colors—even movement for video formats.

Communities are social systems that are highly relevant to the collective action called for in dealing with climate change. But the collective action required for climate change adaptation involves our individual consumption of limited shared resources that are considered as public goods—water, food, and energy. This situation is called a social dilemma. From a psychological perspective, a social dilemma is when members are asked to give up something for the good of everyone, that is, to minimize their use of the public goods. This is easier said than done in terms of climate risk adaptation because for real people reducing consumption of energy is not just the abstract notion of "energy" but includes reducing your use of gasoline, cooking gas, bath water, bringing baskets to the grocery, not eating meat, limiting your gadgets, and use of internet. Studies show that in such cases, there is a notion of "diffused responsibility" and a strong tendency towards "social loafing"—when individuals choose not to participate in the collective action because of self-interests, placing the responsibility instead on others, thinking the effect of not doing anything will not be noticeable. The social dilemma is whether the individual will choose to be cooperative or selfish.

How can risk communication strategies address this dilemma? If our messages are also crafted so that we do not just pass on risk information without a frame of reference, but present these messages so that the interests of each individual "self" is a frame that is overlapping with the interests of others in the community, we can help build a collective goal for the community. It is the sense of identity as a Filipino that made "It's more fun in the Philippines!" such a successful tourism communication strategy.

17.1.1 Understanding Disaster Risk Reduction

Under the Sendai Framework for Disaster Risk Reduction, the first priority for achieving its seven global targets for 2015–2030 is understanding disaster risk in all its dimensions. This includes not only understanding the features of hazards in our environments but also the characteristics of people, including vulnerabilities and capacities, values, and experiences.

The notion of disaster and disaster risk as consequences of interactions between people and their natural and built environments is not at all new and partly traces its roots to the emerging field of environmental psychology during the 1950s and 1960s. The discipline was formed out of interest in gaining a better understanding of the physical features of the environment in which behavior occurs, and also out of some uncertainty with the social relevance and ecological validity of research conducted mainly in a laboratory, defined as a "non-environment." The trend eventually developed into the "contextual revolution" in psychology, with its main emphasis on a transactional-contextual approach (Bonnes and Bonaiuto 2002, 28–30).

In the 1970s, increasing human populations and vulnerabilities spurred more research in natural disasters across different disciplines, each with their own particular interests, such as loss reduction for geography and emergency preparedness in sociology. Notably, the common element in these interests was the recognition that disasters and disaster risks are the consequence of the interaction between the natural, social, and built environments. In the mid-1970s, the different disciplines combined efforts to assess the existing research on natural hazards, thereby mixing perspectives from climatology, sociology, economics, engineering, geology, law, meteorology, psychology, public policy, and many others (Peek and Mileti 2002, 512–513). The result was the establishment of the interdisciplinary approach to disaster risk reduction (DRR) that is now an important feature of the Sendai Framework.

In the Philippines, disaster risk reduction became part of disaster management only from 2005 onwards. The Climate Change Act (CCA), enacted by Congress in 2009, was followed a year later by the Philippine Disaster Risk Reduction and Management (PDRRM) Act. These two laws enable the integration of disaster risk reduction and climate change programs and initiatives particularly in relation to capacity building of communities. There are different views as to how integration can be made, but DRR may be more easily grasped by communities because of its clear benefits to both individual members as well as the whole community. Hence, one strategy is to employ community-based DRR as a leverage for climate adaptation programs (Mercer 2010). Using disaster risk reduction as a way of adaptation can be achieved by strong community participation in identifying their vulnerabilities and exposure to a variety of hazards. Such as seen below, in a Maribojoc tourist spot—the SAVIMA Mangrove boardwalk tour.

Risk assessments that are based on such inputs can produce warning information that are more likely to lead to community compliance because they are perceived as meaningful and consistent with the beliefs and value systems of local folk. In the case of Typhoon Yolanda, one painful lesson was that the warning system may have underestimated the strength of the typhoon and also used a locally unfamiliar term (i.e., storm surge) as a signal for which appropriate actions could not be discerned. This led to delayed and non-adaptive responses not just by the community but even by the local DRRM structure.

Successful risk reduction measures often utilize mixtures of technological and psychosocial designs. PAG-ASA still uses number coding for typhoon alerts but has

Fig. 17.1 Photo of the SAVIMA Mangrove boardwalk tour, a Maribojoc tourist spot

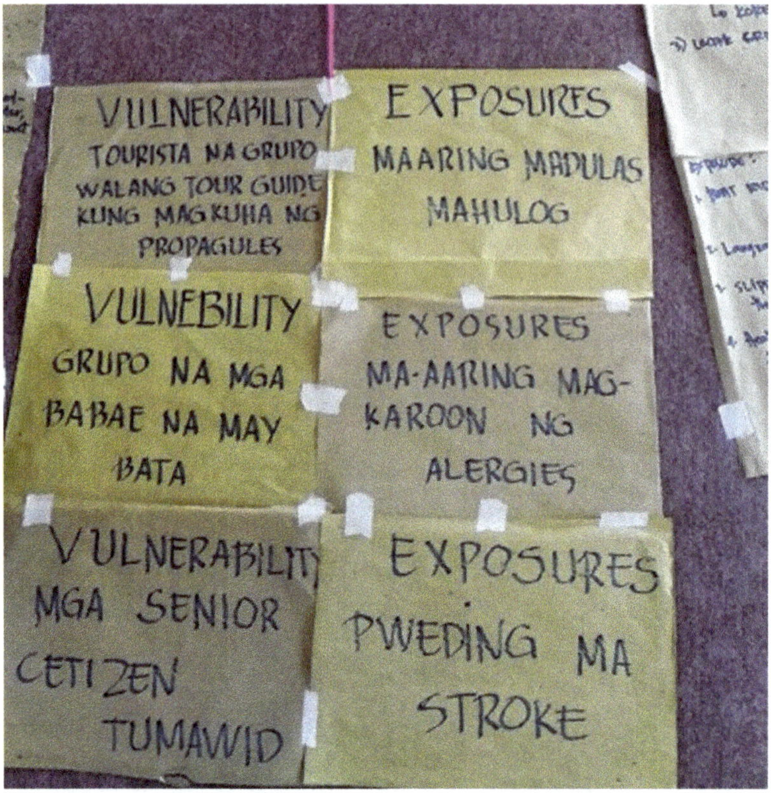

Fig. 17.2 Photo from a DRRM workshop in Maribojoc held in 2018 as part of a UP Emerging Interdisciplinary Research (EIDR) project on Learning Tourism Destinations which produced vulnerability and exposure maps by and for the community, which they used on their own initiative to set up just a few months later their own warning signs

enhanced their warnings with affect-laden color codes and visual and/or vivid text-based descriptions of possible impacts, such as tree branches breaking and roofs flying off. But climate change adaptation requires not only collaboration of experts from different disciplines but also continuous partnering with communities who can contribute their own knowledge and share their approaches to decision-making. Effective risk communication thus allows individuals and communities to learn efficient and acceptable risk reduction behaviors that can transform into adaptive daily habits.

While more researchers and scientists are gaining an increased understanding of disaster risk, and developing more valid, more precise risk assessments, we are yet to increase efforts for the public to understand disaster risk from their own perspective. To communicate disaster risk and influence public behavior towards the goal of disaster risk reduction and resilience, and ultimately, climate change adaptation, we urgently need to create warning–response systems that are based on real risk scenarios.

17.2 Models of Communication

17.2.1 People-Environment Framework in Communication

The human nervous system is probably the most amazing communications system. It allows various parts of the body to receive stimuli from the environment and send messages about the stimuli to the brain, traveling along intricate networks formed by tiny units called neurons. These neurons create the structure of the nervous system and organize its functions based on "connectivity." This connectivity enables information processing through diverse communication patterns, including convergent, divergent, parallel and feedback-loop circuits, to produce analog and/or digital codes based on the electrical and/or chemical information going through the network (Goldstein 2013).

It is worthy to note that the nervous system, especially the brain, is a malleable entity. It can adapt to special circumstances—when the eyes go blind, the neurons previously assigned to process visual stimulation begin to create new connections with the nearby neurons of the auditory cortex. In this manner the system is learning to behave differently—it compensates for the loss of the visual sense by strengthening another sense, in this case, hearing. This capacity for learning and adaption is called neuroplasticity (Goldstein 2013).

The end purpose of the nervous system is to allow people to perceive their environment through specialized sensing mechanisms—visual, auditory, olfactory, gustatory, tactile, and kinesthetic. Communication begins with the detection of stimuli in various forms of energy from the environment which are received by the sensory organs of the body. These stimuli are encoded into neural signals which are processed along sensory pathways, also called nerves, in the peripheral nervous system.

The processed information is analyzed further and interpreted in the central nervous system in the cortex of the brain to produce our perceptions of the environment. For example, the brain analyzes the patterns of light waves sent from the eyes into edges, shapes, colors, and movements. The mind interprets these patterns into recognizable and meaningful objects such as mango trees, speeding cars, dark clouds, old houses, or crowds of people. To perceive is somewhat a way to "make sense" of what is happening, to know what is out there so that we can respond appropriately to stimuli around us.

Interestingly, rapid developments in science have allowed for the design and creation of equipment that can mimic or extend the workings of our sensory systems. We have invented telescopes, amplifiers, cameras, seismometers, smoke detectors, etc.—essentially what we call remote-sensing devices because these devices are outside our bodies. Disaster warning systems often rely on such devices placed strategically on the sides of mountains, at the bottom of oceans, on the ground, or in space, to monitor the environment for hazard risks. These devices can detect signals that can be accessed and then analyzed by trained experts such as meteorologists, geologists, and volcanologists. Such experts can interpret the data to determine if hazard risks are developing and if the risks warrant the issuance of warnings to alert the public.

Unfortunately, ordinary people may often not perceive the same meanings in the warning systems that experts have developed according to what they themselves perceive from their special remote sensing devices. Such a gap in perceived meanings between experts and the ordinary public stems not from differences in sensing capabilities, but from diversity in interpretation based on personal knowledge and experiences. Hence, the gap of meanings is likely to result in a failure in communication because scientists interpret the data from their technical equipment using their technical expertise from their technical experiences. To develop more effective warning systems, the gap of meanings must be reduced not only by enabling the public to be more knowledgeable about hazards and hazard risks but, importantly, also by encouraging scientists to be more aware of people's attitudes and experiences of disasters and hazard risks. The alternative to "encouraging scientists" to understand people, is to add another feature in our DRRM communication system that can serve as a "connector" between the scientists and the public. This would be social scientists and artists working together with the natural scientists.

This type of collaboration already has begun, of course, particularly in developing measures of risk assessment. Following a people–environment framework that views disasters as products of transactions between people and their environments, it is argued that public perceptions of hazards and disasters have very personal meanings derived from individual experiences and values. These meanings often contrast with the objective and analytical descriptions of hazard risks provided by hazard experts. More often than not, these experts define disaster risk in terms of measured probabilities of negative consequences associated with the physical properties of hazards. The choice of risk measures may in fact be subjective even for some consequences that are assumed by many to be highly objective, such as the number of fatalities.

In traffic research, the effects on traffic behavior are observed to vary greatly when providing statistics such as the number of traffic deaths per kilometer, compared to when using the number of traffic deaths per year. Vehicular deaths per year are meaningless as a "risk warning" to a person who has no choice but to drive or take a jeep every day to work. This can also be illustrated by the lack of effect on the behavior of pedestrians of a warning sign found in local streets indicating "*Huwag tumawid, Nakamamatay*" (translated roughly as "You may die if you cross here"), because there is no other alternative for crossing and/or the warning itself is not credible because you see many people crossing but you see none of them are dying.

Similarly in mining, the appropriate risk measure from a governance point of view may be deaths per million tons of coal, given the benefits obtained from coal. But from the miner's point of view, the more relevant risk measure would be deaths per thousand heads employed since he/she could be one of those deaths (Crouch and Wilson 1982 in Slovic 2010). In other words, choice of descriptions of disaster risks must consider the perspective of the people and must be clearly relevant to their specific contexts.

While DRRM continues to evolve with more risk information from experts and stakeholders' sharing their experiences with particular types of disasters, climate change adaptation is much more complicated. The concept of "climate change" in itself is already highly relative—"more frequent" earthquakes, "stronger" typhoons, and "hotter" summers are all highly dependent on other factors such as the age and gender of the person, or the location of the community. Additionally, the debate on the role of human activities in global warming continues and involves scientific data from both sides that cannot be ignored (Shafer 2017). Attributions to climate change for particular experiences with hazards can therefore be ambiguous—if events such as increased drought can be believed to be a natural occurrence, or to be even religious in nature, the urgency of planning specific actions to reduce risk tends to diminish, or at least becomes subject to disagreements among members of a community. These issues can be addressed by collecting clear information about the nature of climate change and climate change risks in a specific locale, in a language that is appropriate to the recipients. For example, rather than simply imposing the concept of climate change, a flipped method can be used so that evidence for climate change can actually be drawn out from the community itself from their own disaster narratives. For CC adaptation, more efforts are required to address these issues of sociodemographic, temporal, and spatial variabilities on such a large scale. More efficient processing of information can be made by adding more "connectors" in their communication models that can help clarify messages passing through to receivers.

Social scientists are important components in these models because their knowledge of behavior can help guide public response (Peek and Mileti 2002, 516). Ordinary people tend to think of risk more subjectively. People's conceptions of risk are usually multidimensional, with emotional components ("dread"), cognitive components ("known/familiar"), and behavioral components ("controllability"). For example, in describing risks associated with typhoons, it may be important to (1) understand what is fearful about wind, (2) find out what do people understand or not understand about typhoons, and (3) what control will people have over typhoon risks—will it be easy or difficult to comply with safety measures? Consequently, disaster risks as determined by natural science experts are often fixed among them, but public perceptions of risk are dynamic and vary considerably across individuals.

While more psychometric techniques are being developed to improve measures of risk perception including disaster risk, there are psychological approaches which no longer focus on risk perception but on combinations of factors that influence how people will comply with safety procedures and warning systems. Some models examine homeostatic motivational processes that involve comparisons between perceived risks vis-à-vis acceptable risks to predict compliance (Mendoza 2004). Other studies examine the social amplification or attenuation of risk through media coverage, interest groups, politicians, and other social institutions (Pidgeon et al. 2003) which can lead to extreme regulation of industries perceived to be associated with disasters. All this information can be incorporated into crafting communication designs and messages.

Fig. 17.3 Risks are assessed in objective ways. This photo refers to objective ways that risks can be assessed and physically described

Fig. 17.4 Risk is personal and social. People perceive risks with different levels of dread and act in different ways. This photo shows people taking the opportunity to have fun in the midst of hazards while others show concern for others

17.2.2 Semiology (Ferdinand de Saussure) and Triadic Semiotics (Charles S. Peirce)

We start this section with a true-to-life narrative.

You have been appointed by the local government to implement disciplinary measures in the mid of a viral pandemic. You tell people to stay home. You tell them that if anyone is on a necessary errand, *they* need to display *their* quarantine pass, don a face mask, and not to cross barangay boundaries. On top of this, you instruct them to observe a 2-m distancing space between persons in any public place. This policy is state-sponsored and you, as a designated executor of this policy, not only implement it as matter function of your office but believe in it wholeheartedly, thus, implement it zealously believing that these actions are the only ways by which the community can contribute in the effort to curve the contagion. There may be initial successes, but as days wear on, the people you see milling around no longer observe this set of policies and just go about their merry ways as if the pandemic does not exist, or so you thought. But they counter, "We need to go out! We need to earn for us to eat!"How would you feel? What do you want to do? Will you be able to understand them? Or will the situation frustrate you so much as to wish violators of regulated health protocols to be shot? Sounds unlikely? Read the article "QC official threatens 'shoot-to-kill' policy for MECQ violators" to know how—when communication shuts down,—even the unlikely becomes real (Talabong 2020).

Risk communication works where the sender of information and the receiver of information reciprocate their signals in a meaningful way. Reciprocate meaning is a two-way communication, and meaningful in that the information is processed, meaning is understood, and an attendant response continues the thread of both communicators' wave of thinking.

The current practice of communication current relies heavily on multi-media—the use of text, images, and sounds—to convey meanings. Communicators or senders of information craft their messages to a target audience by way of printed materials or moving images that are designed so that the intended meanings will not be lost in the communication process. The "sign," as indicated in the field of semiology, stands for the idea being communicated. It is the marriage between the word or sound (the *signifier*) and the concept for which it stands for (the *signified*) (Smith et al. 2005. 228–229). The semiotic theory model shows a reliance on the intersection of the sender's and the receiver's realm of experience to effectively communicate. The content represented by the sign from the sender matches the content received from

Fig. 17.5 Semiotic Model based on Ferdinand de Saussure (referred to as semiology), the sign relation is dyadic, consisting only of a form of the sign (the signifier) and its meaning (the signified)

the sign by the receiver, and the channel of communication transmits the information with integrity. If experiences do not match, or the channel distorts the messages, misinterpretation is very likely to happen.

An inclusive manner by which participants can craft their own narratives based on given images is espoused by Charles Sanders Peirce called Sign Theory of Semiotic or the Triadic Semiotics, "is an account of signification, representation, reference, and meaning. Although sign theories have a long history, Peirce's accounts are distinctive and innovative for their breadth and complexity, and for capturing the importance of interpretation to signification" (Atkin 2013). The "interpretation" made by the receiver is a very important factor in communication. Communicators endeavor to engage the potential receivers of information in crafting communication materials to gain insights into process of an audience's creation of meaning. For example, participatory workshops can be directed at surfacing individual or collective risk concerns. Using available resources—such as cartons, yarns, clay, stones, etc., participants in a disaster mitigation workshop can transform a flat surface into a properly scaled topographic representation of an area that residents in a community can mark with individually or collectively perceived hazards. One model of implementing a participatory workshop is outlined in the "Participatory three-dimensional mapping for disaster-risk reduction" (Gaillard and Maceda 2009). In summary, participants were made to construct three-dimensional maps (Divinuvo [Eastern Samar], Masantol [Pampanga], and Dagupan [Pangasinan]) of their community. They represent the different sectors and age brackets in the community. They were directed to mark livelihood areas, public and private lands, and the areas that they perceived to be hazardous or risky. They then discuss the reasons for their perceptions of risks and agree on recommendations to mitigate the risks. The generated information were then input into GIS hazard maps that can be used as references by local government officials and scientists for structural interventions and crafting local policies that aim to mitigate the risks to the communities. Indigenous knowledge is a good resource of authentic information.

Bottomline? We just can't expect an audience to understand us if our language and our tone do not match their temper. Engage with our audience and learn their ways. Talk the way they talk, walk the way they walk, and they ease up on conversing mode. Once they get comfortable, there is better chance for us to influence their thinking and their actions.

17.3 Disaster Risk Communication

17.3.1 Disaster Risk and Related Factors

"Disaster risk is expressed as the likelihood of loss of life, injury or destruction and damage from a disaster in a given period of time" (UNISDR 2015). The hazards that cause the risks will have to be properly managed to mitigate the disastrous effects of the hazards when they do occur. Notably, disasters happen as a result of natural functions of the earth—can be volcanic eruptions and earthquakes, and human-influenced—such as floods and fires. Disaster risk relates highly with environmental forces, as it does with human habitation and its attendant activities. Whether natural or human-induced, people had to be informed about these hazards and how much risk they are exposed to.

The duality of disaster risks—ranging from being *wholly naturally occurring* on one end to being *human-propelled* on another—contributes to polarizing the issue of climate change. Climate change, appearing with cyclic regularity in earth's history to many, predestines its occurrence, and so espouses a do-nothing attitude. For many others, human activity impacts the earth's climate and so human intervention must be directed to mitigate its ill-effects. What point of view will you take? It can only do us good to take a proactive role towards the latter.

Communicating risks may be hampered by many constraints—especially the financial kind—and access to resources such as staff, domicile, and mobility. However, there are ways we can get by. Seek out partners such as corporations and government units. Include them in our information collaterals. Make them own up to our advocacy. Appeal to their altruistic sense. The energy that we will devote to convincing partners or sponsors should extend to our development of communication materials for our various audiences. Communicate with sharp and direct words. Speak with their language and tone. Will the bearer of the message be a support group, a partner? Or an implementing arm of the government whose aim is to impose or regulate? Or an educator who seeks to explain and inform? Communicating the presence of hazards, and risks to population often means prescribing certain activities including evacuations, for example, to pre-identified places that do not match the perceived levels of safety and basic comfort for the population. An evacuation center may be decrepit and over-populated, and worse, perceived to be prone to sex predation, Instructions may be met with apathy, or worse, resistance. Are these perceptions real or imagined? If real, then physical adjustments have to be made. If largely imaginary, communicating the benefits of evacuation over inaction to the population is essential.

Risk communication can take the form of static materials, moving images, or can be a face-to-face exchange. It may be a poster, infographic spread, video, or movie. It can be a storyline framed in "komiks" strips, or an emotional message contained in a painting or a public art. They can be delivered by people, "the emergency response worker rallying a community to evacuate in the middle of the rising flood. It is the community representatives sitting down with industry to discuss the siting and operation of a hazardous waste incinerator. Risk communication involves people from all walks of life—parents, children, legislative representatives, regulators, scientists, farmers, industrialists, factory workers, and writers. It is part of the science of risk assessment." (Lundgren and McMakin 2013). Communication is a process, and it is not concluded with any one device.

Considering its vast array of stakeholders and the diversity of perspectives involved, communicating risks will have to take into account different value systems, but part of risk communication is to also develop in people a common value of safety.

Risk information does not matter, if people are willing to take risks. Such systems may have been learned by people through legal instructions or cultural practices. It is worthwhile to take stock of signs and signals that are observed as a matter of legal abidance. Observing these should provide context in the process of continually creating signs and symbols that are more meaningful to specific receivers of information. It is often kidded about that Filipino are culturally happy people that even in adverse situations, humor permeates social exchanges.

As a visual communicator, let me illustrate a successful use of "irony" spiced with humor in communication material. To those who experienced the energy crisis of the 1980s, *Asiong Aksaya* is a character in a comic strip illustrated by national artist Lauro "Larry" Alcala. Ironical in Asiong's predisposition to expend unnatural amounts of energy—gasoline, light, etc., but in the end leaves him the victim of his own wastefulness (Valino 2002). As a vehicle for messages on energy conservation, it is an idea out-of-the-box. The impact of Asiong Aksaya having saw print from 1976 to 1984 is not lost to many. Asiong Aksaya was easily assimilated into the popular lexicon as a pejorative—even an endearing term to chastise a wasteful friend.

Communicating through a filter that will serve to modulate the channel of communication may be through *lenses* or *frames*. "To break through the communication barriers of human nature, partisan identity, and media fragmentation, messages need to be tailored to specific medium and audience, using carefully researched metaphors, allusions, and examples that trigger a new way of thinking about the personal relevance of climate change, Framing—as a concept and area of research—spans several social science disciplines. Frames are interpretive storylines that set a specific train of thought in motion, communicating why an issue might be a problem, who or what might be responsible for it, and what should be done about it" (Nisbet 2009, 14). The Philippines need to up the ante on climate change discussions, and in doing so, there are *frames* worth considering, for isn't climate change associated with pollution, deforestration, overpopulation, and overconsumption, among other human-induced problems? Wasn't growing food crops over forest preserves a contested issue? Isn't the virtue of protecting natural resources opposed to commercial forestry and mining interests. Isn't water impounding and damming to serve the needs of sprawling metropolis anathema to preserving and protecting ancestral land and culture? These may be presented as polar opposites but in the end, the question is what should be done about them? From whose lenses are we framing the issues?

17.3.2 Communicating Disaster Risk

17.3.2.1 Standards in Texts and Graphic Requirements in Information Systems

"Recognizing that international uniformity of road signs, signals, and symbols and of road markings is necessary in order to facilitate international road traffic and to increase road safety,

Standards in text and graphic requirements for information systems were developed," thus states the preamble of the Convention on Road Signs and Signals originally crafted in Vienna on November 8, 1968, with 56 subscribed member countries by July 1, 2007. This underscores the need for signs and signals that transcend political boundaries. Before this agreement, countries appropriate their own colors and shapes for road warning and information systems, often influenced by their own historical development in road and transportation systems, and external influences via trade or imperial expansion. The STOP sign was not always Red. It was also Yellow (Miller 2016).

In a shrinking global landscape, this leads to different interpretations of signs, and thus, it was imperative for countries to subscribe to an internationally accepted standard. The Vienna Convention of 1968 attempted to address the issue of standardizing road signs and signals ranging from road structural details, graphics, texts, and colors of road safety and information systems. Other supplementary agreements, such as the European Agreement, introduced additional details to the implementation of the standards (Vienna Convention, part 6).

Despite the prevalence of this standard on road signs and signals, there are marginal deviations that are culturally influenced. To many, a red traffic light always means STOP, a yellow means WARNING, and a green light GO. Not so clearly in many areas in Japan.

Drive around Japan long enough and you'll probably run into one of the country's mythical blue traffic lights. Elsewhere around the island, you'll find "go" signals that are decidedly teal, turquoise, and aqua. "Is this signal broken?" you might wonder. "Did some overworked light-monger install the wrong bulbs?" The answer, it's in the Japanese language…while it may appear that Japan uses blue traffic lights, the government assures us it's actually just a very blue shade of green—green enough to satisfy international regulations, blue enough to still be called "ao" (Spektor 2016).

17.3.2.2 Survey of Legislated and Accepted Conventions in Hazard and Risk Communication

The world's governments offer rich resource of standards relating to health, work, transportation, building, and different areas that form the built environment. However, an institution bearing a very strong influence in the design and manufacture of goods is not government but a private institution.

The ANSI or the American National Standard Institute is a "federation formed by standards writers and users, that manages the voluntary standards system in the United States" (ANSI, "What is ANSI"). ANSI is not a state institution, yet the design, performance specifications, and marking of safety eye and face products, and millions of other safety equipment all over the world subscribe to their specifications. The specifications were not decided at random. The standards were based on a long legacy of industrial practice and scientific investigations involving perceptions and user experience. This brings to the fore ANSI-prescribed safety colors that have roots in industrial America that are internationally observed as well today.

Colors can be distinctively differentiated by persons with normal sense of vision. To those with impaired visions, discriminating on color differences may be a problem. In the grayscale version of the safety colors where only color *value* (defined as the lightness or darkeness of a color) is shown, the colors (because they appear grayed out) do not appear distinct. This highlights the case of "colorblind" people who cannot accurately discriminate between the hazard colors that are of similar "value," Elements other than *hue* and *value* must be taken into account in making signs, especially the ones that have widespread use. The shape of the sign, the integration of diagonal, vertical, and horizontal lines, and the thickness of these lines can come into play. If the design is going to be used mostly for a specific regional or cultural group, it will be more appropriate to use symbolic meanings that the population can relate to, even while carrying over some of the characteristics of the "standard" signs. Symbols and colors can have very strong cultural meanings. Their use, when appropriately used and made by the community itself, will have wider acceptance and following.

17.3.2.3 Language and Symbols Associated with Specific Cultural and Demographic Groups

Language and the cultural meanings associated with symbols make designing for a general audience problematic. Language appears to be a barrier in the full appreciation of the risks associated with issued warnings in the case of supertyphoon Haiyan in the Mideastern Philippines. Even though the hardest-hit areas received early warnings, the weather service and other officials later admitted that the victims were unfamiliar with the term "storm surge." The last deadly storm surge in Tacloban had hit in 1887, more than a century before Haiyan. In a country with scores of regional languages, the government also did not have local terms to be able to communicate the phenomenon to everyone. After the disaster, the government agency worked with linguists to craft simpler meteorological terms to ensure the danger posed by typhoons, floods, landslides, and other adverse events would be fully understood by all (Why Super-Typhoon Haiyan was so Deadly 2018).

The style by which language is embedded in texts is by itself a powerful conveyor of meaning. This may be seen in corporate logotypes where the Brand is written in a way that makes it unique and easy to recall. The logotypes of Coca-Cola and San Miguel Beer come to mind for being so ubiquitous and do not need illustrating. But language-in-text combined with image is yet more powerful. In designing symbols, the shapes and colors need to correspond in meanings for the symbol to be received clearly. For a better assessment of the acceptability of design symbols, the concept of a "cultural palette" was proposed in the book Handbook in Visual Communication, to assist the designer in the development of culturally sensitive symbols (Smith et al. 2005, 119–122). The book states that a cultural palette is developed using a qualitative method involving a series of interviews and a collection of images compiled to an image bank. A panel of experts then determines which of the images are good (appropriate and inoffensive) and those that are poor (inappropriate and offensive) symbols and colors. The final step is to create the palette of symbols and colors that provide a range of culturally sensitive graphics at the same time identifying those that are insesitive and inappropriate and thus should be avoided.

17.4 Storytelling as Basis for Disaster Risk Communication

17.4.1 Narratives and Personal Experiences as Guides for Creating Meaningful Messages

Stakeholders' participation and interaction with those crafting the communication material or those who are managing the risk adds a more authentic tone in communicating risks. Workshops, focused-group- discussions (FGDs), and surveys are often used but there is no substitute for a face-to-face conversation with people in a local area sharing their own stories, imbued with their hopes, aspirations, and fears, in dealing with their perceived risks given the hazards in their areas. The least effective but most often used form of stakeholder participation is the formal hearing or public meeting, for which the organization sets a time and place for the audience to present formal testimony, which is transcribed and used later in the risk management process (Lundgren and McMakin 2013. 227–250).

In the World Bosai Forum 2019, held in Sendai, Japan that I attended and participated in with colleagues from the university, I recall a high school student (now a professional) sharing her experience as a victim of the earthquake that shook Japan that fateful day of March 11, 2011—alternatively referred to as the 3.11 earthquake, the Great Sendai Earthquake, and the Great Tohoku Earthquake. Unfortunately, there was nothing humanly great about the earthquake as it was followed by a devastating tsunami that in combination laid waste to great part of that Pacific coast near the city of Sendai. Worse, it disabled the Fukushima Nuclear Plant laying radioactive waste in the area and causing fear and panic not only in Japan but other parts of the Pacific Basin. According to the student, she was witness to this carnage and loss of family members, but also witness to the resiliency of the human spirit to prevail against odds, and to aid others in dire need of help. I remember her relating how she was adopted by a foreign family—for her safety and education, and the care and kindness she received from elders and her peers. She made a promise to herself in not failing to narrate her experience on all possible occassions, and deliver her message of the need to prepare incessantly, so that people will pay attention to hazards, and will always be prepared. Her appeal to the intellect and to the emotion cannot be eclipsed by mere technical warnings, or warnings consisting of threats, or appeals based on fictional stories.

17.4.2 Hazard Maps and Individual Vulnerability Profiles

Individuals may take the initiative in learning about hazards. They may initiate learning for their personal benefit or for informing others. A good learning resource for this kind of study are hazard maps that are usually prepared for specific types of hazards such as typhoons, floods, volcanoes, and faultlines or combinations of related hazards, (i.e., typhoons and floods, volcanoes and faultlines). A hazard map is prepared using cartography conventions, so it should not be hard to learn and to understand them. However, the differentiated use of legends, indicating the different hazards' levels of intensity or severity, are not the same which may hamper learning.

To the users of computers and mobile devices, an application called Hazard-HunterPH (URL: https://hazardhunter.georisk.gov.ph/) was launched in July 2019 where users of the platform can click on any part of the Philippine map and check on that area's assessment in terms of susceptibility to ground shaking, lahar, flood, tsunami, etc., and its proximity to known volcanoes and faultlines. This takes care in part to understanding the technical aspects of the hazards. The significance may extend to the individual level if users can relate the hazards to their own perceptions of vulnerability so that flood, for instance, is not just a measurement in inches but a factor of "*dread* " to the individual.

In an example, acceptable risk levels to flood may be different for Individual A from Indivdual B due to factors in height or competency in swimming, despite the fact thay they are similarly exposed. The result of this assessment can have many uses. One concept that I can use as an example is the creation of individial avatars (representing one's personality type) that will start with a color, shape, and size, even in the general likeness of the *user* (the individual). By learning and implementing actions that will lessen the risks to the avatar—or say increase in its survival index, the colors, shapes, or even size may change in a desired way, Depending on the care (or abandonment) that the owner provides for the avatar, changes occur in a combination of ways. This way, the user may not only *learn* about good practices in risk mitigation but may *practice* them as well. This is just an example of a way by which current technologies and communication platforms, and the way people—especially the young ones—adopt technology and appropriate them for custom use. The avatar example is similar in a way to earlier trends, adopting *pet rocks* a generation ago, that was followed by the *tamagochi* craze, and then there was *Picachu*! This time, their pet is their avatar and they have to train their avatar the survival tricks neccessary to increase their life stream. They have to be provided with tasks that they need to do, and when fulfilled, the avatar is enhanced.

A presentation delivered by Dr. Tali Sharot in her TEDx Cambridge produced talk "How To Motivate Yourself To Change Your Behavior." gives us a good idea on some ways of motivating people to be guided to a line of action. In the video, she indicated that age groups react to warnings differently. Learning from mistakes tend to be low at young age, raises up until age forty, and again declines in older years. This also translates to heeding warning signs (Sharot [n.d] 6:10–7:15). This perhaps explains in part why young people pick up smoking in teenage years despite warning signs on the label due to their low level of risk aversion.

Dr. Sharot further adds that people want to take in information that they want to hear more than information that they don't. People feel better when they are presented with an improving image of themselves than when they're presented with a negative image of themselves (Sharot, 07:15 -08:15). On the subject of individual avatars proposed earlier, the opportunity given to users in enhancing their avatars will correspondingly help them acquire knowledge of risk management best practices. The improvement in the avatars may just rub on to the users.

17.5 New Approaches and Creative Design Considerations in Developing Risk Communication Materials for Specific Hazards and Target Groups

17.5.1 Targeted vs Standardized Disaster Risk Communication Campaigns

Disaster Risk Communication cannot rely on standard and templated communication materials alone. This must be supplemented with materials that are designed specifically to target audiences. Targeted communication refers to the process of identifying a particular group or an audience that is assessed according to set of demographic criteria that my be based on the answers to the following types of questions:

- What does the audience do for a living?
- What is its income level?
- What is its typical age?
- What is the gender of the typical audience?
- What is the education level of the typical audience?
- Do they have children?

Classifying the audience to a set defined by these attributes provides the communicator with a direction in delivering an appropriate message or solution to the intended audience. It may appear that in risk communication, information disseminated to the public need not discriminate on the type of audience since the information is of technical nature. This is not so. Without a target audience, the message will be lost in the sea of diversity in the mental and emotional states of a "blind" audience. As in the advertising industry, a good practice is to develop different—even short (ranging in seconds) materials that are specifically designed for the different audiences. The best practice in this case is to conceptualize a design program that will map out for whom the communication material is intended for, and the overall message that the communication wishes to deliver, with a projection of a line of action that the receiver will undertake. For these materials to be successful, they are usually designed along the principles of *dominance, frequency*, and *duration*. This is referred to as a campaign, short for marketing or advertising campaign.

Campaign is indicated by "the groups of advertising messages which are similar in nature. They share same messages and themes placed in different types of media at some fixed times. The time frames of advertising campaigns are fixed and specifically defined." (Advertising Campaigns 2019). Following this indicator, risk communication campaign then is a series of messages through media channels that share a single idea and theme. A communication campaign targets specific objectives using strategic visual content customized to the mindset of the target audience. It is creatively directed so that the content—visual, auditory, and at times tactile, typically does not require explanation to give context, relying instead on the impact of the overall experience to drive meaning. Campaigns may include graphic design, moving images, interactivity, music and sound effects, and other sensory devices to aid in the delivery of the message.

Interactivity in communication campaigns in particular is a relatively developing mode of information delivery system with increasing web- delivered and web-accessed content. It allows the receiver to respond making feedback tracking immensely more immediate than the traditional broadcast media such as television and radio. Interactivity means that the receiver may also share online content as well as collaborate in the creation of the content itself. This allows developers to design messages that entice the viewers to participate in the creation of content giving rise to information systems that are crafted by the participation of the crowd themselves and the messages are more meaningful among them. Visuals may also be in the form of moving images, and as claimed that a visually supported presentation is more persuasive than an unattended one by 43% (Vogel et al. 1966). (Many websites claim to an incredible visuals processing at 60,000 more times than text alone but these are largely unsubstantiated.) The use of moving images and sound will undoubtably increase the value of user experience, translating this to increased use and exposure to the messages. Constant exposure to the message will embed messages subconsciously for easier recall and learning consisting "of the acquisition of a new pattern of response (or behavior) without consciousness of the keys (stimuli, contingencies, or rules) that initiate or regulate it" (Partido 2012).

17.5.2 Using Dynamic and Multisensory Designs for Disaster Risk Communication

Multisensory learning approach is often used to teach children with learning disabilities. The approach appeal to a particular child's area of sensory learning strength, sometimes called a learning style. This approach suggests that when students are taught using techniques consistent with their learning styles, they learn more easily, faster, and can retain and apply concepts more readily to future learning. Most students, with a disability or not, enjoy the engaging variety that multisensory techniques can offer (Logdon 2019).

The approach may also be applicable to communicating risks to children and young adults. A General Education (GE) course in the University of the Philippines Diliman named DRMAPS (Disaster Risk Mitigation, Adaptation, and Preparedness Strategies) applies multisensory techniques and has effectively enhanced class participation, understanding of technical principles, integrating and working with team members, and creating communication materials on hazards and risks.

Visual components often include a mix of visual teaching methods and strategies such as using:

- Text and/or pictures on paper, posters
- Illustrations in aid of narrative
- Creation of mind maps
- Using maps and organizing information pertaining to hazards
- Student-created art and images
- Creative work in a video container

Auditory techniques include:

- Recording ambient audio sounds in aid of narrative
- Creation of music, speaking rhymes

Kinesthetic methods include:

- Modeling hazard related scenes using their bodies with minimal props
- Games in acting out scenes flashed on screen
- Role playing (acting) in their video recording for their creative communication material

Multisensory techniques enable students to use their personal areas of strength to help them learn. In the DRMAPS context, students come from diverse disciplines and when grouped in teams, they can contribute to the team effort by drawing on their own creative and academic interest. Current multisensory techniques can be further enhanced by the introduction of new technologies in the digital area. 3D modeling, motion capture and virtual reality technologies have advanced to a degree that the total cost of their operation is within hobbyist level. Outside of creating content, these technologies can also ease creating programs that aid training. In immersive authoring systems, virtual reality developers build applications from within the virtual environment, an approach similar to visual programming. Programming elements are represented with 3D virtual objects, and direct 3D manipulation is used to assemble virtual scenes and specify object behaviors and attributes (Lee, Kim & Billinghurst, 79).

A digital sandbox is one example to this. This technological contraption that can easily be DIY (do-it-yourself) and is a very good way of making young people (but not exclusively so) learn about the nature and effects of climate change. "An augmented reality (AR) sandbox is a 3D, interactive, dynamic educational tool to help understand mapping, topography, and watersheds. It was created by Oliver Kreylos,

Fig. 17.6 Student of DRMAPS "role-play" by acting out and using a small amount of props to simulate a disaster. The class then process the image and seek out the highlights of the role-play in conjuction with the disaster. Depicted in this photo is an event of volcanic eruption

Peter Gold, and M. Burak Yikilma, and uses a motion sensor and computer software to map contour lines and colors based on the real-time shape of the sand" (Augmented Reality Lesson 2019). The sandbox, for instance will show water virtually fillilng up the lower cavities of the sandpit if the sand mounds are reduced in height, or colors on the slopes show the possiblity of collapse as they are built up to a steep angle.

17.6 Conclusion

17.6.1 Designing Risk Communication Materials

What is the content and form, then, of an ideal communication material? There are models of communication that aid the communicator in understanding the basis of content creation considering the interplay of the components in a communication environment. There are software and digital technologies that aid the creation of media. These can integrate shapes, colors, and sounds, and to some extent, engage the tactile and olfactory senses. There are ways of making the receiver of information interact with the media to suit individual learning styles. By way of conclusion, the following is a proposed set of guides in crafting communication for an audience:

Study the audience. Collect information, observe audience attributes, and involve the audience as much as possible in crafting the communication material. Use local narratives, colors, and shapes, that the audience can relate to, and avoid use of the same to the ones that they dislike. Translate technical terms to words that relate to the audience's experience. Check what information channels are available to them, and what they enjoy using the most. This channel, at the least, ushers a path for a relaxed communication exchange.

Adhere to standards when it is required. The set of standards for signs and signals exists for public order. Where standards are binding, observe them. But beware, some accepted standards and conventions are only influenced by industry and are not established by law. In this case, the so-called standards can be modified and translated to symbols that are closer to the audience's realm of experience.

Use technologies that can work in the area. While the current level of technologies facilitates the authoring of creative media, the communication process can only work if the audience can receive and use the material as designed. Check communication delivery systems such as available transmission and projection systems, access to the internet, space, facilities, etc., to verify that the communication material will not be distorted by way of delivery. Note that the creation is not totally dependent on delivery. For example, a series of printed posters may be made using advanced imaging technologies but are not technology dependent in their display, or a complex topography map is produced through 3D printing but does not require special tools to manipulate it to a useful information media.

Allow the audience to interact with media. Stimulation through different senses—including movement—and interaction with the media and content facilitates learning. For instance, using digital media, engaging the audience in appreciating the communication media to the point where they may inject new information materials establishes their "stakeholder" posture and allows multiple points of data entry. The traffic navigational map *Waze* has made good use of crowdsourcing, some modified forms of user engagement—such as adding a layer with data vetting points —can be used for hazard maps. On the other hand, using manual or analog approach, the "stakeholders" may be engaged in creating murals or public art where they can impart their hopes and aspirations, as well as their most dreaded fears. These artworks need not be cast for posterity, but are continuously augmented, improved, or reconstructed, to better reflect their sentiments at any time.

An example of a successful interactive campaign fueled by contributions from viewers is the Department of Tourism's "Its More Fun in the Philippines." Launched by adman and erstwhile Secretary of the Department of Tourism Ramon "Mon J" Jimenez, the campaign features tourism destinations in the Philippines peppered with human interest as he declared "that his strategy will tap the growing power of social media. 'Believe in the beauty of your country. Sell it at every turn. Sell it on Facebook, on Twitter, on Multiply'" (Albano 2020).

To this day, the "Its More Fun in the Philippines" site (https://www.itsmorefuninthephilippines.com/) receives viewer-contributed content from both local and foreign visitors. Photos and accompanying texts of their visits to different Philippine destinations are self-uploaded to the website for all the world to see. Without much curation effort, the site self-updates and self-promotes. Something similar but on the hazard side can be crafted. How about "Hazards Reveal In the Philippines" or "Beware: Extreme Fun" web portals to crowdsource "discovery" of hazards?

Hazards are poised like the Sword of Damocles of the ancient Romans, ever-present and looming, and can strike without warning. Some have been hanging around so long they have transmuted to a sight of awe—that mixed feeling of dread and relish. It is akin to staring down the mouth of a volcano, looking up at a jagged precipice, or being buffeted by giant waves—the kinds of experiences that exhilarate tourists. It is up to the communicator to craft how the population can relish in the delights the "hazards" bring but to prepare them for the havoc they can cause as well. The hazards may not be tamed, but the risks can be reduced.

Plan for a communication campaign. Planning campaigns that are audience-focused and time-sensitive ensures the appropriate use of resources. Campaigns are tactically short and focused but strategically protracted and extensive. Latch onto current trends, but explore original creations. Disrupting the mainstream but keeping on track of delivering the correct science behind the message is a creative challenge.

Hazards that suddenly become real—such as the COVID-19 pandemic—required immediate action and emergency policies that did not afford long discourses. Issues of pressing economic, political, and social concerns that are easily headline banners tend to bury issues of great importance but are creeping in slowly and with great impact, such as climate change. Climate change is such an encompassing topic that it dissolves into the background, even if easily, we can point a finger at climate change as the main culprit of the ills of our time. Take the earthquakes and volcanic eruptions, for example, or the typhoons and floods, even energy needs—and include the COVID-19 pandemic for good measure, Headline hogging events that have effectively sidelined the issue of climate change, and yet these events can relate to climate change. Presentations, items of text or video, or fora that discuss how all these things interrelate are so scarce and inconspicuous. It is time to see climate change as the stress that's pulling apart the threads \of human existence. We may feel so small in the grandness of nature—especially when it starts acting up in ways that scares us, but cope with it we must. Climate change is the big window that opens to the grand vista of the world unfolding. The frames in this big window are our channels in communicating small chunks of information so we will not overload our audiences even as we deliver and receive information with urgency and at great speeds.

References

Albano Lou (2020) 'Its more fun' with Mon Jimenez," GMA News Online. April 30, 2020. https://www.gmanetwork.com/news/specials/content/145/it-s-more-fun-with-mon-jimenez/

Advertising Campaigns - Meaning and its Process (2019) Management study guide. https://www.managementstudyguide.com/advertising-campaigns.htm. Courses. November 24, 2019

Atkin Albert (2013) Peirce's theory of signs. The Stanford Encyclopedia of Philosophy (Summer 2013 Edition), Edward N. Zalta (ed.). https://plato.stanford.edu/archives/sum2013/entries/peirce-semiotics/. Accessed Novermber 24, 2019

"Augmented Reality (AR) Sandbox Lesson (2019) Resources for teachers. USGS science for changing world, July 25, 2019. https://www.usgs.gov/science-support/osqi/youth-education-

science/augmented-reality-ar-sandbox?qt-science_support_page_ related_con=1#qt-science_support_page_related_con
Barnard Malcolm (2008) Approaches to understanding visual culture. New York: Palgrave
Bergstrom Bo (2008) Essentials of visual communication. Laurence King Publishing Ltd., pp 221–222
Bonnes M, Bonaiuto M (2002) Environmental psychology: from spatial-physical environment to sustainable development. In: Bechtel R, Churchman A (eds) Handbook of environmental psychology. John Wiley & Sons, NY, pp 28–54
Climate Change Act of 2009 (RA 9729). https://www.officialgazette.gov.ph/2009/10/23/republic-act-no-9729/
Commission on Audit: Disaster Management Practices in the Philippines: An Assessment, 2017. https://www.coa.gov.ph/disaster_audit/doc/National.pdf
Cvetkovich G, Earle TC (1994) Culture and risk. In Psychology and Culture, eds. Walter Lonner and Roy Malpass. Boston, MA: Allyn & Bacon
Gaillard Jean-Christophe, Emmanuel A Maceda (2009) Participatory three-dimensional mapping for disaster risk reduction. Community-based adaptation to climate change. Chap. 8. http://elti.fesprojects.net/RESOURCES/Community_Based_Adaptation. pdf#page=111
Goldstein E Bruce, James Brockmole (2016) Sensation and perception. Cengage Learning
International Safety Equipment Association (2019) What is ANSI? https://safetyequipment.org/isea-standardsstandards-faq/what-is-ansi/
Kleinhesselink RR, Rosa EA (1994) Cognitive representation of risk perceptions: a comparison of Japan and the U.S. J Cross-Cult Psych
Lee Gun, Gerard Kim, Mark Billinghurst (2005) Immersive authoring: what you experience is what you get (WYXIWYG). Commun ACM 48:76–81
Let's DOH it (How we did it) (2013) Knowledge Centrum. http:// commmedia.aijc.com.ph/comm_media/knowledge/knowledge_best_lets_doh_it.html
Logsdon Ann (2019) How to make multisensory learning materials. Verywell Family. https://www.verywellfamily.com/make-multisensory-teaching-materials-2162327
Lundgren RE, McMakin AH (2013) Risk Communication [electronic resource]: a handbook for communicating environmental, safety, and health risks, 5th edn. Wiley, Hoboken, N.J.
Mendoza AC (2004) Filipino Drivers' intentions to commit traffic violations: a social psychological perspective. Unpublished dissertation, UP Diliman
Mercer J (2010) Disaster risk reduction or climate change adaptation: Are we reinventing the wheel? J Intern Devel 22(2):247–264. https://onlinelibrary.wiley.com/doi/ abs/https://doi.org/10.1002/jid.1677
Miller Aaron (2016) The reason traffic lights are red, yellow, and green. Thrillist. October 6, 2016. https://www.thrillist.com/cars/nation/traffic-light-colors-history
Mirzoeff N (ed) (1998) Visual culture reader. Routledge, London
National Disaster Risk Reduction and Management Council (2013) Final Report re Effects of Magnitude 7.2 Sagbayan Bohol Earthquake. Quezon City. http://ndrrmc.gov.ph/attachments/article/1330/FINAL_REPORT_re_Effects_of_Magnitude_7_2_Sagbayan_Bohol_ Earthquake_15OCT-04NOV2013.pdf
Nisbet Matthew C (2009) Communicating climate change: why frames matter for public engagement. Sci Policy Sustain Develop 51. https://doi.org/10.3200/ENVT.51.2.12-23
Partido N, Juan P (2012) Unconscious learning. Encycl Sci Learn. https://doi.org/10.1007/978-1-4419-1428-6_1067
Peek LA, Mileti DS (2002) The history and future of disaster research. In: Bechtel R, Churchman A (eds) Handbook of environmental psychology. John Wiley & Sons, NY, pp 511–524
Pidgeon Nick, Roger E Kasperson, Paul Slovic (eds) (2003) The social amplification of risk. Cambridge University Press
Simonet S, Wilde GJS (1997) Risk: perception, acceptance, and homeostasis. Appl Psychol Intern Rev 46:235–252
Slovic Paul (1987) Perception of risk. Science 236(4799):280–285

Smith K, Moriarty S, Barbatsis G, Kenney K (eds) (2005) Handbook in visual communication: theory, methods, and media. Lawrence Erlbaum, Mahwan, New Jersey, London

Sharot Tali (n.d.) How to motivate yourself to change your behavior. TEDx Cambridge Video. 16:48. https://www.tedxcambridge.com/talk/how-to-motivate-yourself-to-change-your-behavior/. Accessed August 25, 2020

Skolnick AA (n.d) Manufacturer of 'Death Cigarettes' Says He's Working to Bring About the Death of Smoking. JAMA. 1992; 267(2):204–207. https://doi.org/10.1001/jama.1992.03480020014003. Accessed August 25, 2020.

Specktor Brandon (2019) This Is Why Japan Has Blue Traffic Lights Instead of Green. The Reader's Digest. https://www.rd.com/culture/heres-japan-blue-traffic-lights/

Talabong Rambo (2020) QC official threatens 'shoot-to-kill' policy for MECQ violators, Rappler. https://rappler.com/nation/quezon-city-official-ludovica-threatens-shoot-to-kill-mecq-violators

Thank you for sharing the fun (2020) Its more fun in the Philippines. https://www.itsmorefuninthephilippines.com/

Thompson Derek (2018) The four-letter code to selling anything. TEDx Binghamton University. YouTube Video. 21:09. https://www.youtube.com/watch?v=6pY7EjqD3QA

UNISDR Global Assessment Report (2015)

Valino Raul (2002) Asiong Aksaya and other classic ad campaigns. Philstar Global. July 16, 2002. https://www.philstar.com/ business/2002/07/16/168446/asiong-aksaya-and-other-classic-ad- campaigns

Vienna Convention on Road Signs and Signals (1968) Consolidated version: https://www.unece.org/fileadmin/DAM/trans/conventn/ Conv_road_signs_2006v_EN.pdf

Vogel DR, Dickson GW, Lehman JA (1966) Persuasion and the role of visual presentation support: the UM/3M Study. Working Paper Series. http://misrc.umn.edu/workingpapers/fullpapers/1986/8611.pdf. Accessed November 24, 2019

Why Super Typhoon Haiyan was so deadly (2018) Phys Org, November 5, 2018. https://phys.org/news/2018-11-super-typhoon-haiyan- deadly.html, Earth section

Leonardo C. Rosete is a Professor at the College of Fine Arts, University of the Philippines Diliman. He served as Chair of the Department of Visual Communication thrice and as Dean of the College of Fine Arts. He serves as a member of the Technical Committee for Fine Arts of the Commission on Higher Education and is the president of the Council of Visual Arts and Design Educators (ArtDEC). Academically trained in the Fine Arts and Urban Planning, he is professionally engaged in creative works for visual communication, product design, exhibition design, and digital media. His international works include designs of Philippine Pavilions in international expositions (Tsukuba '85 and Vancouver '86)

Aurora Odette Mendoza is Professor of Psychology and former Chair at the Department of Psychology, University of the Philippines Diliman. She has served as Director of the UP System Office of Admissions, Board member of the Psychological Association of the Philippines, and is currently member of the Commission on Higher Education Technical Committee for Psychology. She has strong interest in the application of psychology in interventions leading to behavior change in diverse settings.

Chapter 18
Tackling Climate Emergency Head-On: An Action Agenda for Policymakers

Antonio G. M. La Viña, Kristoffer B. Berse, Juan M. Pulhin, Alfredo Mahar A. Lagmay, and Kirsten Lianne Mae C. Dedase

Abstract The climate emergency in the Philippines requires decisive action now, not later. The realities on the ground show that the country's food, water and health sectors are highly vulnerable to the impacts of the changing climate. The experiences of communities in select river basin ecosystems and coastal communities demonstrated how the climate emergency has stretched local resources, challenging the capacities of both local and national governments in managing current and future climate risks. These highlight the need for multi-disciplinary responses and whole-of-government approaches in responding to the crisis. In this light, this chapter presents a nine-point action agenda to effectively address the country's greatest existential conundrum.

Keywords Policy recommendations · Call for action · Climate emergency · Philippines

As stated in the introductory chapter, this book is about how the Philippines can address the climate emergency. The chapters in Part 1—"Sectorial Realities of Climate Emergency in the Philippines," take a closer look at climate-vulnerable sectors and communities in the Philippines. This is followed by the chapters in Part 2—"Local Experiences from Selected River Basin Ecosystems and Coastal Communities"—which share documented experiences at the local level of how the climate emergency is being experienced and addressed. Finally, the chapters in Part 3—"Multidisciplinary Responses, Prospects and Challenges"—assess some of the responses that are being implemented in the country. Each chapter provides a "Summary for Policymakers" that includes recommendations to policymakers, national and local decision-makers, and the general public. We synthesize all these recommendations in

K. B. Berse (✉) · A. M. A. Lagmay · K. L. M. C. Dedase
University of the Philippines Resilience Institute, Quezon, Philippines
e-mail: kbberse@up.edu.ph

A. G. M. La Viña · K. B. Berse · A. M. A. Lagmay
University of the Philippines Diliman, Quezon City, Philippines

J. M. Pulhin
University of the Philippines Los Baños, Los Baños, Philippines

K. B. Berse et al. (eds.), *Climate Emergency in the Philippines*, Disaster Risk Reduction, https://doi.org/10.1007/978-981-99-7804-5_18

this final chapter by identifying actions necessary to successfully address the climate emergency.

18.1 Be a Leader in Global Climate Negotiations

In the 27th Conference of the Parties (COP 27), held in Egypt last November 2022, the Philippines sent a 29-member delegation. While the Philippines demanded climate justice from developed countries, some of the supposed steering leaders were not able to complete their attendance at the conference for one reason or another. Not every person in the delegation is tasked to be a negotiator; however, the absence of significant figures does not necessarily help in making a powerful entrance into the negotiations.

The climate summit highlighted several issues that need to be addressed such as (1) resource mobilization for loss and damage, (2) climate finance, and (3) worsening greenhouse gas emissions. The concerns raised were also similar to the priority items for negotiations of the Philippines per DENR Secretary Maria Antonia Yulo, who was also the appointed Chairperson of the delegation.

In terms of resource mobilization for loss and damage, the participating countries reached a compromise of establishing a loss and damage fund, and technical assistance leading to installation of more early warning systems, aiming to further capacitate the vulnerable countries. On the other hand, the countries have yet to devise operational measures for the use of the fund; hence, they have to wait until COP28 before they agree upon the recommendations of the Transitional Committee.

On climate finance, concerns about the $100 billion annual commitment of the developed countries arose. Most countries at the summit were frustrated that the pledged amounts were not met for the previous years. The developing countries also raised that finances remain to be heavy on loans more than grants, resulting in more disadvantaged and highly leveraged countries.

Lastly, greenhouse gas emissions have not met the targets from the Glasgow Climate Pact as rooted in the Paris Agreement. While there are countries who have taken action to prevent the rise of temperatures, more countries should still enhance and further strengthen their plans to narrow the variance between actual and targeted NDCs. To be able to achieve the target of net zero emissions by 2050, non-compliant countries should be pressured to stick to their commitments.

COP 27 brought to the forefront nature-based solutions. In COP26, several countries committed $12 billion to be allocated to forest protection and projects up to year 2026. As of COP27, it was reported to have reached a 22.25% utilization rate.

In the upcoming COP28 in Dubai, these issues should continue to be lobbied for, if not defended intensely. The delegates that the Philippines will be sending are expected to bolster their negotiating powers and bring an extra 2% to their game to ultimately succeed in the fight for climate justice.

18.2 Strengthen Institutional Structures for Disaster Risk Management and Climate Action at the National and Local Levels

In the Philippines, the Office of Civil Defense (OCD) is mainly tasked as the secretariat of the National Disaster Risk Reduction and Management Council (NDRRMC). Despite its mandate to bring forth changes in the disaster risk management and national civil defense programs, OCD remains bounded in terms of authority and resources as an attached agency to the Department of National Defense (DND).

In the implementation of disaster-related activities, command and control is imperative to effectively carry out its mission and establish legitimacy to increase compliance among local government units (LGUs). However, the rank of Undersecretary (Usec) remains to be the highest rank in OCD; hence, the Usec will still need to seek the approval of someone in the upper echelon to make critical decisions and mobilize resources. Due to this, some time-sensitive concerns may not be decided upon with a sense of urgency.

The disjoint of the government agencies hinders proper accountability in terms of executing disaster management at the local level. In some cases, LGUs may find it difficult to identify the point person from the national level due to overlapping policies. For example, based on the law, the National Economic and Development Authority (NEDA) serves as the vice-chairperson to lead the disaster rehabilitation and recovery; however, the NDRRMC also approved the National Rehabilitation and Recovery Planning Guide designating the OCD through its the Rehabilitation and Recovery Management Service to handle the implementation of recovery efforts.

Strengthening the national focal agency for DRM means creating a dedicated agency that can effectively manage disaster risks in an era of #ClimateEmergency. It can be in the form of a standalone line agency (i.e., a new Department) or an authority attached to the Office of the President, as has been proposed in various versions in both houses of Congress. In a study commissioned by The Asia Foundation in 2015, the establishment of an OP-attached agency was found to be the most feasible option as it can address current concerns pertaining to OCD's authority, capacity, and resources, without the burden of a bloated bureaucracy and exponential financing. It was also considered to be most acceptable to most stakeholders and more realistic as its creation may not be via legislation but through the mere issuance of an executive fiat by the President (Berse, 2015).

In any case, capacity building must be a priority of the country's national focal point. To be more strategic and efficient in the use of resources, the agency may focus only on DRM functions, while OCD retains its civil defense responsibilities under the DND. This bifurcation of functions can pave the way for both the new agency and OCD to efficiently carry out their respective mandates. The national focal agency should also have a clear alignment with the current institutional arrangements at the local level and must be able to leverage local capacities and technologies. In Chap. 11 of the book, it is recommended that there should be an inventory of all available technology pertaining to DRM, which can be of help to all LGUs. These

LGUs and their personnel must be trained on how to use the available technology to mitigate and respond to climate change emergencies and calamities, for that is the very purpose for which these technologies were created.

Chapter 16 proposes initiatives on climate education that could be part of capacity building. The author recognizes the importance of conducting training of trainers who will initiate multidisciplinary conversations across disciplines, at national and local levels. Concrete outputs could be curricular and education modules and strategies that can guide teachers, not only in explaining the science of climate change but also the significance of the arts and humanities as a springboard for action and transformation. Education, awareness-raising, and risk communication must be an integral component of the focal agency's functions.

18.3 Scale up the Adoption of Ecosystem-Based Climate Action and Disaster Risk Reduction

Nature remains to be the best shield against the effects of climate change. Conservation of natural resources and ecosystems from forests to oceans not only provides sustainable solutions in reducing risks but also beneficial in terms of establishing long-term resiliency.

In Luzon, for example, the Sierra Madre mountain range has historically saved countless Filipinos from being hit severely by typhoons coming in from the Pacific Ocean. The 540-km mountain range running from Cagayan to Quezon Province showcases 14 protected areas which helps in weakening typhoons prior to its landfall.

Soft engineering and infrastructure solutions do not always work. For instance, in cities or urban areas, it is a common practice to dredge the rivers to alleviate flooding woes. However, this practice is only a band-aid solution. In fact, there are studies pointing out that dredging is not only costly and risky but may also cause flood intensification due to soil erosion and feasible siltation. As a result, it might decrease the stability of nearby structures, heightening their vulnerability and exposure.

Instead of dredging, an NBS approach would look into the protection of the uplands, where the remaining forests are mostly situated. Addressing concerns through nature-based solutions would focus on the rehabilitation of damaged land while integrating sustainable livelihoods for the forest communities.

Nature-based solutions are emphasized throughout the book, especially in Chapters 3, 5, 6, 7, 8, and 9. Thus, in Chapter 3 on the coastal environment, the authors recommend that policymakers must modify or craft better laws and guidelines that can improve the management of the coastal zone, including considering the concepts of longshore currents and geomorphic compartments in studying coastal erosions for Environmental Impact Assessments, minimizing the impact of piers and ports by rethinking the design and placement of these structures (e.g., building piers and ports on stilts should take precedence over solid-based structures to ensure the continuity of longshore transport), improving the easement guideline in light of rising sea levels and ongoing erosion wherein the current guideline can be set as the minimum easement to which the projected rates of sea level rise and coastal erosion will be added, and adopting a rolling easement that will require coastal structures to be perched on stilts. They also propose that: (a) newly accreted lands such as river deltas and spits should be claimed by the government and set aside as greenbelts or parks that can buffer against waves and sea level rise; and (b) legal prohibitions be set on the removal of beach vegetation and forest for newly developing lands as they provide the first line of coastal defense on open sandy coasts, where mangroves do not thrive.

The authors of Chapter 5, which tackles agriculture and food insecurity, have similar recommendations. They emphasize how a risk-based agricultural planning is strategic for climate adaptation to be effective because of the high vulnerability and risks faced by the agriculture sector to climatic impacts.

This is rooted in the concept that impacts of climate change are widespread and responses to the risks should be systematically organized. Thus, they propose the following actions: (a) establishing risk transfer mechanisms such as crop insurance to protect the livelihoods of farmers; (b) integrating traditional knowledge and practices with advances in science and technology to improve monitoring, forecasting and crop advisories and practice can help to build a system of early warning systems for our farmers; (c) examining place-based vulnerability and risks due to changing precipitation patterns, increasing temperature, and increasing and/or intensifying extreme climate events, coupled with continuous environmental resources degradation. From a landscape framework, vulnerability and risk analysis of local areas is imperative as bases for a risk-based strategic adaptation plan to ensure food security for the country.

In responding to water insecurity resulting from the climate emergency, Chapter 6 proposes that Congress enact a new legislation that recognizes the threat of climate change on our country's water supply, rationalizes existing economic regulations and the financial aspects of providing water services to Filipinos to make potable water more accessible and affordable, reviews the various legal frameworks that govern the water resources sector in the country, and urgently addresses the sector's governance problems. The authors also propose budgetary support for research and development in the areas of water desalination, water filtration, water purification, water recycling, rainwater harvesting, and wastewater treatment, among others, and for the construction and maintenance of water sewerage systems and wastewater treatment plants all over the country as part of the government's infrastructure program.

These policy recommendations that mainstream nature-based solutions in disaster risk reduction efforts are also reflected in the case studies in Part 2 of the book. Thus, in Chapter 7 on Manila Bay, it is recommended for DENR and BFAR to collaborate closely on the inventory and GIS-mapping of aquaculture ponds with fishpond lease agreements (FLAs), and their respective status (i.e., active/productive); pursuing a science-advised mangrove restoration roadmap (i.e, site-specific, map-based, modeled using robust inputs, including various climate change scenarios), considering the food security mandate of BFAR, while optimizing the forest extents and the services (e.g., biodiversity, nutrient and sediment fluxes, trophic functions, etc.), as well as climate resiliency function, that these coastal ecosystems provide;

and, with other national agencies in the economic cluster of the cabinet, agree on a short-, medium-, and long-term development roadmap that ensures conservation, enhancement, and protection of the non-negotiable areas of key coastal and upstream ecosystems.

The other case studies in Part 2—in Cagayan De Oro River, Davao, Tacloban, Cebu, and Iloilo—also illustrate this approach, with a common recommendation that highlights the importance of scientific data and analysis in making development decisions for the river basins, coastal regions, and island ecosystems that the authors investigated. The case studies also highlight the need for an integrated approach to natural resources and environmental management.

18.4 Enhance Inter-agency Collaboration for Early Warning, Anticipatory Action, and Timely Response

Whole-of-government coordination is highly promoted by the OECD when it comes to responding to disasters. On the other hand, the government agencies remain lacking in this area, in lieu of the recent disasters that ravaged the country. The shortfall in coordination often results in the misalignment of plans and strategies not only in disaster response (i.e., evacuation, damage assessment) but also in disaster preparedness (i.e., early warning, forecast, and prediction). More often, these only trickle into more problems on the ground level. Efficient communication in the government, from national to local, will surely aid the country in the future, especially when compounding disasters become more prominent.

Due to the lack of coordination and clear delineation of responsibilities among government agencies, they tend to lack accountability and find ease in pointing fingers against each other. When this happens, it becomes more difficult to establish proper monitoring, evaluation, accountability, and learning (MEAL) mechanisms in responding to the communities. In order to efficiently progress in disaster management, these should be done collectively and collaboratively.

In Chapter 9 of the book, which is on the Davao River Basin, the authors propose the following policy recommendations in order to enhance collaboration for early warning and response: (a) allocate budget and resources to level up the state of science and technology for planning and in retaining service agencies that conduct monitoring, evaluation, interpretation of data to inform decision-makers; and (b) establish state-of-the-art information and communication management and sharing system that is user-friendly, open access, and even interactive

The authors of this case study also emphasize that national agencies and the local governments must work with the local communities in the lowlands and the indigenous communities in the uplands using community science as a tool to increase awareness and co-stewardship. This is also to ensure the sustainability of programs meant to address the impacts of flooding and related extreme events.

In Chapter 13, on hazard mapping, the authors recommend that policymakers at the local level must collaborate with the academe to address gaps in technical knowledge on climate change and natural hazards. They push for engaging the academe for technical assistance on climate and disaster risk assessments, land use planning, and the implementation of regular demographic surveys on top of mandated censuses.

18.5 Prioritize the Construction or Retrofitting of Evacuation Centers and Other Critical Infrastructure to Ensure People's Safety

Evacuation centers and other critical infrastructure are imperative when disasters hit communities to ensure people's safety and security. However, most evacuation centers in the country are not permanent; many are situated in nearby basketball courts, churches, and schools. In terms of providing long-term shelter, these evacuation centers remain inadequate. Due to this, there have been calls to establish a National Evacuation Center Investment Program (NECIP), aiming to prioritize the construction or retrofitting of evacuation centers and other critical infrastructure. The UP Resilience Institute has partnered with Agap Bata and ASSURE to lobby this proposal in Congress with the goal of aiding susceptible communities against the lack of permanent evacuation centers.

In building evacuation centers, the location must be checked vis-a-vis its hazard exposure (i.e., landslides, floods). There have been cases (e.g., Typhoon Ulysses and Yolanda) where evacuation centers turned out to be non-functional due to their susceptibility to certain hazards such as landslides and floods. To avoid this, multiple and localized hazard assessments must be taken into consideration in planning the construction of these relocation sites. After all, these are supposed to provide security to affected communities during disasters and minimize casualty.

Permanent evacuation centers are also encouraged to be multi-purpose and serviceable to the community in the absence of natural hazards. In that way, investing in these will also be sound from an economic vantage point. Moreover, these evacuation centers can be used for several community activities, be it for social or livelihood purposes.

Chapter 8 which is a case study of the Cagayan de Oro River Basin has several good suggestions on designing flood control infrastructures. Chapter 10, which looks at the Tacloban experience post-Yolanda, proposes reviewing mass relocation as a strategy to protect residents in coastal communities and looks into contractor-driven housing and interagency coordination. Mass relocation sites lack social facilities and economic opportunities, further deteriorating survivors' quality of life in a climate emergency. Areas for relocation do not allow agriculture-related activities, thus hindering residents' resilience at the onset of climate emergency.

18.6 Institutionalize Interlocal Cooperation throughout the Whole Disaster Cycle

LGUs in the country have significantly progressed in increasing their capacity to address the issues of climate change through proper disaster risk management. However, , the resources that most LGUs have, depending on their class/level, remain limited. An example of this would be the COVID-19 pandemic wherein the response relied heavily on the capacity of local governments. The pandemic highlighted how some LGUs were able to cover the necessities of all their constituents, while others had to make ends meet with their tight budget. While the national government can be relied on, the reality still stands that it would be impractical for them to deal with several disaster areas at once.

With limited resources, it is also paramount for LGUs to explore other sources for funding to ensure continuity and sufficiency of funding their needs, most especially in the context of the anticipated compounding and complex disasters. The LGUs can look into inter-LGU cooperation, wherein the richer LGUs aid the disadvantaged ones, mostly classified under fourth to sixth-class municipalities in the Philippine setting. Aside from that, the LGUs should also tap the CSOs, the private sector, and the academe not only for monetary support but also for in-kind and technical assistance. These mechanisms for partnerships are set forth both in RA 10121 and the Local Government Code of 1991.

In Chapter 12, which is a case study of Daanbantayan, Cebu, the author highlighted the importance of the Local Government Code which provides for the preparation of Comprehensive Land Use Plans (CLUPs) among all LGUS. However, it is important to strengthen implementation as it is quite alarming to know the percentage of LGUs which have not complied with the preparation of their CLUPs, and that even some highly urbanized cities themselves have not complied with the passage of their respective CLUPs. This shows how constrained the LGUs are in coming up with this very important piece of document, wherein DRR-CCAM strategies and measures to address the impacts of the climate emergency are supposed to be integrated. The national government must mandate and provide resources for the conduct of seminars, training, and activities at the provincial level, to make the planning process manageable, where guided actual preparations of CLUPs can be conducted.

In Chapter 14, on local planning, the authors demonstrate the process on how to formulate the CLUP with a climate and disaster risk assessment (CDRA) in the Science City of Muñoz (SCM), Nueva Ecija. They emphasize the importance of informing stakeholders on how climate change increased their risk and vulnerability to floods. Mainstreaming climate change and disaster risk information into the CLUP ensured that stakeholders considered these impacts in their city plans, which are necessary to create measures that are truly adaptive to predicted climate change impacts. Anticipatory planning for future hazards that are predicted to be bigger than those in the historical record is also important.

The authors highlighted the importance of communicating disaster-related information in a way that could be understood by the local community and encouraging stakeholders to get immersed in the planning of their community. As shown in the case study, this capacitated them with better knowledge to become more effective leaders. This shows how formulation of local plans is technical and participatory and results in ownership of response actions and development plans for the community.

The authors also point out how essential it is to establish a partnership with the academic community and other relevant agencies that can supply science-based information to local government as the basis for policy formulation. Local knowledge and community participation must be given equal importance to ensure a whole-of-society approach.

18.7 Legislate an Open Data Policy to Improve Climate and Disaster Risk Assessment

A clear policy on Open Data to effectively respond to the #ClimateEmergency is necessary if we are to meet our international commitments under the Sustainable Development Goals and the Sendai Framework for Disaster Risk Reduction. The latter, in particular, emphasizes the need for science-based and whole-of-society approaches to adapt to the adverse impacts of climate change, which can only be possible if relevant data is made available and accessible to all. Having an Open Data Policy for climate change and disaster-related data will free data that are kept in silos and endow leaders and the public alike with accurate and timely information for decision-making to perform effective and efficient courses of action toward a more resilient nation.

Accessible and unrestricted data is pivotal in contextualizing and tailor-fitting certain disaster risk parameters to specific communities. Open data also spurs more engagement and involvement from individuals and communities at the local level. The significance of open data was highly emphasized throughout the book. In Chapter 13, for example, the authors tackled the Open Data Kit approach. Meanwhile, in Chapter 14, the authors exhibited its significance in land use planning, more specifically the integration of variables related to climate and disaster risk.

At the policy level, open-access data has been integrated into several government plans such as the Philippine Development Plan (2017–2022) and the Philippine Open Government Partnership Action Plan. In fact, the government has already established an Open Data Philippines (ODPH) Portal with the intended purpose of being a data repository for government-funded institutions. On the other hand, the ODPH portal remains to have limitations. Datasets, for example, still cannot be aggregately downloaded. Moreover, it should be feasible to convert these datasets from one format to another for ease of use. Lastly, making the data publicly available should entail that these can be freely and unconditionally utilized, delimiting only to the attribution of proper citation.

To address the above-mentioned limitations, it is only necessary to legislate an Open Data Act. This will institutionalize transparency and accountability in the government agencies into providing the public with accurate and useful data to enhance their participation and responsiveness. Moreover, data sharing is also critical in terms of resolving climate change issues, most especially in the generation of geo-climatic data.

18.8 Incorporate Integrated Natural Resources and Environmental Management in Climate Change Adaptation and Mitigation

The fragmented approach to the cultural, economic, and natural capital results in an ineffective and inefficient management and optimization of natural resources. In tackling the climate change crisis, there should be a paradigm shift from an industrial to an inclusive environmental approach. Doing this entails ensuring the protection of the country's ecological resources and looking at the root causes of environmental degradation while achieving economic progress.

There are gaping trade-offs among economic growth, environmental sustainability, and social systems. Due to this, an integrated approach to development should be embedded and standardized into various institutional frameworks. Paramount to this is the engagement of relevant stakeholders, from the government to its citizens. A clear delineation of responsibilities when implementing environmental policies will not only make people accountable for possible negligent actions but also increase their engagement to the future plans and strategies.

The importance of an integrated approach is emphasized in the case studies included in Part 2 of this book. For example, in Chapter 11 on the experience in Gigantes islands, the authors propose building the resilience of small island communities to climate change and disasters by government addressing the layers of marginalization that exacerbate their vulnerabilities. Thus, the chapter discusses the Small Island Community Resilience Framework that local governments may adopt in their plans, policies, and activities. The multi-pronged framework includes (a) risk assessment, (b) capacity building, (c) addressing isolation, (d) resource management, (3) community education, and (f) mainstreaming disaster risk reduction and climate change adaptation into local plans and policies.

In a similar vein, Chapter 10 which is a case study on Tacloban, it is proposed that community risk assessments be conducted, their results integrated into development plans, and climate information and services are communicated to local communities in the local language.

18.9 Declare a Nationwide, Whole-of-Government, Whole-of-Society Climate Emergency Now

The declaration of a climate emergency must be done now. The declaration would set forth the prioritization and integration of the Philippine climate agenda into various policies. The call would also prompt awareness and a sense of urgency on the pressing issues of climate change.

The call for the nationwide climate emergency is supported by different stakeholders, including the Climate Change Commission's National Panel of Technical Experts (NPTE), CSOs, and the academe. In November 2020, the Philippine House of Representatives adopted House Resolution 535, declaring a disaster and climate emergency across the country. This was authored by Congressman Joey Salceda, the representative of the second district of Albay, as supported by the UP Resilience Institute.

Mechanisms in promoting mitigation and adaptation with the end goal of achieving climate justice should be prioritized and adopted by legislative bodies both at the national and local scales. Activities under mitigation include shifting from brown economy to green and blue economy which means protecting lands and forests through forest restoration and implementing climate-sensitive agricultural systems. Meanwhile, measures under adaptation include enhanced policies for land use, constructing climate-adaptive infrastructures, and effective management of water assets (i.e., supply and reserve). In achieving climate justice, the Philippines also has to be institutionally prepared for shifts in strategies and modifications in mechanisms to align with the global framework.

Consistent with a climate justice perspective, Chapter 15 addresses the problem of inadequate climate financing. The authors state that there is a clear and urgent necessity for a climate finance policy framework for the Philippines where resilience needs are efficiently and effectively identified, based on a comprehensive climate and disaster risk assessment, loss and damage assessment both at the national/sectoral and local levels based on vulnerability. Addressing institutional fragmentation on climate finance in the Philippines is extremely urgent due to the worsening impacts under a climate emergency, because the longer this institutional fragmentation persists, the greater the loss and damage will be for the Filipino people, especially poor communities in vulnerable local governments. Indeed, the Philippines, being a highly vulnerable developing country, must have a climate finance policy not only for the purpose of achieving climate resilience for its own people but also to systematically access climate funds (both for adaptation and mitigation) provided under the Paris Agreement as a nationally intended strategic move.

Given the emergency nature of the climate challenge, risk communication is critical. Chapter 17 proposes concrete standards in texts and graphic requirements, accepted conventions, language, and symbols for specific groups, and provides concrete applications of communication strategies and new approaches. The authors suggest the honing of communication skills so that disaster managers become spontaneous, effective communicators.

As risk communicators, disaster managers must consider the platform of their message, whether it is through the media, communal, face-to-face, broadcast, or the internet, using a formal tone or injecting humor, among others. Educating them on how to use tools in risk communication and knowing how the population relates to these will most likely reduce the discrepancies between the expected and the actual results in communication initiatives.

Antonio G. M. La Viña Ph.D. is Professorial Lecturer at the University of the Philippines College of Law and National College of Public Administration and Governance (NCPAG), former Executive Director of the Manila Observatory, former Dean of the Ateneo School of Government, and was the lead negotiator for the Philippines in various international climate change issues

Kristoffer B. Berse Ph.D. is Associate Professor and Dean of NCPAG, University of the Philippines Diliman and previously served as Director for Research and Creative Work of the University of the Philippines Resilience Institute (UPRI).

Juan Pulhin Ph.D. is Professor and former Dean of the College of Forestry and Natural Resources, University of the Philippines Los Baños and a Lead Author of the United Nations Intergovernmental Panel on Climate Change (IPCC).

Alfredo Mahar Francisco A. Lagmay Ph.D. is a Professor at the National Institute of Geological Sciences, University of the Philippines Diliman, and concurrently serves as Executive Director of UPRI and Director of the UPRI Nationwide Operational Assessment of Hazards (NOAH) Center.

Kirsten Lianne Mae C. Dedase was a former senior researcher and project manager at UPRI, serves as a lecturer at NCPAG, University of the Philippines Diliman, and currently works as a sustainable banking specialist for an international NGO.

Printed in the USA
CPSIA information can be obtained
at www.ICGtesting.com
CBHW071441020624
9450CB00005B/312

9 789819 978038